From Cells to Societies

Springer
Berlin
Heidelberg
New York
Barcelona
Hong Kong
London
Milan
Paris
Tokyo

Physics and Astronomy   ONLINE LIBRARY

http://www.springer.de/phys/

Springer Series in Synergetics

http://www.springer.de/phys/books/sssyn

SSSyn – An Interdisciplinary Series on Complex Systems

The success of the Springer Series in Synergetics has been made possible by the contributions of outstanding authors who presented their quite often pioneering results to the science community well beyond the borders of a special discipline. Indeed, interdisciplinarity is one of the main features of this series. But interdisciplinarity is not enough: The main goal is the search for common features of self-organizing systems in a great variety of seemingly quite different systems, or, still more precisely speaking, the search for general principles underlying the spontaneous formation of spatial, temporal or functional structures. The topics treated may be as diverse as lasers and fluids in physics, pattern formation in chemistry, morphogenesis in biology, brain functions in neurology or self-organization in a city. As is witnessed by several volumes, great attention is being paid to the pivotal interplay between deterministic and stochastic processes, as well as to the dialogue between theoreticians and experimentalists. All this has contributed to a remarkable cross-fertilization between disciplines and to a deeper understanding of complex systems. The timeliness and potential of such an approach are also mirrored – among other indicators – by numerous interdisciplinary workshops and conferences all over the world.

Stuttgart, June 2002   Hermann Haken

A. S. Mikhailov   V. Calenbuhr

From Cells
to Societies

Models of Complex Coherent Action

With 109 Figures

 Springer

Professor Alexander S. Mikhailov
Fritz-Haber-Institut
der Max-Planck-Gesellschaft
Faradayweg 4—6
14195 Berlin, Germany

Dr. Vera Calenbuhr*
Joint Research Center
Directorate for Science Strategy
European Commission
1049 Brussels, Belgium

* This book reflects the personal views of the author, and not the institution for which she is working.

Library of Congress Cataloging-in-Publication Data

Mikhailov, A. S. (Alexander S.), 1950-
From cells to societies : models of complex coherent action / Alexander S. Mikhailov, Vera Calenbuhr.
     p. cm. -- (Springer series in synergetics, ISSN 0172-7389)
   Includes bibliographical references and index.
   ISBN 3540421645 (alk. paper)
   1. Self-organizing systems.
   I. Calenbuhr, Vera, 1960- II. Title. III. Springer series in synergetics (Unnumbered)

Q325 .M54 2002
003'.7--dc21   2002066998

ISSN 0172-7389

ISBN 3-540-42164-5 Springer-Verlag Berlin Heidelberg New York

Springer-Verlag Berlin Heidelberg New York
a member of BertelsmannSpringer Science+Business Media GmbH

http://www.springer.de

© Springer-Verlag Berlin Heidelberg 2002
Printed in Germany

The use of general descriptive names, registered names, trademarks, etc. in this publication does not imply, even in the absence of a specific statement, that such names are exempt from the relevant protective laws and regulations and therefore free for general use.

Typesetting by authors and LE-TEX GbR, Leipzig
Cover design: *design & production*, Heidelberg
Printed on acid-free paper   SPIN: 10567648   55/3141/YL - 5 4 3 2 1 0

Foreword

This book, written by two well-known scientists, represents an excellent addition to the Springer Series in Synergetics in several ways. It shows how by rather simple models we can gain remarkable insights into the behavior of complex systems. At the same time it demonstrates the progress made in this interdisciplinary field. While in the early days of Synergetics, the self-organized coherent action of atoms in the laser – a physical device – was in the foreground of interest (cf. my book *Synergetics: An Introduction* (Springer, Berlin, Heidelberg, New York 1977)), the coherent action of nerve cells got into the focus of research, as is witnessed by the book by P. Tass in this series (P. Tass, *Phase Resetting in Medicine and Biology* (Springer, Berlin, Heidelberg, New York 1999)). In these books the elements were disturbed by noise. Now, in the present book by Mikhailov and Calenbuhr, the self-organized coherent action of otherwise *chaotic* elements is studied and important as well as surprising results by Kaneko, Mikhailov and others are presented. Let me mention just another highly interesting problem treated in this book: the coherent interaction of tens of thousands of reactions going on in biological cells. But other phenomena, such as the formation of swarms of fish or the collective behavior of ants, are also modelled.

These are just a few examples of the many fascinating subjects dealt with in this book that relate to many disciplines under unifying aspects. I am sure that the readers of this book, who may be graduate students, professors or scientists of a variety of disciplines, such as physics, chemistry and biology, will read this book not only with great pleasure, but will find it highly stimulating for their own studies in their respective fields.

Stuttgart, July 2001   *Hermann Haken*

Preface

This book is devoted to the discussion of functional self-organization in large populations of interacting active elements. The possible forms of self-organization in such systems range from coherent collective motions in the physical coordinate space to the mutual synchronization of the internal dynamics, the development of coherently operating groups, the rise of hierarchical structures, and the emergence of dynamical networks. Such processes play an important role in biological and social phenomena. Remarkably, they can however be found already in the relatively simple mathematical models which we are going to discuss. These basic models may be ordered into a sequence, progressing from the non-differentiated collective dynamics, characteristic of cell populations and swarms, to the hierarchically organized dynamics of many interacting networks, which is the property of societies.

The aim of the book is to provide an elementary introduction to the collective dynamics. We have chosen a series of models from physics, biochemistry, biology, sociology and economics, and will systematically discuss their general properties. Our approach follows the tradition of theoretical physics. Though we shall deal with mathematical problems, our emphasis will not be on logical constructions and rigorous proofs. Since the book covers a wide range of topics and touches many applications, it was practically impossible to provide a complete bibliography. The included references and suggestions for further reading reflect the subjective preferences of the authors. Generally, only references related to applications and modifications of the particular models considered are given.

The book can be viewed as a continuation of the two volumes of *Foundations of Synergetics* that were previously published by one of the authors (A.S.M.) in two Springer-Verlag editions. This book is, however, independent from these two volumes and familiarity with them is not needed. We expect that this book will be useful both for students and researchers, interested in applications of the complex systems theory in a variety of disciplines, from physics, chemistry and biology to informatics and social studies.

Writing this book was a long project. We are grateful to many friends and colleagues who supported us in this endeavor. One of the authors (A.S.M.) would especially like to thank the Department of Physical Chemistry of the Fritz Haber Institute of the Max Planck Society and its director G. Ertl for providing excellent conditions for the creative work. The new book is partly based on a lecture course given by this author at the Technical University of Berlin. At various project stages, financial assistance has been provided by the Thyssen and Volkswagen Foundations, by the Humboldt Foundation and by the Peter and Traudl Engelhorn Foundation in Germany.

We express our deep gratitude to H. Haken for his critical reading of the manuscript and valuable comments. We are pleased to acknowledge discussions with W. Alt, R. Beckers, H. Bersini, A. Coutinho, J.L. Deneubourg, L. Edelstein-Keshet, B. Hess, J. Hudson, K. Kaneko, M. Kaufman, Y. Kuramoto, D. MacKernan, O.E. Rössler, K. Showalter, W. Ebeling, L. Schimansky-Geier, E. Schöll, G. Sonnino, J. Stewart and F.J. Varela. Our special thanks to M. Hildebrand, M. Ipsen, S.C. Manrubia, P. Stange and D.H. Zanette, who have greatly contributed to the emergence of this book.

Berlin, Brussels,   *Alexander S. Mikhailov*
July 2001   *Vera Calenbuhr*

Contents

1. Introduction

The phenomenon of life has long puzzled physicists. The difficulty was not that biological organisms and processes are complex (physics routinely deals with complex phenomena), but why they are so much different from anything encountered in the inorganic world. In his book *"What is Life?"* published in 1944 Schrödinger wrote: "From all we have learned about the structure of living matter, we must be prepared to find it working in a manner that cannot be reduced to the ordinary laws of physics. And that not on the ground that there is any 'new force' or what not directing the behavior of the single atoms within a living organism, but because the construction is different from anything we have yet tested in the physical laboratory" [1]. He went further to point out that, from the thermodynamical viewpoint, all living systems are open. Together with the flux of energy passing through it, an open system may export entropy and thus maintain its order or even progress into a more ordered nonequilibrium state.

The analysis by Schrödinger has stimulated intensive theoretical research on open nonequilibrium systems. The works by Turing [2], Prigogine [3], Haken [4] and others have revealed that such systems can organize themselves and spontaneously build complex spatial and spatiotemporal patterns. It was found that this behavior is not restricted to biological organisms, but is also observed in physical or chemical systems. Indeed, the operation of any laser is based on such principles. Chemical systems, such as the Belousov–Zhabotinsky reaction or catalytic surface reactions, exhibit an impressive variety of wave patterns which are similar to excitations in the cardiac muscle or in the animal retina. Rich nonequilibrium pattern formation is observed in semiconductors and systems with gas discharges.

A parallel line of theoretical development was to study complex behavior in relatively simple dynamical systems. In 1963 Lorenz has discovered a model with just three ordinary differential equations where stochastic dynamics in the absence of applied noise was found [5]. In the same year Sinai showed that even the motion of two balls in a billiard is intrinsically chaotic [6]. Since that time, studies of chaotic nonlinear dynamics have flourished.

An important role was played by the progress of computer technology, which has made possible extensive numerical simulations of complex nonlinear phenomena. This progress has also allowed us to begin systematic analysis of evolutionary processes. The studies of "artificial life" were thus initiated by the Santa Fe group.

Subsequently, these different research directions have merged to produce what is today known as the interdisciplinary theory of complex systems. At present, it represents one of the most rapidly developing scientific fields. The number of publications on complex systems is growing almost at an exponential rate, with new journals being established and many meetings being organized. Introductions, dealing with various aspects of complex systems, are already available [7–20].

In this book we focus our attention on a particular class of problems in the theory of complex systems. Roughly speaking, they can be described as related to *social* life. In other words, we shall consider typical kinds of collective behavior in large populations of interacting active agents. The discussion will be based on a sequence of simple mathematical models which illustrate the principal properties of collective behavior. We have decided to order this sequence in such a way that our analysis begins with primitive non-differentiated populations, which are only capable of coherent collective motions and can therefore be characterized as swarms. We proceed then to the models where individual elements already have internal dynamics and consider such phenomena as mutual synchronization, spontaneous formation of coherently operating groups and development of the hierarchical organization. Finally, the emergence and the evolution of dynamical networks are analyzed. As used in this book, the words "swarm" and "society" refer to generic forms of collective dynamics. Thus, financial markets may sometimes behave as swarms and the immune system of an animal can be viewed as building a primitive society.

When collective social or biological dynamics is investigated, the first question is how to describe individual elements of a population. Even when such elements represent single macromolecules, organella or biological cells, they are so complex that to incorporate all internal complexity into a model is impossible. Fortunately, in most cases the elements interact not fully expressing their complexity. Therefore, they can be described as *automata* with a limited repertoire of responses and relatively simple effective internal dynamics. In Chap. 2 we give examples of the automata description.

The distinguishing property of reaction–diffusion systems is that their elements interact (or react) only locally in space, while their motion through the medium is essentially passive, i.e. has a diffusive nature. A classical example of this situation is provided by chemical reactions in weak aqueous solutions, where active molecules perform random Brownian motion and react only by direct collisions. In contrast to this, many biological organisms, even small bacteria, are able to *actively move* through the medium. Therefore, the motion of these organisms is no longer controlled just by external random forces acting on them and leading to the diffusive motion. Instead, they can go to predefined destinations or behave, by changing the direction and the magnitude of their velocity, in response to the signals received from other population members. Hence, physical motion in space and time becomes directly controlled by the agents forming the populations; it is based on communication between the agents. Some phenomena related to active motion are considered in Chap. 3.

The active motion is not necessarily based on the generation of forces propelling a body through the medium. Nonequilibrium systems allow rectification of noise, i.e. its transformation into directed physical motion. Much attention has recently been paid to the studies of ratchets, which represent simple mechanical devices endowed with this property. Navigation of intelligent particles in fluctuating media provides another variant of active noise-induced motion. As we show in Chap. 4, the collective dynamics of some systems *makes use* of noise.

An intrinsic property of any living organism is its *age*. Depending on their age, animals and insects can reproduce, switch between different kinds of food or behavioral patterns, and eventually die. The events experienced at an earlier age can also influence the later behavior. Treating age as an individual internal coordinate of an agent, we show in Chap. 5 that the collective dynamics of such populations is similar to the coherent motions of swarms. However, this dynamics takes place in the internal "demographic" space of the population, rather than in the physical coordinate space.

In addition to its age, a biological organism has other variables specifying its internal states. Some forms of behavior, described as motions along certain internal coordinates, can be very simple. For instance, the behavior of a firefly is that it periodically emits light. Active agents with cyclic individual behavior can be modelled as periodic oscillators. Communication and interactions between such agents can lead to the emergence of synchronous collective behavior of the entire population. *Mutual synchronization* of collective dynamics is discussed in Chap. 6.

A more refined kind of coherent collective dynamics consists of the formation of synchronously operating groups (or clusters) of elements. Dynamical clustering, considered in Chap. 7, is typical for populations of elements with chaotic individual dynamics. It can already be viewed as the spontaneous development of a *functional structure* in an initially uniform population, with different functional roles played by different self-organized groups.

The next step in the direction from swarms to societies is the appearance of *hierarchical organization*. This is discussed in Chap. 8. In particular, we show that coherently operating groups which develop through dynamical clustering are often hierarchically ordered, so that this system can be viewed as a "glass". Moreover, examples are presented that show how enslaving and a hierarchy of effective interactions between structures can develop.

The dominant structures of a society are networks formed by individuals and their coherently operating groups. These *dynamical networks* emerge, evolve and interact. Through the formation of various dynamical networks, a society can organize itself to perform certain required functions. It can also respond to environmental variations by appropriately changing its network structure. Illustrations of such behavior, using simple mathematical models, are given in Chap. 9.

2. The Games of Life

Life is so complex and diverse that the mere possibility of its mathematical modeling is sometimes denied. Nonetheless, there are many successful models of biological and social systems. The common property of many such models is that they treat an individual organism, a cell or an animal, as an *automaton*. An automaton is essentially a 'black box' which receives information and generates an output, choosing it from a fixed repertoire. Hence, it is described by specifying a set of mapping rules that must be applied to generate responses for various combinations of incoming signals.

In mathematics, the automata are often assumed to be discrete. However, automata with gradually varying input and output signals and a continuous temporal evolution may also be considered. Moreover, automata with stochastic dynamics are possible.

When an automaton which mimics the individual behavior of a biological organism is constructed, a population of such automata can be considered whose members communicate by generating, sending and receiving signals. We shall formulate and discuss in this chapter several characteristic examples of communicating automata. The first example refers to excitable media, such as the cardiac tissue, where the cells communicate via electric signals. Following Wiener and Rosenblueth, we show how simple operation rules of excitable automata lead to complex patterns of rotating spiral waves. In the next section our attention is focused on the collective amoebae *Dictyostelium discoideum*. The properties of such individual amoebae are similar to those of excitable cells, though they communicate by sending chemical signals. The important new feature is that the amoebae can slowly move through the medium in response to the received chemical signals. Then, we consider automata models that reproduce collective motions of fish schools. The last section deals with ant societies modelled as automata populations.

2.1 Excitable Media

Investigations of excitation waves have a long history. Already in 1876 Romanes saw in his studies of jellyfish (medusae) that the tissue of their umbrella was excitable and conducted contraction waves [21]. Following his observations, Mayer cut in 1908 a complete ring of the tissue from the circumference of the medusa's bell. He found that, by appropriate stimulation, he could induce a contraction wave which started to travel around the ring and continued to circulate for hours or days [22]. It was soon realized that the heart muscle had similar excitable properties. Persistent circulation of contraction waves was observed in 1913 by Mines in the rings cut from the auricle of fish [23] and in 1914 by Garey, who used the rings from the ventricle of large turtles [25]. When registration of the electrical cardiac activity became possible, electrical waves causing local muscle contractions were further observed.

In 1925 Lewis [24] gave a detailed description of excitation waves in the heart. Under normal physiological conditions, a new excitation wave is generated approximately each second by a group of cells forming the sinus node. This concentric wave spreads over the heart and is responsible for its normal beat. The propagation velocities of the cardiac excitation waves were measured, showing that they remained approximately constant. It was found that the excitation of a tissue element is followed by an interval of time (called the refractory period) during which the muscle refuses to respond to electrical stimulation. It was also found that, in some pathological regimes, such normal wave propagation pattern is replaced by a wave that rapidly circulates inside the heart.

Despite experimental successes, no theoretical model of excitable media was available at that time. The electrochemical mechanisms, underlying excitation cycles in single cardiac cells, were understood only much later. Nonetheless, in 1946 a mathematical theory was formulated [26] that explained the principal observed properties of excitation waves, and described the circulation of such waves and the formation of rotating spiral waves. This theory was proposed by the mathematician Norbert Wiener together with the cardiologist Arthuro Rosenblueth.

The idea of Wiener was simple: if we do not know the detailed mechanisms of single cells, we can still try to model them as *automata*. The biological excitable media, such as the cardiac tissue, are then described as networks formed by such idealized connected excitable elements. Each element can be found in one of three different functional states of *rest*, *excitation* and *recovery* (the refractory state). The arrival of an external perturbation at an element in the rest state triggers its transition into the state of excitation. The element stays a short time in the excited state and then goes into the refractory state of a fixed duration. In this state, the element does not respond to any external perturbations. Only when the refractory time has expired and the element has returned into its original rest state, can it again undergo excitation. The excitation cycle of a single element is shown schematically in Fig. 2.1.

An element that is currently occupying the excited state induces excitation of all neighboring elements that are in a state of rest. Thus, application of a local perturbation to an array of such elements would produce a propagating excitation wave. Though the system actually consists of individual elements, Wiener and Rosenblueth preferred to treat it as a continuous medium. In the

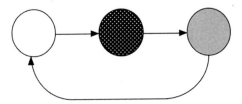

Fig. 2.1. The schematic representation of a cycle of an excitable element. Arrival of a supercritical perturbation initiates a transition from the rest state (white) into the state of excitation (dark gray). After that the element goes into the refractory state (light gray) and then returns to the rest state

continuum approximation, propagating excitation waves represented traveling fronts of excitation that were followed by a refractory zone. Each small element of the curve, specifying the excitation front, moved with time in its normal direction at a constant velocity. The front followed the boundaries of the medium and was orthogonal to them.

The excitable medium could contain obstacles. These might have been produced by cutting a hole or creating a region not accessible for excitation[1]. Wiener and Rosenblueth considered the motion of excitation waves around obstacles with different shapes. Since in their theory the wave could not separate itself from the boundary of an obstacle, this implied that the wave must indefinitely circulate around the obstacle, yielding a spiral. Figure 2.2, reproduced from the original article [26], gives an example of two spiral waves rotating in opposite directions around two small holes of equal sizes. Note that only the excitation fronts are shown here.

For numerical simulations, the Wiener–Rosenblueth model can be reformulated in terms of arrays of locally connected automata (or *cellular automata*). Below, we present the cellular automata model of excitable media which was proposed in 1986 by Zykov and Mikhailov [27] (see also [28]). Let us consider a two-dimensional square lattice occupied by excitable elements. Each lattice site is specified by a pair of coordinates i and j ($i, j = 1, 2, 3, \ldots, N$). The state of an individual element at a discrete moment of time n is described by two variables ϕ_{ij}^n and s_{ij}^n. The first of these variables, ϕ, represents the integer *phase* of an element. The zero phase ($\phi = 0$) corresponds to the rest state. The phases in the interval $0 < \phi < \tau_e$ correspond to excited states (τ_e is the duration of the excitable period). The refractory states correspond to the interval $\tau_e < \phi < \tau_e + \tau_r$, where τ_r is the duration of the refractory period. The second variable s specifies the *accumulated signal* received by the element from its neighbors. If this variable

---

[1] The obstacles for propagation of excitation waves in the heart muscle are typically formed by blood vessels spanning the muscle. After an infarct, small regions of the cardiac tissue can die, thus also forming regions where excitation is not possible.

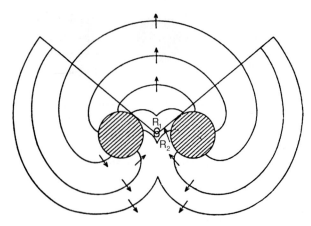

Fig. 2.2. Two counter-rotating spiral waves. (From [26])

exceeds a certain threshold h, the element goes from the state of rest into the first excitable state and its cycle is thus initiated.

The accumulated signal of the element (i, j) is obtained by summation of signals coming from its neighbors. It is assumed that such signals are produced by any of these elements in the excitation state. To take into account temporal accumulation, we additionally assume that the signal at the next time moment includes also a fraction $g < 1$ of the signal which the element had one time step earlier. Hence, we have

$$s_{ij}^{n+1} = gs_{ij}^n + \sum_{k,l} C(i-k, j-l)\theta_{kl}^n, \tag{2.1}$$

where the coefficients C determine the range of spatial summation. If only the signals coming from the element itself and its nearest neighbors (including the diagonally located elements) are summed, this corresponds to the choice $C(i-k, j-l) = 1$, if $|i-k| \le 1$, and $|j-l| \le 1$ and $C(i-k, j-l) = 0$ otherwise. The variables θ are defined here as $\theta_{kl}^n = 1$ if the respective element (k, l) is currently in the state of excitation, i.e. $0 < \phi_{kl}^n < \tau_e$, and $\theta_{kl}^n = 0$ otherwise.

The updating laws for the phase variable are

$$\phi_{ij}^{n+1} = \begin{cases} 0, & \text{if } \phi_{ij}^n = 0 \text{ and } s_{ij}^{n+1} < h \\ 1, & \text{if } \phi_{ij}^n = 0 \text{ and } s_{ij}^{n+1} > h \\ \phi_{ij}^n + 1, & \text{if } 0 < \phi_{ij}^n < \tau_e + \tau_r \\ 0, & \text{if } \phi_{ij}^n = \tau_e + \tau_r \end{cases} \tag{2.2}$$

Thus, the element operates like a clock: the cycle is started if the threshold h is exceeded and then lasts for the time $\tau_e + \tau_r$, returning the element to its original rest state.

Simulations of this model can easily be performed. When a group of elements in the middle of the system is initially excited, this gives rise to a concentric excitation wave that spreads over the medium. At a large distance from its center, this excitation wave becomes flat. If we take such flat wave and erase half of it, the remaining half will evolve into a steadily rotating spiral wave, as shown in Fig. 2.3. Varying the parameters, one can reproduce in this model all basic phenomena observed in excitable media (see, e.g. [28])

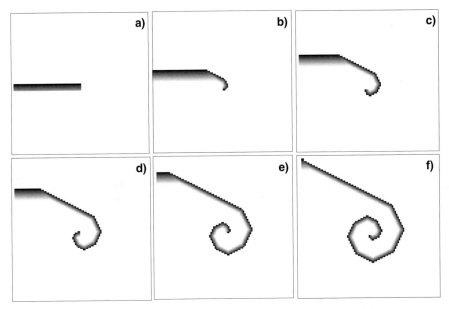

Fig. 2.3. Development of a spiral wave. Simulation using the cellular automata model with $h = 3$, $g = 0.5$, $\tau_e = 3$ and $\tau_r = 5$; the size of the medium is 100×100. Six consecutive frames (**a–f**), separated by 10 time steps each, are displayed

We can also introduce obstacles and inhomogeneities into this model. The simplest way of doing this is to define an additional array h_{ij}, specifying individual threshold values for all elements. Setting $h_{ij} = h$ outside obstacles and assigning high threshold values to the nodes inside the obstacle, we can make this region inaccessible for the excitation. Randomly distributed inhomogeneities can be introduced by assuming that each lattice site is excitable only with some probability p, i.e. that $h_{ij} = h$ with probability p and $h_{ij} = \infty$ otherwise. Figure 2.4 shows the results of a simulation for the medium where $p = 0.8$. We start with the same initial condition as in Fig. 2.3. The inhomogeneities lead to the appearance of irregular wavefronts and to the breakup of waves, so that a complex pattern of many rotating wave fragments will be eventually established. Similar behavior is actually observed in the heart muscle under fibrillation conditions.

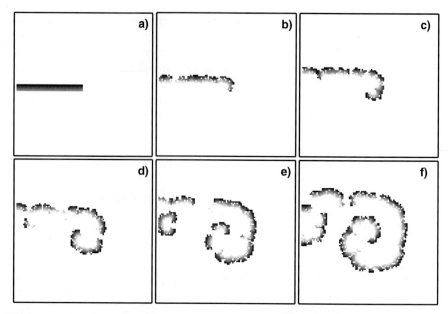

Fig. 2.4. Wave patterns in a randomly inhomogeneous excitable medium. Simulation for the cellular automata model of size 100×100 with the parameters $h = 2.5$, $g = 0.5$, $p = 0.8$, $\tau_e = 3$ and $\tau_r = 5$. Six consecutive frames (**a–f**), separated by 10 time steps each, are displayed

2.2 Collective Amoebae

The slime mold *Dictyostelium discoideum* represents an unusual example of a biological system that can exist in two different forms, either as a population of microorganisms or as a macroorganism consisting of thousands of individual cells. This property makes such amoebae an extremely appealing object in theoretical investigations of biological morphogenesis. Mathematical models reproducing the main aspects of their collective behavior have been constructed and are discussed in this section. When nutrients are abundant, the amoebae form a dispersed cloud with a low cell density. In this state the activity of individual cells is not correlated and they behave like any other population of microorganisms. If, however, the nutrients become scarce, a collective structuring transition begins in this population. The individual amoebae condense into a macroorganism, where contacts between single cells are established. This macroorganism develops later a three-dimensional fruit body, looking similar to a small mushroom, and produces spores. Under natural conditions, the spores would be spread by the wind over a large area. When they land in a region where the nutrient concentration is sufficiently high, they give rise to a new dispersed population. Thus, from the evolutionary viewpoint, the structuring transition allows the amoebae to avoid extinction in an environment that suddenly becomes low in nutrients and to colonize new potentially rich area.

Especially interesting are the early stages of this transition, leading to the formation of a condensed structure. Observations show (Fig. 2.5) that first the patterns of concentric rings and rotating spirals appear. Later, slow motion of cells towards the centers of such patterns begins and the amoebae rearrange themselves into a system of thin dense streaks.

Fig. 2.5. Aggregation pattern of *Dictyostelium discoideum*. *Left*: initial stage (cell density is essentially homogeneous, and white regions correspond to moving amoebae, marking the position of cAMP wavefronts; in darker regions, the amoebae are stationary). *Middle* and *right*: developed cell streams approximately 60 and 120 min later (dark regions in the photograph are depleted of amoebae). The size of the field shown is 1.5 cm × 1.5 cm. (From [32])

These phenomena look similar to the wave patterns in the excitable media considered in the previous section. The analysis reveals that single amoebae may indeed be viewed as excitable elements with the states of rest, excitation and recovery (cf. Fig. 2.1). The difference, however, is that interactions between individual cells are not based on the propagation of electrical signals, as in the cardiac tissue. Instead, chemical intercellular communication is realized in this system. In their excited states, individual cells release into the surrounding medium a special chemical substance (cyclic-adenosine 3,5 monophosphate - cAMP) that diffuses through the medium and can reach other cells. If the local concentration of this substance exceeds a certain threshold and a cell is currently in the state of rest, its excitation cycle is initiated.

The collective chemical activity of the cellular population leads then to the appearance of concentration patterns formed by traveling waves and the development of chemical gradients in the medium. Now, another property of collective amoebae becomes significant. These microorganisms can move themselves through the medium by following the chemical gradients (such phenomena are known as *chemotaxis*). Namely, they tend to move towards the regions with higher concentrations of the chemical mediator cAMP. Moreover, the active motion takes place only when the amoebae are in the states of excitation.

Today, quite detailed models of these processes are available. As an illustration, we consider below only a greatly simplified mathematical description proposed by Kessler and Levine [29]. In this model, each amoeba is represented by an automaton, whose internal dynamics is characterized by the phase variable ϕ that obeys the same evolution law (2.2) as in the case of excitable cardiac cells. However, the communication between the cells is organized differently.

Each cell m is characterized by its spatial location $\mathbf{r}_m$. It is convenient to introduce the cell variable θ_m, such that $\theta_m = 1$ when the cell is excited $(0 < \phi < \tau_e)$ and $\theta_m = 0$ otherwise. In the excited state, the cell releases a chemical activator that diffuses through the medium. The evolution of the activator concentration $u(\mathbf{r}, t)$ is then described by the equation

$$\dot{u} = -\gamma u + D\nabla^2 u + q \sum_m \theta_m(t)\delta(\mathbf{r} - \mathbf{r}_m). \tag{2.3}$$

Here D is the diffusion constant and q is the rate of release of the activator by an excited cell. The first term on the right-hand side describes the decay of the activator that should be present in order to prevent the accumulation of this chemical substance in the system. The cycle of an individual cell begins when the local activator concentration exceeds the threshold h.

Though the mechanism of communication between the automata is now different and involves diffusion rather than direct exchange of electrical signals, this system possesses all the typical properties of an excitable medium. It supports the propagation of excitation waves. As an example, Fig. 2.6 shows a steadily rotating spiral wave in this medium [29]. Here we have 2000 cells that are randomly distributed over the medium. The parameters are $\gamma = 0.5, D = 1, h = 1, \tau_e = 2, \tau_r = 20$, and $q = 150$. This simulation was performed using the discretized version of the diffusion equation (2.3) on a lattice of size 100×100 with no-flux boundary conditions. The black boxes in Fig. 2.6 contain cells; inside the boxes with black borders the activator concentration exceeds the threshold $(u > h)$.

The amoebae can move by following the gradients of the activator concentration. To incorporate this behavior into the model, we assume that the coordinates $\mathbf{r}_m$ of the cells may change with time. The rule of the cell motion was formulated in [29] for the discrete version of the model.

For a cell located at a given moment of time n at the lattice site (i, j), the activator gradients in the horizontal and vertical directions are computed as

T = 66.96

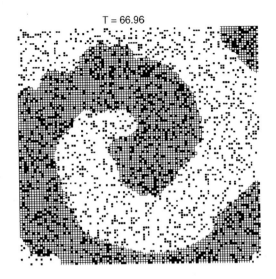

Fig. 2.6. A spiral in the medium of size $L = 100$ with no-flux boundary conditions. The black boxes contain cells, and the boxes with black borders have $u > h$. (From [29])

$$u_x = \frac{1}{2}(u^n_{i+1,j} - u^n_{i-1,j}), \quad u_y = \frac{1}{2}(u^n_{i,j+1} - u^n_{i,j-1}). \tag{2.4}$$

If this cell is in the excited state and at least one of the gradients exceeds the motion threshold h', the cell attempts to move. It tries to jump into the neighboring site with the higher activator concentration along the direction of the gradient, exceeding the threshold. If both gradients are larger than the threshold, the cell tries to jump into the diagonally located cell. However, the jumps actually take place only if the target site is empty. When it is currently occupied, the cell remains at its old location. This rule is applied at each moment of time for all cells in the population. However, the cells that have already moved during their excitation cycles would not move again until the new excitation cycle begins. Hence, each cell can move only once per chemical signal.

Figure 2.7 shows evolution of the spatial distribution of cells on the lattice of size 400×400 for the model with the same parameters as in Fig. 2.6 and the motion threshold $h' = 2$. In this simulation, concentric spreading wave patterns were generated by a pacemaker placed in the center of the lattice. This pacemaker periodically triggered excitation waves at time intervals of $T = 30$. The cells thus moved in the radial concentration gradients produced by such spreading waves.

Initially the cells were randomly distributed over the lattice (Fig. 2.7a). The concentration gradients produce, as could be expected, the radial motion of the cells towards the center. However, this motion is not uniform. Instead,

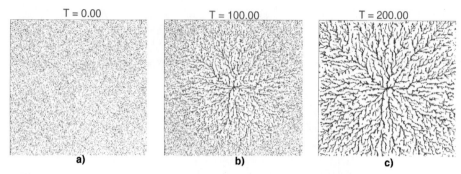

T = 0.00   T = 100.00   T = 200.00

a)   b)   c)

Fig. 2.7. The streaming instability. Simulations for the medium of size $L = 400$ with a central pacemaker. Only the cells are depicted. Subsequent frames **(a–c)** correspond to moments $T = 0, 100$ and 200, respectively. (From [29])

the moving cells arrange themselves into a radial network of dense streams (Fig. 2.7b, c).

Note that in order to obtain the stream patterns shown in Fig. 2.7 the rule of the cell motion was slightly modified, allowing the cells to pile up in the central lattice sites (and thus lifting the restriction of the single-site occupancy here). The real biological cells actually clump in the center of the stream pattern, producing a three-dimensional structure.

The instability, responsible for streaming, has been explained [30–32] by taking into account that in the regions with higher population density the effective excitability of the medium is increased and this leads to a local increase in the wave velocity. A propagating wave therefore becomes curved and acts as a 'lens' focusing the flow of cells.

Similar radial stream patterns are obtained in the simulations if, instead of a pacemaker, the medium contains a rotating spiral wave. A different behavior is, however, found in this case in the central region, where the core of the spiral is located. The cell density in this region becomes depleted and the amoebae form a dense ring rotating around the core. The spirals develop in this system from the wave breaks that occur when, by a fluctuation, the local cell density is strongly decreased and an analog of a hole (cf. Sect. 2.1) is created.

Spontaneous emergence of stream patterns is indeed observed in the experiments with the collective amoebae *Dictyostelium discoideum* (see Fig. 2.6). The theories, picturing this system as a population of automata that move in the collectively created chemical field, provide a good qualitative description of the first stages of the aggregation process. At later stages, the cell density inside the streams is so high that direct contacts between the moving cells are established. When this occurs, other interactions between the cells are expressed and the system slowly acquires the properties characteristic of a multicellular organism.

2.3 Fish Schools

Fish move over many miles in close formations that are called schools. About 5000 fish species school throughout their whole life, and twice the number school as juveniles. The sizes and structures of schools are very different over time and among species. Nevertheless, many typical patterns of such collective behavior are known. While feeding or resting, the fish assume nearly random orientations. When the group is moving, the fish show a highly parallel orientation. If attacked by predators, the schools respond by various structural transformations (fountain effect, split effect, waves of agitation, etc.).

In contrast to most mammal herds, many fish schools have no hierarchical structure. Herring and mackerel are typical species that form schools without leaders. They demonstrate that a highly coordinated group behavior is possible without the domination of some individuals over others. The harmony of action of such schools does not result from the commands of a leader, but emerges from the cooperative interactions of their individual members.

Mathematical models of fish schools are constructed by viewing each fish as an automaton that processes information received from other individuals and responds by changing its motion according to a set of simple behavioral rules. The models should include approximately the same information that is available to a real fish as the end result of its perceptual and cognitive processes. In this section we present and discuss the automata model of fish schooling developed by Huth and Wissel [33–35]. As a simple example we consider below the two-dimensional version of this model; it can easily be extended to include the vertical dimension.

Each fish i in the population of size N is characterized by its position $\mathbf{r}_i(t) = \{x_i(t), y_i(t)\}$ and its velocity vector $\mathbf{v}_i(t)$ at time t. The time is made discrete by dividing it into short intervals Δt. The model consists of a set of rules that determine the new positions $\mathbf{r}_i(t + \Delta t)$ and the new velocities $\mathbf{v}_i(t+\Delta t)$ of all the fish at the next time moment, if their current coordinates and velocities are known.

Several general notations will be used below. We denote as $r_{ij} = |\mathbf{r}_i(t) - \mathbf{r}_j(t)|$ the distance between fish i and fish j. The angle between any two vectors $\mathbf{a}$ and $\mathbf{b}$ is denoted as $\angle(\mathbf{a}, \mathbf{b})$.

We first consider interactions between a pair of fish. These interactions strongly depend on the distance. If the neighbor is close ($r_{ij} < r_1$), the fish tries to avoid the collision (repulsion). At intermediate distances ($r_1 < r_{ij} < r_2$), the influenced fish tries to swim in the same direction as its neighbor (parallel orientation). If the neighbor if far away ($r_2 < r_{ij} < r_3$), the fish swims towards its neighbor (attraction). Finally, if the distance is so large that the other fish is already out of the visual range ($r_{ij} > r_3$), any interaction is absent and the fish performs random motion (search). These regions are schematically shown in Fig. 2.8a.

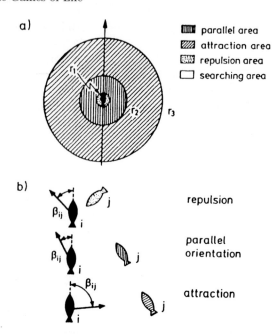

Fig. 2.8. (a) Ranges of the basic behavior patterns. The reaction of the black fish depends on the range in which a neighbor is positioned. (b) Basic behavior patterns. The arrow indicates the new swimming direction of the black fish i and thus shows its behavior reaction to its neighbor j. (From [34])

The behavioral rules in all these four regions must be specified. We do this by introducing the angle β_{ij} by which fish i will turn under the exclusive influence of fish j. In the *repulsion* region $(r_{ij} < r_1)$ fish i turns in such a way that it would move perpendicularly away from the swimming direction of fish j (Fig. 2.8b). If $\mathbf{w}$ is a vector that is orthogonal $(\mathbf{w} \cdot \mathbf{v}_j = 0)$ to the swimming direction of fish j, the angle β_{ij} of fish i represents the smallest turn needed to move in the direction of vector $\mathbf{w}$, i.e.

$$\beta_{ij} = \min\{\angle(\mathbf{w}, \mathbf{v}_j)\}. \tag{2.5}$$

In the region of *parallel orientation* $(r_1 < r_{ij} < r_2)$ the fish wants to move parallel to its neighbor:

$$\beta_{ij} = \angle(\mathbf{v}_i, \mathbf{v}_j). \tag{2.6}$$

In the *attraction* region $(r_2 < r_{ij} < r_3)$ the fish turns towards its neighbor to perform an approach movement:

$$\beta_{ij} = \angle(\mathbf{v}_i, \mathbf{r}_j - \mathbf{r}_i). \tag{2.7}$$

The behavior in the *search* region $(r_3 < r)$ does not depend on the motion of other fish and will be described later.

A fish may simultaneously have several other fish located within the inter-action range and we must therefore specify how the fish mixes the influences of its neighbors. Since the fish must continuously monitor the motions of its neighbors in order to respond to them, it seems natural to assume that it cannot keep track of too many other fish if they are within the interaction range. If the number N_0 of fish inside the interaction region $r_{ij} < r_3$ exceeds a threshold K, the fish chooses K of them following the *priority rule*. Namely, it chooses to respond only to the motion of its K neighbors that have the smallest angular deviations

$$\delta_{ij} = |\measuredangle(\mathbf{v}_i, \mathbf{r}_j - \mathbf{r}_i)| \tag{2.8}$$

from its swimming direction. This means that the fish prefers to respond only to what it sees inside a certain front cone.

A fish mixes the influences of the selected neighbors by taking the average of the influence angles. Since the animal behavior is always also exposed to random influences, the actual turning angles φ_i are only *statistically* deter-mined in this model. The probability distribution of a turning angle φ_i is

$$p(\varphi_i) = \frac{1}{\sqrt{2\pi}s} \exp\left\{-\frac{1}{2s^2}\left(\varphi_i - \frac{1}{K}\sum_{\text{selected } j}\beta_{ij}\right)^2\right\}. \tag{2.9}$$

Here the parameter s controls the statistical dispersion of the turning angle.

The new absolute velocity g_i of fish i is also statistically determined by the Gaussian distribution

$$p(g_i) = \frac{1}{\sqrt{2\pi}s_v} \exp\left\{-\frac{1}{2s_v^2}(g_i - \bar{v})^2\right\}, \tag{2.10}$$

where the mean velocity is given by

$$\bar{v} = \min\left\{v_{\max}, \frac{1}{K}\sum_{\text{selected } j} v_j\right\}. \tag{2.11}$$

We have taken into account that a fish cannot exceed a maximal speed $v_{\max}$.

Now we have all the components needed to determine the new velocity vector $\mathbf{v}_i(t + \Delta t)$ of fish i. It will be given by

$$\mathbf{v}_i(t + \Delta t) = g_i \mathbf{D}(\varphi_i)\mathbf{d}_i(t), \tag{2.12}$$

where $\mathbf{D}(\varphi_i)$ is the matrix of rotation by angle φ_i and $\mathbf{d}_i(t)$ is the unit vector of the swimming direction at time t, i.e.

$$\mathbf{d}_i(t) = \frac{\mathbf{v}_i(t)}{|\mathbf{v}_i(t)|}. \tag{2.13}$$

The new position of fish i at the next time increment $t + \Delta t$ is then given by

$$\mathbf{r}_i(t + \Delta t) = \mathbf{r}_i(t) + \mathbf{v}_i(t + \Delta t)\Delta t. \qquad (2.14)$$

Additionally, we must specify the fish behavior when the number K_3 of fish within the interaction region $r_{ij} < r_3$ is less than K. In this case all neighbors are selected and K is replaced by K_3 in the above rules. If *no* fish is present inside its interaction region ($K_3 = 0$), fish i performs a random search:

$$p(\varphi_i) = \frac{1}{2\pi}, \; p(g_i) = \frac{1}{\sqrt{2\pi} s_v} \exp\left\{ -\frac{1}{2s_v^2}(g_i - v_i)^2 \right\}. \qquad (2.15)$$

Extensive computer simulations of this model have been performed by Wissel and co-workers ([33–35]). The typical model parameters are (all lengths are measured in units of BL, the body length): $K = 4$, $s = 5°$, $s_v = 0.2$ BL/s, $v_{\max} = 1.3$ BL/s, $\Delta t = 0.5$ s. Schools of size $N = 20$ were usually chosen. At the beginning of a simulation, the fish are set randomly in a starting area of 4.5 BL×4.5 BL. Figure 2.9(RF1) shows a highly polarized school with strong cohesion that is obtained by taking relatively small values of the interaction radii $r_1 = 0.5$ BL, $r_2 = 2$ BL, $r_3 = 5$ BL. When these radii are increased ($r_1 = 2$ BL, $r_2 = 4.5$ BL, $r_3 = 7.5$ BL), a weakly polarized school with a loose cohesion is found [Fig. 2.9(RF2)].

Fig. 2.9. Typical appearance of two simulated fish schools with strong (RF1) and loose (RF2) cohesions. (From [34])

Large observational data is available for the schools formed by different fish species. A characteristic structural property of real fish schools is their internal dynamics. The individuals of a school continually change their relative positions. There is no fish that always swims in the vanguard, but various specimens take the lead. Their leadership lasts from fractions of a second to several seconds, after which they fall back to the middle or even the rear, to reach the front later. The dynamics of these changes of positions can be quantified by the leader time, defined as the average time which a fish spends at the top of school. Figure 2.10 shows the frequency distributions of the leader time in the experiments and in the simulations. The merging of two fish schools has also been observed: when two schools meet in open water, the new swimming direction appears to be approximately the resultant vector of

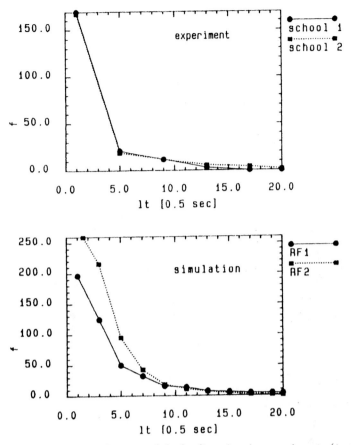

Fig. 2.10. Frequency distributions of the leading time in experiments (*top*) and in simulations (*bottom*). Experimental data is for the fish *Gathapogon elongatus* [79]. In the mean, a fish spends 1.3 to 1.4 seconds at the top of the school. Simulation data is shown for highly polarized (RF1) and weakly polarized (RF2) schools with the same parameters as in Fig. 2.9. (From [34])

the two original tracks. Figure 2.11 illustrates that the simulated fish schools indeed reproduce this behavior.

The model was further modified by introducing rules for the reaction to obstacle, food, gradients, and so on. According to the wall avoiding rule, if a fish comes too near a wall, it turns and aligns parallel to the wall. Figure 2.12 presents the simulations of a school that ran into a cylindrical obstacle. The school splits into two subgroups that swim around the cylinder. If the obstacle is relatively small, both groups merge again into one school beyond it. Large obstacles result in the separation of groups and the formation of two new schools. This form of collective behavior is often observed and is known as the fountain effect. It is used by fish to avoid predators.

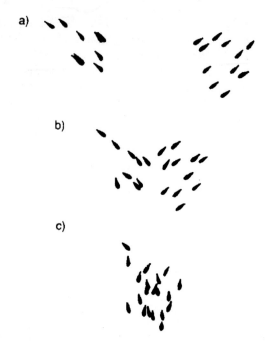

Fig. 2.11. Simulation of the merging of two schools. The same parameters as for the school RF1 in Fig. 2.9. Snapshots at subsequent moments are shown. (From [34])

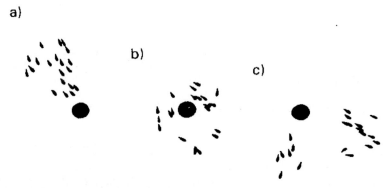

Fig. 2.12. Simulation of the splitting of a school caused by an obstacle. The black disk indicates the obtacle. Figures (**a**), (**b**) and (**c**) correspond to three subsequent moments. The same parameters as for the school RF1 in Fig. 2.9. (From [34])

Another interesting situation is the behavior of feeding schools. If a fish has found a food cloud, it loses its polarization and each fish snaps at food on its own (Fig. 2.13). This behavior is reproduced, if a feeding rule is added to the model. According to this rule, a fish increases its dispersion parameter s and reduces its maximal velocity v_{max}, when it is surrounded by food. The increase of s raises the level of individual behavior of the feeding fish and, in addition, causes the neighbors to have only weak influence on the behavior of this fish. After the food patch has been consumed, the school reforms.

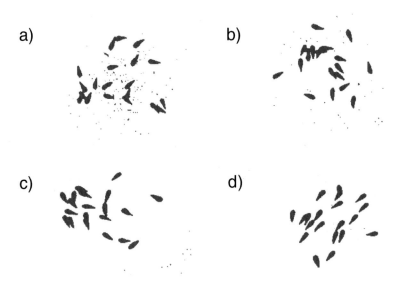

Fig. 2.13. Simulation of a school in a food cloud. The small scattered dots represent food. Figures (**a**), (**b**), (**c**) and (**d**) correspond to four subsequent moments. The same parameters as for the school RF1 in Fig. 2.9. While the school is feeding, it is unpolarized. When most of the food has been consumed, the fish reform into a polarized school. (From [34])

Besides food patches, other external stimuli may affect the movement of a school. It is known that schools can use temperature, oxygen and salinity gradients for their orientation. This phenomenon is reproduced by the model if, when a fish detects an improvement of the abiotic conditions, it tends to maintain its swimming direction and slightly increases its speed. Figure 2.14 shows the effect of such a gradient rule added to the model. The school follows the region of the best abiotic conditions, even if this region has a form of the curved belt.

Fig. 2.14. Simulation of a school that swims along a region with better abiotic conditions (gray). Three subsequent snapshots are superimposed. (From [34])

2.4 Insect Societies

Insect societies exemplify the full range of different levels of organization up to tightly and coherently built animal communities of up to millions of individuals. The social insects include all ants and termite species and the more highly organized bees and wasps. They have attracted interest because of their ecological dominance on the one hand and because of their remarkable organization on the other. They are often considered as a superorganism because they employ social design to solve ecological problems ordinarily dealt with by single organisms [36].

Experimental studies have revealed that social insects in general and ants in particular show a stunning degree of division of labor and efficiency of execution of tasks: there are workers that engage in brood care; damage to the nest is rapidly discovered and repaired; endangered individuals communicate their situation to others and provoke collective defense responses; foragers that have discovered a food source communicate their finding to nest mates and invite and guide them to the food source. Communication plays a central role in all these cases. Hoelldobler and Wilson [55] distinguish not less than twelve different functional categories of responses as a consequence of a received message. Chemical communication, in which so-called pheromones – odorous chemical substances – are emitted by one individual and perceived by another is the principal communication mechanism.

Complex behavioral patterns are related to recruitment in foraging. Many, if not most, of the 12 000 known ant species use collective foraging strategies, which are generally based in one way or another on chemical trails. The spectrum of possible recruitment strategies ranges from very simple ones in which a relatively small number of individuals is involved to very complex ones involving several millions of individuals. In the simplest case, an ant that has found a food source returns to the nest laying a chemical trail and invites a nest-mate using chemical and tactile signals to follow to the food source (tandem running). In intermediary strategies, the chemical trail itself suffices to cause individuals in the nest to follow the trail to the nest. After a successful food find, these individuals return to the nest laying pheromones, thereby reinforcing the chemical trail. If there is no more food available or if there is crowding at the food source, individuals return without laying pheromones. The trail is thus reinforced or weakens as a function of the quality of the food source and regulates the recruitment process (mass recruitment). It can be shown that the dynamics of a recruitment strategy using trail pheromone can help ant societies to solve logistic problems such as finding the shortest way to a food source [41, 42]. The most impressive foraging strategies based on chemical communication can be found in the colonies of army ants, one of the sub-families of ants known as the *Dorylinae*. They have very large colonies of up to 20 000 000 individuals. Some army ant- (and termite-) species have been observed to form large trails followed by thousands of individuals and extending over 100 m. These trails sometimes display multi-lane traffic, where the inner lanes carry the nest-bound traffic and the outer lanes carry the outward traffic. The order on the trail is characterized on a short length-scale (perpendicular to the trail axis) and is repeated along the axis of the trail [50]. Moreover, the flanks of the trail are guarded (Fig. 2.15) by large soldiers protecting the busy traffic [51].

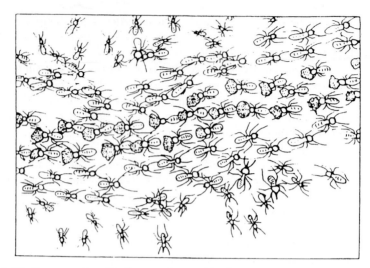

Fig. 2.15. Structured traffic flow as observed in the termite species *Lacessititermes lacessitus*. (Redrawn from [51])

The division of labour and the successful execution of tasks can be partly attributed to the caste structure of an ant society. There are at least two castes of reproductive and non-reproductive individuals, i.e. the *queen* and the *workers*. In many cases, there are several worker castes that are morphologically distinct. While it is true that the queen exerts a remarkable control on the activities of a colony through, for example, the emission of pheromones, it is unlikely that she can control the multitude of tasks requiring orientation and the coordination of a large number of individuals such as in the case of foraging. A major step forward in the understanding of how complex collective behavior can result from individual behavior is due to Oster and Wilson [54], who analyzed the structure of ant societies from the viewpoint of optimal control theory and who were able to show that the extreme ergonomic efficiency of ant societies comes about as a result of the manner in which problems are tackled: typically, small groups of individuals work on the same problem. After one problem has been solved, the groups switch to the next task and so on. That is, a task is carried out in a parallel way. Oster and Wilson showed that a series of parallel solutions is more reliable and thus less energy-wasting than having the same series of chores carried out by one individual. This approach sheds some light on how social behavior may have evolved and it explains why societies have the structure they have. What is not addressed in this approach is how a certain behavioral pattern is formed. Another approach focusing on this latter question was developed in the late 1980s and early 1990s by Deneubourg and co-workers. The basic idea is that individuals communicate information of local conditions such as the presence of food, building material or an enemy to their nest mates. The perception of such a signal, mostly chemical, triggers behavioral responses, including the emission of signals to other nest mates. In this view, ants are seen as automata that display a limited number of behavioral responses. Collective behavior patterns result from the amplification of communication signals through positive feedback. This view sharply contrasts with the traditional biological approach, in which individual complexity is necessarily at the root of collective complexity. This would have entailed that the observation of a different collective behavior automatically elicits the search for a different individual behavior [38].

One of these models that is considered below deals with the spontaneous formation of foraging trails. The model is based on the release of pheromone from an idealized trail and on the behavior of an individual ant. The behavioral algorithm determines how an ant orients and moves in a scent field. In the following we shall first describe the orientation mechanism and then the equations that govern the dynamics of scent fields.

The basis of the behavioral algorithm is a stimulus–response relation specifying the behavior of an ant in a scent field. Hangartner has shown experimentally that ants can follow chemical trails using a mechanism called *osmotropotaxis*: an individual perceives the concentration of a scent trail with the tips of its antennae and tries to minimize the concentration difference by turning towards the higher concentration [52]. Calenbuhr and Deneubourg suggested modeling an individual ant as a discrete automaton whose position is specified by its coordinates $x(t)$ and $y(t)$ and whose orientation is given by the angle $\theta(t)$, defined as the angle between the body axis of the animal and the normal of the trail axis (Fig. 2.16). Translating the general notion of osmotropotaxis and taking time to be discrete, the movements of the automata at time step t are described by the following algorithm [43]:

(1) Perception of the pheromone concentrations C_l (left) and C_r (right) with the tips of the antennae and determination of the concentration difference:

$$\Delta C = C_l - C_r. \tag{2.16}$$

(2) Change of direction according to the rule:

$$\Delta\theta = F(\Delta C)\Delta t. \tag{2.17}$$

(3) Making one step forward:

$$\Delta x = v\Delta t \cos\theta(t) \text{ and } \Delta y = v\Delta t \sin\theta(t). \tag{2.18}$$

The new position and orientation of the individuals (x, y, θ) at time $t + \Delta t$ is then given by

$$x(t + \Delta t) = x(t) + \Delta x \text{ and } y(t + \Delta t) = y(t) + \Delta y, \tag{2.19}$$
$$\theta(t + \Delta t) = \theta(t) + \Delta\theta. \tag{2.20}$$

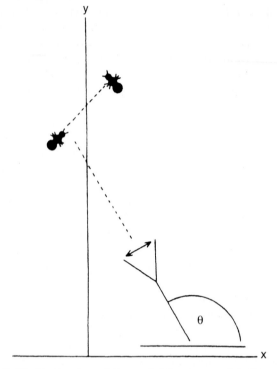

Fig. 2.16. Illustration of the trail-following model. (From [43])

Here, v is the absolute speed, which we assume to be constant, and Δt is the interval between time steps.

The perception–reaction function is chosen in the form:

$$F(\Delta C) = \text{sign}(\Delta C)\frac{\theta_{\max}|\Delta C|^n}{|\Delta C|^n + a^n}. \tag{2.21}$$

Hence, the ants turn towards the higher concentration and the turning angle is proportional to the concentration gradient between the two antennae. Moreover, there is a maximal behavioral response $\theta_{\max}$ that will be reached nonlinearly. By adjusting the parameter n this response can be tuned from smooth behavior to threshold-like stiffness. The larger the value of n the more the behavioral response resembles the Heaviside step function. The parameter a determines the characteristic concentration gradient at which the nonlinear saturation sets in. Functions of this type are often found in biological systems that show cooperative effects [58].

The pheromone is deposited on the ground and then evaporates into the air, where diffusion takes place. The local pheromone concentration $c(\mathbf{r},t)$ in the atmosphere near the surface satisfies the diffusion equation

$$\frac{\partial c}{\partial t} = D\nabla^2 c + q(\mathbf{r},t), \tag{2.22}$$

where $q(\mathbf{r}, t)$ describes the sources of the pheromone substance. Since the pheromone is released through evaporation only at the ground, all these sources are located on the horizontal plane $z = 0$.

In reality, the pheromone concentration in the air close to the ground fluctuates because of the turbulence in the atmosphere. This can be taken phenomenologically into account by replacing the calculated concentration field $c(\mathbf{r}, t)$ with $c(\mathbf{r}, t)(1+\varepsilon\xi(\mathbf{r}, t))$, where $\xi(\mathbf{r}, t)$ is white noise of unit intensity and ε is a parameter that specifies the mean amplitude of fluctuations.

The ants sense the pheromone concentration in a thin air layer close to the surface ($z \to 0$). If an ant has coordinates x, y and orientation θ, the concentration difference ΔC between its left and right antennae will be determined by the concentration gradient along the direction that is orthogonal to the ant orientation, i.e. $\Delta C = l(\mathbf{k} \cdot \nabla c)$, where l is the distance between the left and right antennae and $\mathbf{k}$ is the unit vector with the components $k_x = \sin\theta$ and $k_y = -\cos\theta$.

Special experiments have been performed, in which the pheromone was initially applied in a fine circular line on a flat solid surface. An ant was then placed at some point and the distance followed on the circular trail was measured. It was found that, with an appropriate choice of its parameters, the above model yields reasonable quantitative agreement with the experimental data [43, 44].

In the ant societies the scent field is collectively produced by the animals that release the pheromone. Therefore, the model must be extended to describe these processes [45]. The build-up of a chemical trail involves a collective decision: individual ants should patrol as efficiently as possible the area around the nest, thereby dispersing themselves. If a food source is found by one of the ants, its exploitation would require the ants to concentrate around the point of the food discovery. The switch from search to exploitation can be achieved by adding a new behavioral rule.

The internal state of the ith automaton, representing an ant, will be described by the variable h_i, taking two values 0 and 1. If $h_i(t) = 0$, the ant is in the search mode. When $h_i(t) = 1$, the ant returns to the nest (homing behavior). We assume that the nest is at the origin of the coordinates and the food source is located at a point with the coordinates x_f and y_f.

At the next time step, $t + \Delta t$, the variable h_i is updated according to the following rule:

$$h_i(t + \Delta t) = \begin{cases} 0, & \text{if } r_i < r_n^{\mathrm{cr}} \\ 1, & \text{if } \tilde{r}_i < r_f^{\mathrm{cr}} \\ h_i(t), & \text{otherwise} \end{cases} . \tag{2.23}$$

Here $r_i = \sqrt{x_i^2 + y_i^2}$ is the distance between the ith individual and the nest, r_n^{cr} is the critical distance to the nest, $\tilde{r}_i = \sqrt{(x_i - x_f)^2 + (y_i - y_f)^2}$ is the distance between the ith individual and the food source, and r_f^{cr} is the critical distance to the food source.

When $h_i(t) = 0$, the behavior of the automaton is described by (2.16)–(2.20), i.e. its motion is controlled by gradients in the chemical landscape. If the concentration gradients are too weak, i.e. $|\Delta C|$ is less than a certain threshold, the individuals perform the search by random wandering.

In contrast to this, if $h_i(t) = 1$, the automaton turns and moves along the direction towards the nest, ignoring any chemical signals. This motion is continued until the nest is reached. Indeed, the individuals of most ant species are capable of finding the nest without chemical cues. Thus, in the homing mode (for $h_i = 1$) we have

$$x_i(t + \Delta t) = x_i(t) + v\Delta t \cos\theta_i, \text{ and } y_i(t + \Delta t) = y_i(t) + v\Delta t \sin\theta_i, \quad (2.24)$$

where the angle θ_i corresponds to the direction towards the nest, i.e. $\theta_i = \arctan(y_i/x_i)$.

When the ants return home after they have found a food source, they lay a chemical trail and therefore modify the chemical landscape. In our model this can be described by assuming that the pheromone source in (2.22) is

$$q(\mathbf{r}, t) = \sum_{i=1}^{N} \Phi h_i(t)\delta(\mathbf{r} - \mathbf{r}_i(t)), \quad (2.25)$$

where $\mathbf{r}_i = (x_i, y_i, 0)$ is the location vector of the ant $i = 1, 2, \ldots, N$ and Φ is the pheromone release rate for a homing ant.

Numerical simulations of this automata model have been performed [45]. In these simulations, a discretized version of the diffusion equation (2.22) has been used.

A group of individuals obeying these rules shows the following behavior (Fig. 2.17). After the individuals are placed at random on the foraging ground they start to move randomly. When an individual finds the food source (indicated by the right bright dot in Fig. 2.17) it returns to the nest (the left bright dot), laying a chemical trail. At the nest the individual stops laying a trail and starts to move according to the signal perceived by its antennae. In most cases an individual will follow its own trail upon arriving at the nest. When the individual reaches the food source again, it returns to the nest, laying a chemical trail, and so on. Other individuals will be attracted by the trail, and randomly choose a direction; either they move towards the food source, where they will start "homing" and lay a chemical trail, or they will move directly towards the nest without laying a trail. After a while the majority of individuals will travel back and forth between the food source and the nest. After 100s a trail is established that is used by most individuals.

The snapshots in Fig. 2.17 are taken (from top left to bottom right) at time steps 0 s, 20 s, 40 s, 55 s, 70 s, and 100 s. The actual model, used in this simulation, was slightly more complex than the one that has been described above. It was assumed that the pheromone was partly emitted directly into

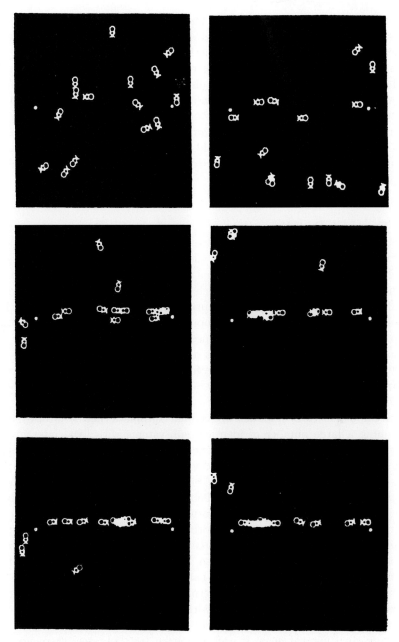

Fig. 2.17. Subsequent snapshots (from *top left* to *bottom right*) showing the gradual formation of a chemical trail. (From [45])

air and partly laid on the ground. The ground pheromone evaporated at a certain rate into the air, where its diffusion took place. The parameter values used in the simulation are given in [45].

As long as there is food, individuals reinforce the chemical trail. When the food source is removed, the trail will disappear and the group of individuals will switch again from the exploitation mode to the search mode, where they walk about randomly to find a new food source. Those individuals that did not reach the trail before the food source was removed are not wasted for the exploitation process, as they could find a second food source and start exploitation there. The model can be tuned to reproduce different details of various recruitment strategies.

2.5 Further Reading

The examples given in this chapter show the usefulness and the power of viewing biological and social organisms as automata: an organism is affected by the outside world (inputs) and it behaves (output) according to its current condition (state) and nature (program). The concept of automata has had a considerable influence on the development of modern computers, robotics, neuroscience and psychology. We will use the concept of automata extensively throughout the book and will discuss different types of behavioral response and the collective behavior resulting from interactions between automata. Below, we provide a short overview of research involving the concept of automata.

An important class of models are cellular automata that describe systems containing a large number of identical subsystems with local interactions. They represent lattices with each lattice site having a finite set of possible states. The states evolve synchronously in discrete time steps according to rules that are identical for each site. Since the pioneering work by Wiener and von Neumann in the 1940s and 1950s, cellular automata (CA) have been widely used to model complex natural phenomena.

In one-dimensional CA the neighborhood of cell i is a line of cells on either side of the site i. Two-dimensional CA can be viewed as checker board of cells. In contrast to the one-dimensional case, there are several neighborhoods that can be considered. In CA with the so-called von Neumann neighborhood the evolution of the state of cell i is determined by the states of four adjacent horizontal and vertical cells. In CA with the Moore neighborhood the evolution of the state of cell i depends on the states of all adjacent eight cells, including the diagonal ones. If each cell have $k = 2$ states, there are 2^{32}, or roughly 4×10^9, different rules for the von Neumann neighborhood and 2^{512}, or roughly 10^{154}, possible rules for the Moore neighborhood. Although it is impossible to test the evolution of a starting configuration for all possible rules, empirical investigations indicate that, roughly speaking, there are four different classes of long-term behavior of one- and two-dimensional CA.

Wolfram has shown that the CA behavior mostly (perhaps always) falls in one of the four classes [68]. They are analogous to the limit points, limit cycles and chaotic behavior found in systems based on differential equations. The fourth class shows complex evolution of spatial patterns retaining the memory of initial conditions.

Originally, von Neumann used CA to investigate the abstract properties of the biological self-reproduction process. If biological self-reproduction is carried out by a biochemical process (i.e. by a machine), then the behavior of this machine should be describable as a logical sequence of steps, i.e. as an algorithm. Von Neumann constructed a self-reproducing "machine" using a cellular array with 29 states per cell [72, 73]. Later, Codd and Langton showed that CA of a simpler structure and with fewer rules are also capable of self-reproduction [67, 69].

An interesting CA is the *Game of Life* formulated by Conway [70]. It consists of a square lattice in which each cell has two states, so that the cell can be either "alive" or "dead". The interactions between neighboring cells are defined as follows. A cell with more than three living neighbors will die at the next time step from overcrowding. A cell with less than two living neighbors will die from isolation. A dead cell that is surrounded by exactly three living neighbors will give birth, i.e. it will be a living cell in the next time step. The Moore neighborhood is here chosen. The Game of Life has become popular because, despite its simple rules, it generates a large number of spatiotemporal patterns, including oscillations and translational motion [71].

Cellular automata have found extensive applications in modeling biological processes. Indeed, the first models of excitable media were based on discrete CA and described abstract excitable media [59], aggregation in *Dictyostelium* [60], artrial fibrillation [61], heart flutter [62] and cardiac arhytmias [63]. Subsequently, Tyson et al. improved such models by considering the rules with larger neighborhoods that give rise to more smooth propagating fronts (e.g. see [65, 66, 74]). Another approach to this problem was proposed by Markus and Hess, who considered lattices with random locations of active sites inside the cells [64]. Other applications of CA to the modeling of biological phenomena include the problem of vertebrate skin pigmentation patterns [75] and DNA sequences [76]. More complex discrete automata using intricate collections of states and interaction geometries are due to Lindenmayer, who modeled the growth of plants [77], and to Kauffman, who modeled gene activity in an evolutionary context[15].

We have shown how collective behavioral patterns in ant societies result from chemical communication and an orientation mechanism, and how these are used to solve logistic problems. Besides recruitment, other collective activities are important for social insects in general and for ants in particular. Social insects build complex nest structures that do not only provide shelter but also serve for heat and moist regulation, keeping food stocks and other functions (for an overview, see [36, 55–57]). What are the rules governing the insects' behavior in building a nest? Deneubourg et al. described a simple automata model to elucidate this question [39]. Their automata move randomly on a square grid and start digging with a certain probability. An individual automaton, that extracts a soil particle, marks the site with a pheromone and the probability of digging near the site increases. As a consequence, the number of digging individuals grows. It is assumed in the model that the growth slows down and the local insect population reaches saturation when a certain pheromone level is achieved.

In a similar way, Deneubourg and co-workers could explain the sorting behavior [40]. The motivation was to understand the well-ordered interior of an ant nest. Eggs are arranged in a pile next to a pile of larvae and a further pile of cocoons. Alternatively, there are cases in which the three different types of piles can be found in entirely different parts of the nest [37]. If one tips the contents of a nest onto a surface, the workers will rapidly gather the brood into a place of shelter and sort it into different piles as before. The model is based on the following algorithm. When the randomly moving automata come across an object the probability of picking it up is the larger, the more the object is isolated, i.e. the less the number of similar objects there are in the immediate neighborhood. When carrying the object, the automata's probability of putting it down is all the greater as there are more of the same in the immediate neighborhood. The authors could show that these rules are sufficient to form separate clusters of two object types (see also [18]).

Automata can generate complex behavior from simple rules. Braitenberg has shown how different structures of circuitry and behavioral responses operating on simple and elementary environmental information can lead to very complex behaviors that even make allusions to psychological phenomena such as fear, timidity, aggression, etc. [46]. Brooks built sophisticated robots using a decentralized mindset approach [47]. The investigations of complex collective behavior in fish schools and insect societies have stimulated general research on populations of cooperative active agents, such as robots. Today, the study of distributed autonomous robotic systems (DARS) has become a rapidly developing field of engineering (see the conference proceedings [48, 49]). See also [16].

3. Active Motion

Many biological swarms represent collectively traveling populations. Each population member can move itself in the coordinate space by generating the needed forces. The coherent collective motion results here from physical interactions and exchange of information between the traveling individuals. Bacterial flows, fish schools, bird flocks and animal herds are examples of such behavior. However, essentially the same kind of behavior is found, for instance, in the traffic flows of vehicles on motorways. On the other hand, similar properties may be observed in populations of small physical particles or localized reaction–diffusion patterns. The mechanisms of active motion in bacteria, animals or cars are complicated and involve complex internal machinery. However, it would be wrong to assume that such high complexity is generally needed to produce the motion. Active motion of particles may spontaneously arise in simple nonequilibrium physico-chemical systems.

At low population densities, interactions between traveling individuals are rare and can be described as collisions separated by long intervals of free motion. In this case, the population can be viewed as a gas formed by self-moving particles. When the density is high, it represents an active fluid whose motion is described by special hydrodynamical equations. Active fluids are systems far from thermal equilibrium and, therefore, various forms of instabilities and pattern formation are possible here. An interesting and practically important example of active motion is provided by the traffic flow on motorways. Though the individual particles – cars with their drivers – are quite complex in this case, their flows can still be described in terms of the hydrodynamics of an active fluid. This theory explains the origin of traffic jams and their properties.

3.1 Elementary Mechanisms of Self-Motion

A car moves because it has wheels and a combustion engine. A large animal moves because it has muscles. When we go to microorganisms, such as bacteria, the question of what is the motor and how can it operate within a single biological cell is nontrivial. The best-studied example of a swimming microorganism is the tiny bacterium *Escherichia coli*, which is only 2 μm long. This bacterium has a tail, called a flagellum, which is only about 150 Å in diameter. As suggested by Berg [80] and proved in subsequent experiments [81, 82], these bacteria swim by *rotating* their flagella. Thus, their propulsion mechanism already comprises a motor, a rotary joint and a propeller.

We see that even at submicrometer scales it is possible to have molecular machines that operate similar to mechanical motors. But such complex machinery is not always necessary to produce self-motion and very elementary mechanisms may already bring it about. Below we consider two systems where self-motion emerges as a result of simple physico-chemical processes. In our examples the motion of small particles involves surface forces sensitive to chemical concentrations and temperature.

At a contact line between a liquid and a solid, capillary forces act. Their magnitude is controlled by a coefficient which can be modified by covering ('coating') the surface with special chemical substances, called surfactants. This coefficient depends on the surfactant concentration and temperature. In 1978 Greenspan has noted that, if a surface is nonuniformly coated with the surfactant, this must lead to a slow creeping motion of a droplet over the surface [83]. Because of the gradient in the chemical concentration of the surfactant, the capillary forces acting on both sides of the droplet do not balance each other and this leads to the appearance of a net force causing the translational motion. A similar effect is expected if the solid surface is uniform, but the droplet contains surfactant and its concentration is not uniform inside the droplet.

Liquid drops on a hot solid surface are often seen to perform rapid irregular skating motions before they finally dry out. This is explained [84] by taking into account that the drops usually contain as impurities some chemical substances that influence their capillary properties. When the surface is nonuniformly heated, evaporation proceeds at different rates in different parts of the droplet and, as a result, the surfactant concentration is not constant across the surface.

It is also known that a solid particle inside a liquid with a chemical gradient can move along the direction of this gradient [85, 86]. This effect takes place because the dissolved chemical substance influences the local surface tension coefficient and thus changes the intensity of the surface forces applied at the liquid–solid interface. A similar effect is observed in the presence of a temperature gradient, since the surface tension coefficient has a strong temperature dependence. In the experiments of Young et al. [87] small air bubbles were seen to slowly drift inside the oil towards a hotter region in absence of any convective fluid flows.

Though in these examples the motion of particles is only due to external gradients, they indicate how elementary self-moving objects can be designed: To induce its motion, such an object should create and maintain a gradient of some surface-active substance in the neighboring area of the medium.

A nice toy, familiar to many parents, actually realizes this idea. You make a small paper boat, attach a piece of soap to its bottom near the stern and put it onto water. You see then that the boat starts to move over the water surface and does so until all the soap is dissolved. The explanation of its motion is simple. The soap is a surfactant decreasing the surface tension of water. Since the piece is attached near the stern of the boat, the concentration of the dissolved soap is higher at the end of the boat and lower at its front. Because of this gradient, the capillary forces acting at the interface between the water and the boat are not balanced and induce the motion.

Further exploring this idea [88], let us consider small light particles floating in a deep liquid layer and generating a chemical substance that diffuses through the liquid and serves as a surfactant. Suppose that each particle represents a solid disk of radius R_0. The surfactant is continuously produced at rate J per unit time inside the particle and released into the liquid through a small hole located at the center of the disk bottom. This surfactant diffuses through the liquid with the diffusion constant D and modifies its surface tension. The capillary coefficient Γ, which gives the magnitude of the capillary force acting per unit length of the solid–liquid contact line, depends as $\Gamma = \Gamma_0 - \mu c$ on the local surfactant concentration c. To prevent accumulation of the surfactant, we assume that it decays at a constant rate γ.

When the disk is at rest, the surfactant distribution around it is symmetrical and the capillary forces are balanced (Fig. 3.1a). However, if the disk moves, the distribution is no longer symmetrical, i.e. the concentration c is lower in the front region and higher in the back regions (Fig. 3.1b). To calculate the capillary force acting on the particle, we need to know the surfactant distribution created by the moving source.

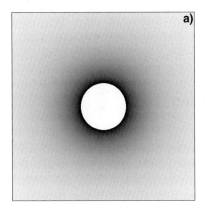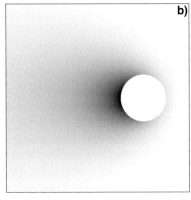

Fig. 3.1. Concentration distributions around stationary (a, $V = 0$) and moving (b, $V = 4$) disks of radius $R_0 = 0.5$; other parameters are $D = 1$ and $\gamma = 0.1$

The concentration distribution created by a moving point-like source is described by the diffusion equation

$$\dot{c} = -\gamma c + D\nabla^2 c + J\delta\left(\mathbf{r} - \mathbf{R}(t)\right), \tag{3.1}$$

where the vector $\mathbf{R}(t)$ specifies the source position at time t. If the liquid occupies half of the space $-\infty > z > 0$ and the source is moving in the horizontal boundary plane $z = 0$, the general solution of (3.1) is given by

$$c(\mathbf{r}, t) = 2J \int_{-\infty}^{t} G\left(\mathbf{r} - \mathbf{R}(t'), t - t'\right) dt', \qquad (3.2)$$

where

$$G(r, \tau) = \frac{1}{(4\pi D\tau)^{3/2}} \exp\left(-\frac{r^2}{4D\tau} - \gamma\tau\right) \qquad (3.3)$$

is the Green's function for the considered diffusion process in the infinite medium.

If the source is moving at a constant velocity V along the x-direction in the surface plane, its coordinates are $\mathbf{R}(t) = (Vt, 0, 0)$. Substituting this into (3.2) and integrating, we obtain the concentration distribution:

$$c(\mathbf{r}, t) = \frac{J}{2\pi D\rho} \exp\left[-\frac{V}{2D}(x - Vt)\right] \exp\left(-\frac{\rho}{2D}\sqrt{V^2 + 4\gamma D}\right), \qquad (3.4)$$

where $\rho = \left[(x - Vt)^2 + y^2 + z^2\right]^{1/2}$ is the distance from the moving source. This distribution is asymmetric. Comparing concentrations at equal distances ρ before and behind the moving source, we see that for $x > Vt$ the first exponent in (3.4) is larger and therefore the surfactant concentration in the front region is higher than in the back region of the moving source.

Capillary forces act at the contact line between the solid and the liquid. The force dF that acts on a contact line element dl is given by $dF = \Gamma dl$. It is directed along the local normal vector $\mathbf{n}$ to the contact line. When a disk-shaped particle floats in liquid, the contact line is just the circular boundary of the disk. The net capillary force, applied to this particle, is calculated by summing the forces acting at all small elements of the boundary, i.e.

$$\mathbf{F}_c = \int \Gamma \mathbf{n} dl . \qquad (3.5)$$

Since the capillary coefficient depends as $\Gamma = \Gamma_0 - \mu c$ on the local surfactant concentration and this concentration varies along the disk boundary, the integral (3.5) does not vanish. The resulting net capillary force acts in the same direction as the velocity $\mathbf{V}$ of the moving disk. Indeed, the surfactant concentration c is *higher* in the back region of the disk boundary and therefore the capillary forces, applied to this part and acting backwards, are smaller than the respective forces, applied to the frontal part of the disk boundary and acting in the forward direction.

The total capillary force acting on the disk is given by the integral

$$F_c = -\mu R_0 \int_0^{2\pi} c(\varphi) \cos \varphi d\varphi, \qquad (3.6)$$

where $c(\varphi)$ is the surfactant concentration on the boundary of radius R_0 at an angle φ with respect to the direction of motion. According to (3.4), it is

$$c(\varphi) = \frac{J}{2\pi D R_0} \exp\left[-\frac{V R_0}{2D} \cos\varphi\right] \exp\left(-\frac{\rho}{2D}\sqrt{V^2 + 4\gamma D}\right) . \tag{3.7}$$

This integral can be approximately calculated when the motion is so slow that the condition $V R_0/D \ll 1$ is satisfied. In this case the capillary force is

$$F_c = \frac{\mu J V R_0}{4D^2}\left[1 + \frac{1}{8}\left(\frac{V R_0}{2D}\right)^2\right] \exp\left(-\frac{R_0}{2D}\sqrt{V^2 + 4\gamma D}\right). \tag{3.8}$$

Since we are interested only in slow motions, we can expand (3.8) in powers of the velocity V, keeping terms up to the third order. Moreover, we assume that the disk radius R_0 is much smaller than the characteristic decay length of the surfactant $L_{\text{dif}} = \sqrt{D/\gamma}$ (otherwise its concentration at the disk boundary would have been already extremely low). As a result, the following expression for the total capillary force acting on the disk is obtained:

$$\mathbf{F}_c = \frac{\mu J R_0}{4D^2}\left\{1 - \frac{L_{\text{dif}}}{8R_0}\left(\frac{V R_0}{D}\right)^2\right\}\mathbf{V}. \tag{3.9}$$

Hence, for small velocities the capillary force acts in the direction of the velocity vector $\mathbf{V}$ and can *induce* the motion.

So far we have neglected the hydrodynamic flows that accompany the motion of our floating particle. These flows would actually somewhat distort the surfactant distribution around the particle, requiring corrections to (3.9). Since our analysis is only intended to provide an illustration, we do not explicitly consider such corrections, which would be small for sufficiently high viscosity of the liquid. There is, however, another hydrodynamic effect which is very important.

When a solid body moves through a fluid, it experiences the force of viscous friction. For slow laminar flows, characterized by small Reynolds numbers, this force is proportional to the body velocity and its magnitude can be estimated, using simple scaling considerations (see [89]). For a floating disk of radius R_0 that moves at velocity V in a liquid with viscosity η, the viscous force $\mathbf{F}_{\text{visc}}$ is approximately given by

$$\mathbf{F}_{\text{visc}} = -\zeta \eta R_0 \mathbf{V}, \tag{3.10}$$

where ζ is a numerical proportionality factor. Note that, in contrast to the capillary force, the force of viscous friction is directed in the opposite direction to the velocity vector $\mathbf{V}$ and thus tends to slow down the motion.

The laws of mechanics say that the acceleration of a physical particle is determined by the total force acting on it, i.e.

$$m\frac{d\mathbf{V}}{dt} = \mathbf{F}_c + \mathbf{F}_{visc}, \tag{3.11}$$

where m is the mass of this particle. Substituting into (3.11) the expressions (3.9) and (3.10) for the respective forces, we obtain a dynamical equation of the form

$$\frac{d\mathbf{V}}{dt} = \alpha\mathbf{V} - \beta V^2\mathbf{V}, \tag{3.12}$$

where the coefficients are

$$\alpha = \frac{R_0}{m}\left(\frac{\mu J}{4D^2} - \zeta\eta\right), \tag{3.13}$$

and

$$\beta = \frac{\mu J R_0^2 L_{dif}}{32mD^4}. \tag{3.14}$$

According to (3.12), the particle slows its motion and comes to rest, if the coefficient α is negative. When this coefficient is positive, the rest state is unstable and the particle accelerates until it achieves the velocity $V_0 = \sqrt{\alpha/\beta}$. If we treat α as a control parameter, it can be thus noted that (3.12) describes the onset of self-motion in our nonequilibrium system.

Using (3.13), we find that the self-motion begins when $\alpha = 0$ or, explicitly, when the rate J of the surfactant release exceeds the threshold

$$J_0 = \frac{4\zeta\eta D^2}{\mu}. \tag{3.15}$$

Remarkably, this threshold does not depend on the disk radius R_0. It is proportional to the viscosity η of the liquid and to the diffusion constant D of the surfactant.

By rescaling time and velocity as $t \to \alpha t$ and $\mathbf{V} \to \sqrt{\beta/\alpha}\mathbf{V}$, (3.12) can be written as

$$\frac{d\mathbf{V}}{dt} = \mathbf{V} - V^2\mathbf{V}. \tag{3.16}$$

Though we have derived (3.16) for a particular nonequilibrium physico–chemical system, it is general. This equation describes the "soft" onset of self-motion, when the rest state of a particle becomes unstable with respect to small perturbations and the particle begins to slowly move through the medium. Such an instability can be interpreted as a supercritical bifurcation of the fixed point $\mathbf{V} = 0$, corresponding to the rest state. Equation (3.16) represents the so-called normal form for this bifurcation. Note that the system is isotropic and the direction of motion above the instability threshold is arbitrary. This means that the bifurcation is accompanied by spontaneous symmetry breaking, similar to the respective behavior near second-order phase transitions (cf. [90]).

Not only physical objects may start to move far from equilibrium. When reaction–diffusion systems are considered, stable localized concentration patterns can appear. These patterns, called *spots*, are usually stationary. However, under certain conditions the stationary spots become unstable and begin to move. The bifurcation to traveling spots in such systems was first described by Krischer and Mikhailov [91]. The normal form (3.16) for this bifurcation in three-component reaction-diffusion systems was later derived [92].

Though the dynamical equation (3.16) is strictly valid only when the velocities are relatively small, it provides a satisfactory qualitative description even far from the bifurcation point and can therefore be used as a good starting point in the general analysis of self-motion phenomena.

3.2 Self-Motion in External Fields

Various external fields can act on self-propagating objects. Gravitational forces act on all physical particles. If a particle is charged, it feels the presence of electrostatic fields in the medium. The forces may also have a capillary origin. In the above-considered example of a floating self-moving particle, externally created gradients of the surfactant concentration would produce additional forces acting on the particle.

In the presence of an external force field $\mathbf{F}(\mathbf{r})$, the dynamical equations of a self-moving particle are

$$\begin{aligned}
\dot{\mathbf{V}} &= \mathbf{V} - V^2\mathbf{V} + \mathbf{F}(\mathbf{r}) \\
\dot{\mathbf{r}} &= \mathbf{V}
\end{aligned}. \tag{3.17}$$

We first consider two-dimensional self-motion in weak external slowly varying fields. Introducing the angle ψ that specifies the direction of motion, we write $V_x = V \cos\psi$ and $V_y = V \sin\psi$. Similarly, we have $F_x = f \cos\Psi$ and $F_y = f \sin\Psi$, where the angle Ψ indicates the local force direction. Using (3.17), we obtain

$$\begin{aligned}
\dot{V} &= V(1 - V^2) + f(\mathbf{r})\cos(\psi - \Psi(\mathbf{r})) \\
V\dot{\psi} &= -f(\mathbf{r})\sin(\psi - \Psi(\mathbf{r}))
\end{aligned}. \tag{3.18}$$

Note that the coordinates $\mathbf{r}$ of the particle in these equations also change with time according to the equation $\dot{\mathbf{r}} = \mathbf{V}$.

If the force is weak ($f \ll 1$), the velocity modulus V is only slightly different from 1, i.e. we can write $V = 1 + \delta V$ where $\delta V \ll 1$. Keeping only the terms of the first order in f, we obtain

$$\begin{aligned}
\dot{\delta V} &= -2\delta V + f(\mathbf{r})\cos(\psi - \Psi(\mathbf{r})) \\
\dot{\psi} &= -f(\mathbf{r})\sin(\psi - \Psi(\mathbf{r}))
\end{aligned}. \tag{3.19}$$

The second of these equations does not explicitly depend on the variable δV. If $f(\mathbf{r})$ and $\Psi(\mathbf{r})$ were constant in this equation, its solution would simply be that the velocity angle ψ becomes equal to the force angle Ψ within time interval $\Delta t \approx 1/f$. During this interval the particle will pass the distance $\Delta L = V \Delta t \approx 1/f$.

If the force field is only slowly varying in space, so that it remains approximately constant over distances ΔL, the particle is able to adiabatically adjust its direction of motion to the local force direction. This means that in this case we have $\psi(t) \approx \Psi(\mathbf{r}(t))$. Moreover, the first of equations (3.19) then yields $V \approx 1 + f(\mathbf{r})/2$.

If we define the force field as $\mathbf{F}(\mathbf{r}) = f(\mathbf{r})\ \mathbf{n}(\mathbf{r})$, where $\mathbf{n}(\mathbf{r})$ is the unit vector specifying its local direction at point $\mathbf{r}$, the above arguments imply that the solution for the velocity of the self-moving particle would be

$$\mathbf{V} \approx [1 + f(\mathbf{r})/2]\ \mathbf{n}(\mathbf{r}), \qquad (3.20)$$

and its trajectory would be determined in the lowest order in f by the equation

$$\dot{\mathbf{r}} = \mathbf{n}(\mathbf{r}). \qquad (3.21)$$

This means that the particle moves at a constant unit velocity by exactly following the force lines. Note that the force here is needed not to produce the motion, but only to direct it. We see that weak slowly varying force fields may be used to efficiently *guide* self-moving particles! The necessary conditions for this behavior are that the force is weak ($f \ll 1$) and the characteristic length scale λ of the field is large ($\lambda \gg 1/f$).

When the force field is strong and/or rapidly varying in space, new effects become possible. Note that, if potential forces are considered, the fields can be expressed as gradients of certain potentials, i.e. one can write $\mathbf{F}(\mathbf{r}) = -\nabla U(\mathbf{r})$. Later in this section we analyze the behavior of self-moving particles in an axially symmetric attractive force field corresponding to a potential well. As a typical example, we consider a Gaussian well $U(r) = -a \exp\left(-r^2/2\rho^2\right)$ of radius ρ and depth a that yields the force field $\mathbf{F}(\mathbf{r}) = -f(r)\mathbf{e}_r$, where $\mathbf{e}_r$ is the unit radial vector and

$$f(r) = a\frac{r}{\rho^2} \exp\left(-\frac{r^2}{2\rho^2}\right). \qquad (3.22)$$

In classical mechanics, the motion of particles in potential fields is accompanied by the conservation of their energy $E = \frac{1}{2}V^2 + U(r)$. Depending on the energy, the particle would oscillate if $E < 0$ in such a potential well, or move along an infinite unbounded trajectory if $E > 0$. The amplitude of oscillations is then controlled by the energy of the particle. On the other hand, if viscous friction and thus energy dissipation are present, the particle slows down its motion until it is found resting at the bottom of the potential well.

The energy of a self-moving particle is not conserved, but also it is not just dissipated. Indeed, using (3.17) we find that

$$\frac{dE}{dt} = V^2(1 - V^2). \tag{3.23}$$

Therefore, the energy grows, if the particle's velocity V is less than 1, and decreases when this velocity exceeds 1. The energy remains constant only if $V = 1$ (though the state with $V = 0$ also corresponds to a fixed energy, it is absolutely unstable for self-moving particles).

As taught in any standard physics school course, under the influence of an attractive radial force f classical particle may rotate with constant velocity V along a circular trajectory of radius r. This radius is determined by the condition that the centripetal acceleration is equal to the acting force (remember that we consider particles of unit mass), i.e. $V^2/r = f$. When self-moving particles are considered, such steady rotation is possible only if $V = 1$. Thus, we obtain an equation that determines the rotation radius:

$$\frac{1}{r} = f(r). \tag{3.24}$$

We see that, in contrast to classical mechanical motion, the rotation radius is not arbitrary, but fixed by the potential field.

The solutions of (3.24) can be found by plotting the functions $f(r)$ and $1/r$. The roots of this equation are then given by the intersections of these two graphs. Figure 3.2 shows this construction for the Gaussian potential force field (3.22) at two different values of the parameter a.

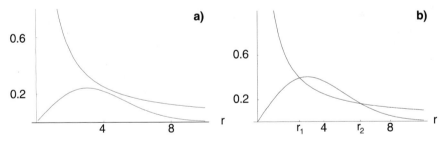

Fig. 3.2. Graphical solution of (3.24) for the force field given by (3.22). **(a)** $a = 1.2$ and **(b)** $a = 2.0$, for $\rho = 3$

When the force is relatively weak (Fig. 3.2a), the graphs do not intersect and thus (3.24) has no solutions. This means that steady rotation of a self-moving particle in such a field is not possible. When the force intensity is increased (Fig. 3.2b), the intersections appear and we find *two* roots of equation (3.24), yielding two different rotation radii. However, only one of these roots corresponds to stable rotation. This can be understood using simple arguments.

Let us consider the motion corresponding to the larger radius r_2 in Fig. 3.2b. If we slightly perturb this motion, by increasing the rotation radius, we see from Fig. 3.2b that the centrifugal force $1/r$ would become larger than the attractive force $f(r)$ and the particle would tend to move further away from the center. On the other hand, if the rotation radius is slightly decreased below r_2, the centrifugal force is weaker than the attractive force and the particle moves closer to the center. Hence, this solution does not correspond to stable rotation.

If we apply the same arguments to the intersection point at radius r_1 in Fig. 3.2b, we see that the situation is then the opposite. On the right-hand side of this point the attractive force is greater than the centrifugal force and tends to return the particle. On its left-hand side, the centrifugal force is larger and brings the particle back to the intersection point. This suggests that such a point may correspond to stable steady rotation.

The suggestion is confirmed by the numerical integration of the dynamical equations (3.17) with the force field (3.22). In our simulations we start with a particle that is initially located outside of the potential well (i.e. at distance $r \gg \rho$) and moves at some angle towards its center. The outcome of the collision depends on the *impact parameter d*, defined as the minimal distance at which the particle would have passed the center if the attractive potential were absent.

When the impact parameter is large, the particle goes through the periphery of the attractive potential and is only slightly deflected by it (Fig. 3.3a). Decreasing the impact parameter, we see that the trajectory becomes more strongly influenced by the attractive potential, but remains infinite (Fig. 3.3b). When a certain critical value of the impact parameter is reached, the particle is captured by the attractive center (Fig. 3.3c). It enters the potential well and, after a few internal reflections, begins to orbit along a circular trajectory inside it. The radius of this stable rotation coincides with r_1 given by the graphical solution of equation (3.24). The capture phenomenon occurs for a range of the impact parameters d and always results in orbiting with the same radius. If, however, the impact parameter is very small (i.e. the particle heads almost directly into the center), it may again pass uncaptured through the potential well (Fig. 3.3d). Note that the roots of (3.22) determine two circular orbits that represent *limit cycles* of the dynamical equations (3.17). The smaller limit cycle is attractive and corresponds to stable orbiting of the captured particle. The larger limit cycle is a repeller of this dynamical system.

This special scattering behavior was first described by Bode [93], who considered the motion of spots in three-component reaction–diffusion models in the presence of axially symmetric inhomogeneities. The equations of self-motion were a little different in his case, i.e. an additional term in the second equation of the system (3.17) was present. However, this difference did not

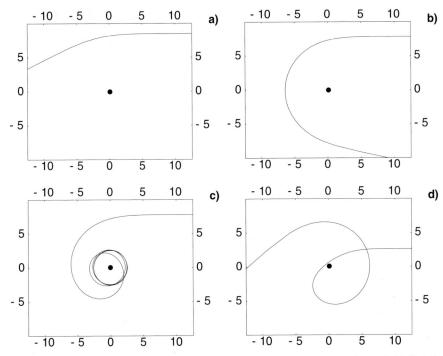

Fig. 3.3. Scattering of an actively moving particle by a central force field (3.22) with $a = 2$ and $\rho = 3$ for four different values of the impact parameter: (**a**) $d = 8.5$, (**b**) $d = 7.84$, (**c**) $d = 7.8$ (stable orbit) and (**d**) $d = 2.62$

play a principal role and the obtained behavior was essentially the same as illustrated in Fig. 3.3.

When the attractive potential goes to infinity at the central point, self-moving particles may fall into the center. If we choose the force field in the form $f(r) = -c/r$ and take $c > 1$, (3.24) has no solutions: the attractive force remains larger than the centrifugal force at all distances from the center. In this case the motion ends in the center of the attractive potential (Fig. 3.4).

Collisions with a repulsive center result in the scattering of self-moving particles, which is generally similar to the respective phenomena in classical mechanics. The difference is that, approaching a repulsive center, the self-moving particle changes its direction of motion without any significant slowing down.

3.3 Hydrodynamics of Active Fluids

Interactions between self-moving particles are possible. The floating particles, considered in Sect. 3.1, produce a distribution of the surfactant concentration around them which influences the motion of other particles entering this

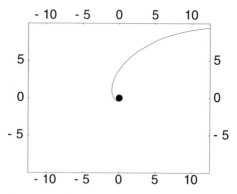

Fig. 3.4. Capture of an actively moving particle by a force center; $f = -c/r$ with $c = 1.05$

region. Simple analysis shows that the resulting interaction should in this case be repulsive. When biological self-moving organisms are discussed, the interactions between them would generally be a consequence of the communication and active response of such "particles" to the received information. The effective interactions may be both attractive or repulsive, depending on the distance between the particles. For instance, when the motion of fish was considered in Chap. 2, we assumed that the fish repel each other when the distance between them is short, and that they attract each other at intermediate separations.

In the simplest model, the collective behavior of an interacting population of N self-moving particles is described by the dynamical equations

$$\dot{\mathbf{V}}_i = \mathbf{V}_i - V_i^2 \mathbf{V}_i - \sum_{j \neq i}^{N} \nabla_i U \left(\mathbf{r}_i - \mathbf{r}_j \right) + \mathbf{f}_i(t),$$

$$\dot{\mathbf{r}}_i = \mathbf{V}_i, \tag{3.25}$$

where $U(r)$ is the potential for binary interactions between the particles and $\mathbf{f}_i(t)$ are random forces acting on the particles because of fluctuations in the medium.

When interactions are repulsive, the particles tend to avoid each other while maintaining their motion. Attractive interactions lead to clustering of self-moving particles (to prevent their indefinite collapse, it should be assumed that the interactions become repulsive at short distances). Eventually, this may lead to the formation of a dense cloud of self-moving particles.

This effect is a direct analog of the condensation phenomena in equilibrium physical systems. When the temperature is decreased, a non-ideal gas of molecules undergoes a phase transition and forms droplets of a liquid. The difference is only that the particles are self-moving and therefore the appearing "drops" travel through the medium even in the absence of any external forces.

The role of temperature is played in (3.25) by the intensity of the random forces. When the attractive interactions are weak, they cannot overcome the stirring effect of the random forces. Increasing the strength of the interactions leads to a phase transition that can be investigated using essentially the same methods that were developed for physical systems in thermal equilibrium. We do not discuss the properties of the phase transition that result in coherent collective motion, and focus below our attention on the theoretical analysis of the condensed self-moving phase.

Our task is to find the equations that would describe the flows of an *active fluid* formed by interacting self-moving particles. Instead of indicating the positions and velocities of each of the particles, we introduce two fields $n(\mathbf{r}, t)$ and $\mathbf{v}(\mathbf{r}, t)$ that specify the local density of particles and their mean velocity at point $\mathbf{r}$ at time t.

Since the particles do not die and are not spontaneously produced, their local density $n(\mathbf{r}, t)$ must obey the conservation equation

$$\frac{\partial n}{\partial t} + \operatorname{div}(n\mathbf{v}) = 0 . \tag{3.26}$$

To construct the evolution equation for the velocity field $\mathbf{v}(\mathbf{r}, t)$, let us consider the motion of a small fluid element $\Delta\Omega$ containing ΔN particles. Suppose that at time t this element is located at point $\mathbf{r}$ and has velocity $\mathbf{V}(t) = \mathbf{v}(\mathbf{r}, t)$. In order to determine the velocity $\mathbf{V}(t + \Delta t)$ of this element after a short time interval Δt we note that, at this later time, the element will be located at a different point $\mathbf{r}' = \mathbf{r} + \Delta\mathbf{r}$, where $\Delta\mathbf{r} = \mathbf{V}(t)\Delta t$. Therefore, its new velocity $\mathbf{V}(t + \Delta t)$ is approximately

$$\mathbf{V}(t + \Delta t) \approx \mathbf{V}(t) + \Delta t \frac{\partial \mathbf{v}}{\partial t} + \Delta\mathbf{r} \cdot \frac{\partial \mathbf{v}}{\partial \mathbf{r}}. \tag{3.27}$$

Hence, the acceleration

$$\frac{d\mathbf{V}}{dt} \approx \frac{1}{\Delta t} \left(\mathbf{V}(t + \Delta t) - \mathbf{V}(t) \right) \tag{3.28}$$

of this fluid element is

$$\frac{d\mathbf{V}}{dt} = \frac{\partial \mathbf{v}}{\partial t} + \left(\mathbf{v} \cdot \frac{\partial}{\partial \mathbf{r}} \right) \mathbf{v} . \tag{3.29}$$

Thus, even if the velocity field is stationary ($\partial \mathbf{v}/\partial t = 0$), the velocity of a fluid element varies with time because it changes its location and comes to regions with different flow velocities.

The particles inside a fluid element are generally characterized by a distribution of individual velocities so that the flow velocity is only a local average of this distribution. However, if the attractive interactions between the self-moving particles are strong enough, they would tend to move with approximately equal velocities so that the velocity $\mathbf{V}$ of a fluid element would also represent a characteristic velocity of the individual particles inside this element. This means that, in the main approximation, the acceleration of the fluid element is again determined by (3.16).

Interactions between self-moving particles lead to the appearance of interactions between neighboring fluid elements. Since we have assumed that at small distances the particles repel each another, it is natural to expect that the particles would tend to leave the regions with high densities and move into more rarefied regions. This effect can be phenomenologically described by introducing a pressure p which is a function of the local particle density, i.e. $p = p(n)$.

Furthermore, interactions between particles would tend to eliminate nonuniformities in the flow velocity. This means that, provided all external fields are absent and the density is constant, small perturbations of the flow velocity should be damped. In other words, the considered active fluid must possess a certain viscosity.

Combining these effects, we arrive at the following evolution equation for the local flow velocity:

$$\frac{\partial \mathbf{v}}{\partial t} + (\mathbf{v} \cdot \nabla)\mathbf{v} = \mathbf{v} - v^2\mathbf{v} - \nabla p + \nu\nabla^2\mathbf{v}, \qquad (3.30)$$

where ν is the viscosity coefficient.

Equations (3.30) and (3.26) describe the hydrodynamic flow of active fluids. They differ from the classical hydrodynamic Navier–Stokes equations only in that they include two additional terms $\mathbf{v}$ and $-v^2\mathbf{v}$ in (3.30). These terms account for the self-motion of active fluids. It should be noted that, when such terms are introduced, the hydrodynamic equations are no longer invariant under the Galileo transformations, i.e. under a transition to any moving coordinate frame. The origin of this asymmetry is clear. Whenever self-motion is considered, the existence of an absolute rest frame is implied. Indeed, for the self-moving particles in Sect. 3.1 such a frame was provided by the thick liquid layer on top of which these particles were floating. For the fish schools discussed in Sect. 3.2 this was the water mass in the sea or a lake, etc.

The active fluid, obeying (3.30) and (3.26), tends to move at a constant velocity $v = 1$ with respect to the absolute rest frame. However, since the fluid is isotropic, the direction of this steady motion remains arbitrary. We start our analysis by investigating the behavior of long-wave perturbations of this steady state.

Let us choose the x-axis along the direction of motion, so that in the steady state $v_x = 1$ and $v_y = 0$ (we consider two-dimensional systems). The fluid density in this state is constant and equal to n_0. We add small perturbations $v_x = 1 + \delta v_x$, v_y, $n = n_0 + \delta n$ and linearize (3.30) and (3.26), obtaining:

$$\frac{\partial \delta n}{\partial t} + \frac{\partial \delta n}{\partial x} + n_0 \frac{\partial \delta v_x}{\partial x} + n_0 \frac{\partial v_y}{\partial y} = 0,$$

$$\frac{\partial \delta v_x}{\partial t} + \frac{\partial \delta v_x}{\partial x} = -2\delta v_x - a\frac{\partial \delta n}{\partial x} + \nu \frac{\partial^2 \delta v_x}{\partial x^2} + \nu \frac{\partial^2 \delta v_x}{\partial y^2}, \qquad (3.31)$$

$$\frac{\partial v_y}{\partial t} + \frac{\partial v_y}{\partial x} = -a\frac{\partial \delta n}{\partial y} + \nu \frac{\partial^2 v_y}{\partial x^2} + \nu \frac{\partial^2 v_y}{\partial y^2},$$

where the coefficient a is defined as $a = p'(n_0)$.

The solutions of this system of linear partial differential equations can be sought in the form

$$\delta n(x, y, t) = \delta n \exp\left[\lambda t + ik_x(x - t) + ik_y y\right],$$
$$\delta v_x(x, y, t) = \delta v_x \exp\left[\lambda t + ik_x(x - t) + ik_y y\right], \qquad (3.32)$$
$$v_y(x, y, t) = v_y \exp\left[\lambda t + ik_x(x - t) + ik_y y\right].$$

Substituting (3.32) into (3.31), we obtain a characteristic equation that determines the complex increment of growth λ of the mode with the wavenumbers k_x and k_y. Solving this cubic equation and keeping only the terms up to the second order in the wavenumbers, we get

$$\lambda_{1,2} = \pm i\sqrt{an_0}k \sin\psi - \frac{1}{2}\nu k^2 - \frac{1}{8}an_0 k^2 \cos^2\psi, \qquad (3.33)$$

$$\lambda_3 = -2 - \nu k^2 + \frac{1}{4}an_0 k^2 \cos^2\psi, \qquad (3.34)$$

where ψ is the angle between the flow direction and the wavevector $\mathbf{k}$, i.e. $k_x = k \cos\psi$ and $k_y = k \sin\psi$.

Thus, all perturbation modes of the uniform flow are damped and, therefore, in the considered active fluid this flow is always linearly stable. The first two modes represent *sound waves* whose damping rate is proportional to the square of the wavevector $\mathbf{k}$ and therefore becomes small in the long-wavelength limit. The third mode corresponding to the root λ_3 is strongly damped even for the uniform perturbations with $\mathbf{k} = 0$. This mode describes a velocity modulation in the direction of motion.

According to (3.33), the sound waves have frequency $\omega = \sqrt{an_0}k \sin\psi$ and speed $s = \omega/k = \sqrt{an_0} \sin\psi$. We see that this speed depends on the propagation direction, decreasing for the sound waves that propagate closer to the x-axis. The speed s vanishes and the respective modes become purely dissipative at $\psi = 0$. This behavior is different from the properties of sound waves in classical isotropic fluids, where the Galilean invariance prohibits any dependence of the sound on the angle between the wavevector and the flow direction.

Besides uniform flows, steady circular flows, or *vortices*, are possible in the considered active fluid. A vortex corresponds to a special stationary solution of (3.30) and (3.26). Introducing the polar coordinates (r, φ) with the origin of the coordinate system in the center of the vortex, we can write this solution in the form

$$\mathbf{v} = v(r)\mathbf{e}_\varphi, n = n(r), \tag{3.35}$$

where $\mathbf{e}_\varphi$ is the local axial unit vector, orthogonal to the radial unit vector $\mathbf{e}_r$.

Because the flow is axial, the conservation equation (3.30) is identically satisfied in this case. Substitution of (3.35) into (3.31) in polar coordinates yields

$$\frac{v}{r}\frac{\partial}{\partial\varphi}(v\mathbf{e}_\varphi) = (1 - v^2)v\mathbf{e}_\varphi - \frac{\partial p}{\partial r}\mathbf{e}_r$$

$$+\nu\left[\frac{1}{r}\frac{\partial}{\partial r}\left(r\frac{\partial}{\partial r}(v\mathbf{e}_\varphi)\right) + \frac{1}{r^2}\frac{\partial^2}{\partial\varphi^2}(v\mathbf{e}_\varphi)\right]. \tag{3.36}$$

We must take here into account that the local unit vectors $\mathbf{e}_r$ and $\mathbf{e}_\varphi$ in the polar coordinate system depend on the coordinates r and φ, so that

$$\frac{\partial}{\partial r}\mathbf{e}_r = 0, \frac{\partial}{\partial\varphi}\mathbf{e}_r = \mathbf{e}_\varphi$$

$$\frac{\partial}{\partial r}\mathbf{e}_\varphi = 0, \frac{\partial}{\partial\varphi}\mathbf{e}_\varphi = -\mathbf{e}_r \tag{3.37}$$

Using these identities to transform the derivatives in (3.36), we obtain

$$-\frac{v^2}{r}\mathbf{e}_r = (1 - v^2)v\mathbf{e}_\varphi - \frac{dp}{dr}\mathbf{e}_r$$

$$+\nu\left[\frac{1}{r}\frac{d}{dr}\left(r\frac{dv}{dr}\right) - \frac{v}{r^2}\right]\mathbf{e}_\varphi. \tag{3.38}$$

Since this must hold separately for the projections on both orthogonal unit vectors $\mathbf{e}_r$ and $\mathbf{e}_\varphi$, we have two equations:

$$(1 - v^2)v + \frac{\nu}{r}\frac{d}{dr}\left(r\frac{dv}{dr}\right) - \frac{\nu}{r^2}v = 0 \tag{3.39}$$

and

$$\frac{v^2}{r} = -\frac{dp}{dr}. \tag{3.40}$$

Remarkably, the first of these two equations does not include the fluid density n. Its solution can be found by introducing the new radial coordinate $\xi = \nu^{-1/2}r$ and writing this equation as

$$\frac{d^2v}{d\xi^2} + \frac{1}{\xi}\frac{dv}{d\xi} + \left(1 - v^2 - \frac{1}{\xi^2}\right)v = 0. \tag{3.41}$$

It must be solved with the boundary conditions $v = 0$ at $\xi = 0$ and $v \to 1$ at $\xi \to \infty$. The function $v(\xi)$, obtained by numerical integration of this equation, is displayed in Fig. 3.5. It is given approximately by $v(\xi) \approx A\xi$ with $A \approx 1.666...$ at $\xi \ll 1$ and behaves as $v(\xi) \approx 1 - (1/2)\,\xi^{-2}$ in the limit $\xi \to \infty$.

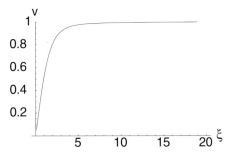

Fig. 3.5. Velocity profile in the rotating vortex

When the velocity field $v(r)$ is known, the local pressure p can be obtained by integrating (3.40):

$$p(r) = -\int_0^r v^2(r)\frac{dr}{r} + const.$$
(3.42)

Hence, the pressure is decreased in the center of the vortex, creating the centripetal force that maintains the circular flow. Using the above approximations for the velocity field near the center and in the far region, we find that

$$p(r) \approx -\frac{A^2 r^2}{2\nu} + const, \text{ for } r \ll \nu^{1/2},$$
(3.43)

$$p(r) \approx -\ln r + const, \text{ for } r \gg \nu^{1/2}.$$
(3.44)

Since pressure is a function $p = p(n)$ of the local density n, equations (3.42)–(3.44) determine the radial density distribution in the vortex.

When an active fluid flows inside a finite region, boundary conditions for the hydrodynamical equations (3.26) and (3.30) should be specified. Since such conditions may depend strongly on the detailed properties of the actual interactions between self-moving particles and a solid boundary, no general form of them can be formulated. An important exception is the case of free boundaries that we discuss below.

A population of self-moving particles condenses into a liquid because of the presence of attractive interactions between the particles. In a similar way to classical fluids, the considered active fluids can form *drops*. The circular shape of the drops is a consequence of the surface tension acting on the open surface. Microscopically, this surface tension results from the same attractive interactions that are responsible for the condensation. In hydrodynamical equations it is specified by a separate parameter Γ, known as the surface tension or capillary coefficient. According to the Laplace law, the surface tension produces a pressure p_s proportional to the local curvature of the surface, i.e.

$$p_s = \Gamma(R_1^{-1} + R_2^{-1}),$$
(3.45)

where R_1 and R_2 are the two principal curvature radii.

The boundary conditions at a free surface are that the velocity **v** is parallel to the boundary and that the pressure $p(n)$ at the boundary is equal to p_s. Note that the last of these conditions (i.e. $p(n) = p_s$) determines the fluid density n at the free surface.

Let us consider a drop formed by a two-dimensional active fluid. Since each element of this fluid tends to move, it is clear that there will be persistent flows inside such a drop. Moreover, at least if the drop is small enough, we can expect its shape to be circular and the flow inside to be steady and axial. But then this flow should be described by the special solution that we have constructed above, i.e. its local velocity is $v = v(\nu^{-1/2}r)$, where the function $v(\xi)$ is determined by (3.41) and the pressure distribution inside the drop is given by (3.42).

At the drop boundary $r = R$ the pressure is determined by the boundary condition as $p = \Gamma/R$ (we now denote as Γ the surface tension coefficient of the *self-moving fluid*, which should not be mixed with a similar notation used in Sect. 3.1 for the surface tension coefficient of a normal fluid). This boundary condition can be used to determine the integration constant in (3.42), and we thus obtain

$$p(r) = \frac{\Gamma}{R} - \int_r^R v^2(r)\frac{dr}{r}. \tag{3.46}$$

Since pressure is a function $p = p(n)$ of the local density, this equation yields the radial density distribution inside a rotating drop. For instance, if this function is linear, i.e. $p(n) = \epsilon n$, we get

$$n(r) = \frac{1}{\epsilon}p(r). \tag{3.47}$$

Integrating this over r from 0 to R, the total number N of particles inside the drop is found:

$$N = 2\pi \int_0^R n(r)r dr. \tag{3.48}$$

As an example, Fig. 3.6 shows the computed radial density distribution in a large rotating population. The density is greatly reduced in the central region of radius $\nu^{1/2}$ and then slowly increases towards the periphery. This means that the rotating population has a ring-like structure.

Inside the ring $R \geq r \gg \nu^{1/2}$, where the main part of the population is concentrated, the density is given by

$$n(r) \approx \frac{\Gamma}{\epsilon R} - \frac{1}{\epsilon}\ln\frac{R}{r}. \tag{3.49}$$

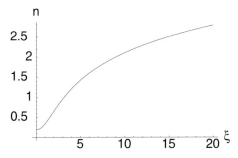

Fig. 3.6. Density distribution inside the rotating vortex; $\varepsilon = 1$ and $n(\xi = 0) = 0.2$

Neglecting the weak logarithmic dependence, the total number of particles can therefore be roughly estimated as $N \approx \pi \Gamma R/\epsilon$. This can be used to estimate the radius of a rotating population of size N as

$$R \approx \frac{\epsilon N}{\pi \Gamma}. \tag{3.50}$$

Hence, this radius is proportional to the number of particles and inversely proportional to the surface tension coefficient Γ that specifies the strength of cohesion inside the population.

There is a largest possible size of a rotating population. Indeed, as follows from (3.46), the pressure in the center of this structure is

$$p(0) = \frac{\Gamma}{R} - \int_0^R v^2(r) \frac{dr}{r}. \tag{3.51}$$

It decreases with the radius R and may become negative. But negative pressures are nonphysical (they would correspond, for instance, to negative densities if the linear dependence $p = \epsilon n$ is assumed). This implies that the radius cannot exceed $R_{\max}$ determined by (3.51) with $p(0) = 0$. Figure 3.7 shows the computed dependence of the dimensionless maximal radius $\xi_{\max} = R_{\max} \nu^{-1/2}$ on the parameter combination $\Gamma \nu^{-1/2}$. It is roughly proportional to the surface tension coefficient Γ.

The above analysis yields only the absolute upper existence boundary for steadily rotating drops. It must be complemented by stability investigations that may well show that the drops actually become unstable even before this boundary is reached. Moreover, one should remember that at very low densities, characteristic for the central region, the hydrodynamical description ceases to be applicable. It can be expected that, when the size of a rotating population becomes supercritical, it would split into two or more smaller groups, each independently rotating.

It is interesting to note that rotating drops and rings are indeed observed in bacterial populations [94, 95]. Figure 3.8 shows a rotating group of bacteria

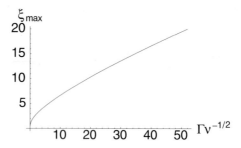

Fig. 3.7. Dimensionless maximal radius $\xi_{\max}$ of a rotating population as a function of the parameter combination $\Gamma\nu^{-1/2}$

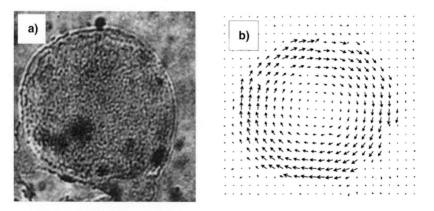

Fig. 3.8. A rotating group of bacteria *Bacillus subtilus* and the corresponding velocity field obtained by digitizing the video recordings. (From [95])

Bacillus subtilus and the corresponding velocity field obtained by digitizing the video recordings.

Of course, our discussion based on the hydrodynamical equations (3.30) and (3.31) can provide only a qualitative explanation of such phenomena. These equations describe the motion of active fluids near the supercritical bifurcation, leading to the onset of self-motion in the system. For simplicity, in the above discussion we have chosen the units of measurement for the time and spatial coordinates in such a way that the velocity of the steady uniform flow is equal to unity. In general notation, the hydrodynamical equation (3.31) will have the form

$$\frac{\partial \mathbf{v}}{\partial t} + (\mathbf{v} \cdot \nabla)\,\mathbf{v} = \alpha\mathbf{v} - \beta v^2\mathbf{v} - \nabla p + \nu\nabla^2\mathbf{v}. \qquad (3.52)$$

It is reduced to (3.31) by the scaling transformations $t \to \alpha^{-1}t$, $\mathbf{r} \to (\alpha\beta)^{-1/2}\mathbf{r}$, $\mathbf{v} \to (\alpha/\beta)^{1/2}\mathbf{v}$, $p \to (\beta/\alpha)^{1/2}p$ and $\nu \to \beta\nu$. The bifurcation, leading to self-motion, takes place at $\alpha = 0$ and the equation (3.52) holds only for small values of the parameter α. The conservation equation (3.30) is invariant under such transformations.

Toner and Tu [96, 97] argued that the most general form of the hydro-dynamical equation for isotropic active fluids, taking into account only the rotational symmetry and keeping only the terms up to the second-order derivatives with respect to coordinates, is

$$
\begin{aligned}
&\frac{\partial \mathbf{v}}{\partial t} + a_1(\mathbf{v} \cdot \nabla)\ \mathbf{v} + a_2(\nabla \cdot \mathbf{v})\mathbf{v} + a_3\nabla \left(|\mathbf{v}|^2\right) \\
&= \alpha\mathbf{v} - \beta v^2\mathbf{v} - \nabla p + \nu_1\nabla^2\ \mathbf{v} + \nu_2\nabla\left(\nabla \cdot \mathbf{v}\right) + \nu_3(\mathbf{v} \cdot \nabla)^2\mathbf{v},
\end{aligned}
\tag{3.53}
$$

where all the coefficients may be arbitrary functions of the squared magnitude $|\mathbf{v}|^2$ of the velocity and of the density n.

Near the bifurcation point, i.e. assuming that the coefficient α is small, (3.53) is simplified: its coefficients become constants and the term with the coefficient ν_3 can be neglected because it effectively has a higher order in the bifurcation parameter α. However, the equation would still contain several additional terms, as compared to (3.52). The terms with the coefficients a_3 and ν_2 do not play a significant role, but the quadratic term with the coefficient a_2 leads to strong modifications of the fluctuation phenomena in this model for systems with spatial dimensionality $d > 2$ [97]. However, its inclusion does not result in any essential corrections to the special solutions for steadily rotating drops that were considered above.

Csahók and Czirók [98] constructed a hydrodynamical description for a concrete case of bacterial motion. The evolution equation for the velocity field had in this case the form

$$
\frac{\partial \mathbf{v}}{\partial t} + (\mathbf{v} \cdot \nabla)\mathbf{v} = -\frac{1}{\tau}\left(\mathbf{v} - v_0\frac{\mathbf{v}}{|\mathbf{v}|}\right) - g\nabla h + \nu\nabla^2\mathbf{v},
\tag{3.54}
$$

where h is the local height of the bacterial colony. It differs from (3.52) only in the terms describing the self-motion of the particles. In the absence of interactions, the particles tend to move at velocity $v = v_0$, whose direction remains arbitrary; they reach this steady velocity within the characteristic relaxation time τ. Using (3.54), analytical solutions for a vortex flow, similar to (3.41) and (3.42), have been obtained by these authors. Numerical simulations of this system were also performed. Figure 3.9 shows an example of a computed flow containing several interacting vortices.

3.4 Traffic Flows

If you fly on a plane and view the traffic from a high altitude, motorways look like narrow rivers where small particles – cars – are flowing. These particles are brought into motion by sophisticated combustion engines and their movements are controlled by humans. At a first glance, the system is

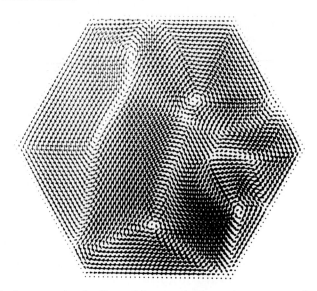

Fig. 3.9. An example of a flow with several interacting vortexes. (From [98])

so complex that its mathematical description seems hopeless. Nonetheless, quite successful phenomenological models of the traffic flows have been proposed whose predictions even quantitatively agree with the actual observed behavior.

In 1955 Lighthill and Whitham published a paper [99] (see also [100]) suggesting that road traffic could indeed be treated in a similar way to fluid flows. This simple theory already describes the development of traffic jams from density perturbations. To construct its equations, we assume that a long one-lane road has no intersections, entrances or exits. Let x be the coordinate along the road. The flow is then described by the local density $n(x,t)$ of cars and by their local mean velocity $v(x,t)$.

Let us consider a short segment of this road lying between the cross-sections x and $x + \Delta x$. The total number of cars at time t inside this segment is given by

$$\Delta N(t) = \int_{x}^{x+\Delta x} n(x,t)dx. \tag{3.55}$$

This number changes because new cars come through the left cross-section x and other cars leave the segment though the right cross-section $x + \Delta x$. Suppose that $J(x)$ and $J(x + \Delta x)$ are the car flows at these two sections, i.e. the numbers of cars crossing them per unit time. Then we can write

$$\frac{d}{dt}\Delta N = J(x) - J(x + \Delta x). \tag{3.56}$$

Note that the flow can generally be expressed as a product of the density of cars and their local mean velocity, i.e. as $J = nv$. Therefore, (3.56) can also be written in the form

$$\frac{d}{dt}\Delta N = n(x)v(x) - n(x + \Delta x)v(x + \Delta x). \tag{3.57}$$

If the flow is continuous, i.e. described by smooth density and velocity distributions, we can use linear approximations:

$$n(x + \Delta x) \approx n(x) + \frac{\partial n}{\partial x}\Delta x,$$
$$v(x + \Delta x) \approx v(x) + \frac{\partial v}{\partial x}\Delta x, \tag{3.58}$$
$$\Delta N(t) \approx n(x)\Delta x.$$

Substituting this into (3.57) we derive the usual continuity equation

$$\frac{\partial n}{\partial t} + \frac{\partial}{\partial x}(nv) = 0. \tag{3.59}$$

Lighthill and Whitham assumed that the flow velocity is directly determined by the local car density and described by a certain known function $v = \overline{v}(n)$. Generally, one can expect that this velocity has its maximum level (e.g. fixed by the speed limit) at low car densities and monotonously decreases with the density n. The cars cannot move ($v = 0$) when their density exceeds some threshold n_{max}. This is confirmed by measurements for real traffic flows. For instance, a good fit with data for the Lincoln tunnel in New York is found by taking $\overline{v}(n) = a\ln(n_{max}/n)$ with $a = 17.2$ mph and $n_{max} = 228$ vehicles per mile [100].

Since v is a known function of n, the continuity equation determines the evolution of the density distribution. It is convenient to write it in a slightly different form, i.e. as

$$\frac{\partial n}{\partial t} + c(n)\frac{\partial n}{\partial x} = 0, \tag{3.60}$$

where

$$c(n) \equiv \overline{v}(n) + n\frac{d\overline{v}}{dn}. \tag{3.61}$$

This equation should be complemented by the initial condition

$$n(x, t = 0) = f(x), \tag{3.62}$$

which specifies the density distribution at $t = 0$. We assume that the considered road section is long and can be treated as infinite, so that the coordinate x varies from $-\infty$ to ∞.

Though (3.60) was obtained by us in the analysis of a particular problem of traffic flow, the same equation arises in a large variety of applications (see [100]). It is known as the equation of *simple* (or *kinematic*) waves.

A remarkable property of this nonlinear equation is that it permits an exact analytic solution for any function $c(n)$ and any initial condition $f(x)$.

First, we take the special case of $c(n) = c_0 = \text{const}$. One can easily check by direct substitution that in this case the solution of (3.60) is

$$n(x, t) = f(x - c_0 t). \tag{3.63}$$

Its interpretation is simple: the density profile shifts with time at a constant velocity c_0, preserving its initial shape. Below, we shall see that in the general case different elements of this profile would move with different velocities, and this would result in the evolution of its shape.

To derive the general solution of (3.60), let us consider a curve Γ in the plane (x, t) that is defined by the equation

$$\frac{dx}{dt} = c(n(x, t)), \tag{3.64}$$

where $n(x, t)$ is a solution of (3.60). Calculating the derivative of n along this curve, we find

$$\left(\frac{dn}{dt}\right)_\Gamma = \left(\frac{\partial n}{\partial t} + \frac{\partial n}{\partial x}\frac{dx}{dt}\right)_\Gamma = \left(\frac{\partial n}{\partial t} + \frac{\partial n}{\partial x}c(n)\right)_\Gamma = 0. \tag{3.65}$$

This means that $n = \text{const}$ on the considered curve Γ. But then, according to (3.64), the curve Γ is actually a straight line with the slope $c(n)$, i.e. it is given by $x(t) = \xi + c(n)t$, where ξ is the starting point of this line at $t = 0$. Along the entire line, the density n is constant and equal to its initial value $f(\xi)$ at the starting point ξ. Hence, we obtain

$$n = f(\xi), \tag{3.66}$$
$$x = \xi + c(f(\xi))t. \tag{3.67}$$

Equations (3.66) and (3.67) constitute a general solution of (3.60) with the initial condition (3.62), written in the parametric form. By varying the variable ξ, we obtain all points (x, n) that together represent the profile $n(x)$ of the density distribution at time t. For example, when $c(n) = c_0$, the second of these equations yields $x = \xi + c_0 t$ and thus $\xi = x - c_0 t$. Substituting this into the first equation, we arrive at the solution (3.63).

The lines defined by (3.67) represent the *characteristics* of the nonlinear partial differential equation (3.60). The characteristics can be viewed as trajectories of certain fictitious moving elements. Since the lines are straight, all such elements move at constant velocities. Their motion starts at $t = 0$ from different points ξ on the x-axis. According to (3.67), the velocity c of an element is fixed by the density $f(\xi)$ at the respective initial point, i.e. $c = c(f(\xi))$. As implied by (3.66), the density remains constant along the entire trajectory.

Hence, the obtained solution allows a simple interpretation. We take the initial profile $f(x)$ and divide it into small elements (Fig. 3.10). In the course

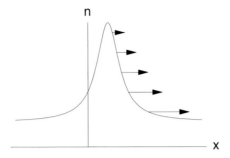

Fig. 3.10. Velocity distribution (arrows) along the density profile

of time, each profile element would move along the x-axis with its own velocity, determined as $c = c(n)$ by the density level in the respective element. To
construct the profile at the next moment t, we should only shift each profile
element in the horizontal direction by a distance $c(n)t$. If c does not depend
on n, each element would thus be shifted by the same distance and the profile
would be simply translated along the x-axis without changing its shape (cf.
(3.63)). Generally, when c depends on the density n, the shape of the profile
changes with time because its different elements move at different velocities.
Note that the velocity $c(n)$, defined by (3.61), differs from the flow velocity
$\overline{v}(n)$. The two velocities coincide only when the density dependence is absent.
The velocity $c(n)$ may be negative even if the flow velocity is positive.

Figure 3.11 shows the computed evolution of an initial density profile
in the system where c is a decreasing function of density n, as typical for
the traffic flows. The initial profile (Fig. 3.11a) has a maximum of the car
density in its center. Because the cars move more slowly where their density
is higher, at the next moment of time (Fig. 3.11b), this maximum becomes
asymmetric: its top part is shifted in the backward direction with respect
to the base. The distortion increases and, at some later moment, the profile
contains a vertical part (Fig. 3.11c). Further evolution leads then to very
strange-looking profiles (Fig. 3.11d) that are not single-valued, i.e. the local
density n takes three different possible values at any position in a particular
interval of the coordinate x.

Clearly, something went wrong with our solution. The density of cars per
unit road section *must* be a uniquely defined quantity! Though the multi-
valued solutions of the evolution equation (3.60) are formally correct, they
are physically unacceptable.

Examining again our derivation, we can notice that, to obtain the differential equation (3.59) from equation (3.57), we had to assume that the density
profile $n(x)$ was continuous, i.e. it did not contain abrupt jumps. However,
discontinuous solutions are not forbidden by the nature of the considered
problem. Suppose that the density profile contains a finite jump at a certain
point X, so that the density levels on both sides of the jump are different:

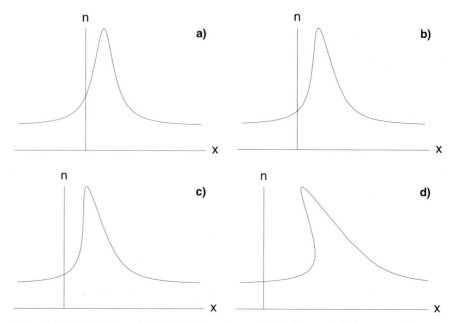

Fig. 3.11. Evolution of the initial density profile (**a**) $t = 0$. Subsequent density profiles at moments (**b**) $t = 2.5$, (**c**) $t = 4$ and (**d**) $t = 10$ are shown

$n(X + \varepsilon) \neq n(X - \varepsilon)$ for $\varepsilon \to 0$. Except for the special point $x = X$, the distribution is smooth and should therefore satisfy at $x \neq X$ the same differential evolution equation (3.60) as the multi-valued continuous solution. Moreover, the initial conditions for both of these distributions are also the same, i.e. given by (3.62).

The discontinuous solution can be obtained from the multi-valued continuous solution by introducing a vertical jump from the lower to the upper solution branches at a certain point X (this jump is shown by the dashed line in Fig. 3.12a). The position of the jump is determined by the condition that the total numbers of particles for both distributions are equal. This implies that the two regions, which are cut by the dashed line from the plot of the continuous multi-valued solution, have equal areas. Removing the parts, corresponding to these two regions, we obtain the profile of the respective discontinuous solution (Fig. 3.12b).

This construction is general and can be applied to any system described by the nonlinear partial differential equation (3.60). The jumps that appear are called *shock waves*. The shocks move with a velocity dX/dt that is determined (see [100]) by the equation:

$$\frac{dX}{dt} = \frac{n_+\bar{v}(n_+) - n_-\bar{v}(n_-)}{n_+ - n_-}, \tag{3.68}$$

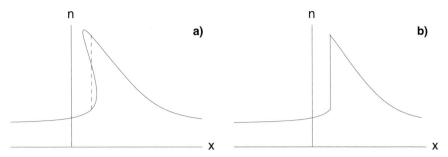

Fig. 3.12. Construction of the discontinuous jump solution

where n_+ and n_- are the densities immediately in front of and behind the shock, respectively,

$$n_+ = \lim_{\varepsilon \to 0} n(X + \varepsilon, t), \quad n_- = \lim_{\varepsilon \to 0} n(X - \varepsilon, t). \tag{3.69}$$

When $n_+ > n_-$ and $\bar{v}(n_+) < \bar{v}(n_-)$, the shock velocity is negative, i.e. it moves up the flow.

Looking again at the construction of the discontinuous solution (Fig. 3.12), we note that n_+ is less than the maximal density in the initial pulse (and in the multi-valued continuous solution). As time goes on, the multi-valued continuous solution becomes increasingly stretched in the vertical direction and the jump is located farther away from the maximum. Therefore, the density n_+ decreases. This means that the jump gets smaller with time and the shock wave eventually disappears.

Summarizing our analysis, we can conclude that shock waves are the typical patterns described by the nonlinear equation (3.60). They develop from finite initial perturbations of the uniform flow. If the initial perturbation has the form of a positive pulse (Fig. 3.11a) and the velocity $c(n)$ is a decreasing function of the density n, the discontinuities (shocks) appear on the rear side of the pulse. They move up the flow and get weaker. The lifetime of the shocks is finite and determined by the magnitude of the initial perturbation. If $c(n)$ is an increasing function of n, the shocks develop in front of the pulse and their velocity, given by (3.68), is positive, and the shocks therefore move *down* the flow. The lifetime of the shocks is finite in this case, too.

When traffic flows are considered, the shock waves are simply *traffic jams*. This was the main result of the study by Lighthill and Whitham [99]. If the density of cars is increased locally, this leads to a decrease in the local flow velocity. Hence, the cars move more slowly in this part of the road. But the drivers up the flow do not know about the slow-down and continue to drive at a high velocity. Thus, they move into a dense group of cars and must slow their velocity appropriately. This process leads to the development of an abrupt jump in the car density. The jump slowly drifts up the flow. In the jam region immediately after the jump, the car density is high and the velocity is low. Once the jam is passed, the cars accelerate and the car density decreases.

Thus, according to the Lighthill–Whitham theory, traffic jams need for their development a local initial increase in the car density. This seems quite reasonable – the jams are indeed often caused by road accidents, temporally preventing the normal flow. Such jams have finite lifetimes and slowly drift up the flow.

Motorways play an important role in modern society and there have been many systematic observations of traffic flows under various road conditions. Surprisingly, these observations show that sometimes traffic congestion develops without any obvious reason from a uniform flow, so that a "phantom traffic jam" appears (e.g. see [101]). To explain this effect, one must go beyond the original Lighthill–Whitham theory.

The problem was addressed in 1993 by Kerner and Konhäuser [102]. They noticed that the assumption of a flow velocity v completely determined by the local density of cars ($v = \bar{v}(n)$) represents a simplification that is not always applicable. Indeed, it implies that a driver can respond without any delays to variations of the driving conditions and immediately adjust the speed to the optimal velocity $\bar{v}(n)$ that corresponds to the current flow density n. A more realistic assumption would be that such an adjustment generally takes a certain characteristic time τ and, therefore, the velocity V of a given car obeys the equation

$$\frac{dV}{dt} = -\frac{1}{\tau}(V - \bar{v}(n)). \tag{3.70}$$

Furthermore, nonlocal effects should be taken into account. A driver is a human being and he or she can look ahead and respond not only to the local driving conditions, but also to the situation developing in front of the car within the visibility region. Two effects can be here distinguished.

Firstly, the driver would probably slow down and even brake if he or she sees that the car is approaching a region with a higher car density. On the other hand, the driver would tend to accelerate if the car density ahead of it is decreasing. This effect can be incorporated into the model by adding a new term, proportional to the gradient of the car density, to (3.70), so that it takes the form

$$\frac{dV}{dt} = -\frac{1}{\tau}(V - \bar{v}(n)) - a\frac{\partial n}{\partial x}, \tag{3.71}$$

where the proportionality coefficient a may depend on the density n.

Secondly, even if the flow density is constant in the visibility region, the flow velocity may vary inside it. It is natural to assume that, in this case, the driver would tend to choose the velocity that would be the mean of the flow velocities in front of and behind him or her. Including this effect, the equation of motion becomes

$$\frac{dV}{dt} = -\frac{1}{\tau}(V - \bar{v}(n)) - a\frac{\partial n}{\partial x} - \frac{1}{\tau_1}\left[V - \frac{1}{2}\left(v(x + \Delta x) + v(x - \Delta x)\right)\right], \tag{3.72}$$

where τ_1 is another characteristic relaxation time and Δx is the visibility radius.

Next, we note that the acceleration dV/dt of a given particle is related to the flow velocity v by (3.29), which can be written for the one-dimensional flow under consideration as

$$\frac{dV}{dt} = \frac{\partial v}{\partial t} + v\frac{\partial v}{\partial x}. \tag{3.73}$$

Combining (3.72) and (3.73), we thus obtain

$$\frac{\partial v}{\partial t} + v\frac{\partial v}{\partial x} = -\frac{1}{\tau}(v - \overline{v}(n)) - a\frac{\partial n}{\partial x} - \frac{1}{\tau_1}\left[v - \frac{1}{2}\left(v(x + \Delta x) + v(x - \Delta x)\right)\right]. \tag{3.74}$$

If the flow velocity only changes slightly within the visibility radius, we can approximately write

$$v(x + \Delta x) + v(x - \Delta x) \approx 2v(x) + \Delta x^2 \frac{\partial^2 v}{\partial x^2}. \tag{3.75}$$

Substituting this expansion into (3.74), we finally obtain the evolution equation for the velocity field:

$$\frac{\partial v}{\partial t} + v\frac{\partial v}{\partial x} = -\frac{1}{\tau}(v - \overline{v}(n)) - a\frac{\partial n}{\partial x} + \nu\frac{\partial^2 v}{\partial x^2}, \tag{3.76}$$

where $\nu = \Delta x^2/2\tau_1$. It should be complemented by the conservation equation

$$\frac{\partial n}{\partial t} + \frac{\partial(nv)}{\partial x} = 0. \tag{3.77}$$

Equations (3.76) and (3.77) provide the basis for the hydrodynamical theory of traffic flows. Generally, the coefficients τ, a and ν may depend on the local car density n. It should be noted that (3.76) was first constructed in 1974 by Whitham [100], but without the last "viscous" term.

Traffic flows are a special case of active fluids formed by self-moving particles. In the previous section we have considered active fluids near the spontaneous onset of self-motion, when the flow velocities are small. Moreover, we were interested in isotropic active fluids where the direction of developing active motion is arbitrary. If traffic flows are considered, the flow is one-dimensional and its direction is prescribed by the road signs. Moreover, the flow velocities are high under normal traffic conditions. Comparing (3.76) and (3.30), we see that the main difference between these two equations is in the first terms, describing the local dynamics of the velocity field. According to equation (3.30), in the absence of interactions a particle tends to move at unit velocity in an arbitrarily chosen direction. According to (3.76), under uniform conditions a driver takes the "optimal" velocity $\overline{v}(n)$ determined by the traffic density.

Stationary uniform traffic flow is described by the solution $n = n_0$ and $v = \bar{v}(n_0) = v_0$ of (3.76) and (3.77). To investigate the stability of this uniform flow, we add small perturbations $v = v_0 + \delta v$, $n = n_0 + \delta n$ and linearize (3.76) and (3.77), obtaining:

$$\frac{\partial \delta n}{\partial t} + v_0 \frac{\partial \delta n}{\partial x} + n_0 \frac{\partial \delta v}{\partial x} = 0, \tag{3.78}$$

$$\frac{\partial \delta v}{\partial t} + v_0 \frac{\partial \delta v}{\partial x} = -\frac{1}{\tau} \delta v - \frac{1}{\tau} b \delta n - a \frac{\partial \delta n}{\partial x} + \nu \frac{\partial^2 \delta v}{\partial x^2}. \tag{3.79}$$

Note that, if the coefficients τ, a and ν depend on the density n, they should be taken at $n = n_0$ in the linearized equations. We have introduced the notation $b = -d\bar{v}/dn$ at $n = n_0$; this coefficient is positive because $\bar{v}(n)$ is a decreasing function of the density n.

The solutions of this system of linear partial differential equations can be sought in the form

$$\begin{aligned} \delta n(x,t) &= \delta n \exp\left[\lambda t + ik(x - v_0 t)\right], \\ \delta v(x,t) &= \delta v \exp\left[\lambda t + ik(x - v_0 t)\right]. \end{aligned} \tag{3.80}$$

Substitution of (3.80) into (3.78) and (3.79) leads to the following characteristic equation:

$$\lambda^2 + \lambda \left(\frac{1}{\tau} + \nu k^2\right) + a n_0 k^2 - \frac{i}{\tau} b n_0 k = 0. \tag{3.81}$$

It has two roots

$$\lambda_{1,2} = -\frac{1}{2}\left(\frac{1}{\tau} + \nu k^2\right) \pm \left[\frac{1}{4}\left(\frac{1}{\tau} + \nu k^2\right)^2 - a n_0 k^2 + \frac{i}{\tau} b n_0 k\right]^{1/2}. \tag{3.82}$$

The uniform flow is unstable with respect to small perturbations if $\mathrm{Re}\,\lambda > 0$ for the modes with certain values of the wavenumber k.

Examining (3.82), we notice that the real part of the root λ_2, corresponding to the minus sign in this equation, is always negative, and therefore the respective perturbation modes are always damped. The situation is different for the modes corresponding to the plus sign in (3.82). If we plot $\mathrm{Re}\,\lambda_1$ as a function of the wavenumber k, we see that $\mathrm{Re}\,\lambda_1$ can be positive for a certain interval of wavenumbers, so that the respective perturbation modes grow with time (Fig. 3.13).

It follows from (3.82) that near the instability threshold we have

$$\mathrm{Re}\lambda_1 \approx \tau n_0 (b^2 n_0 - a)k^2 - 2\nu\tau^2 a n_0 k^4, \tag{3.83}$$

$$\mathrm{Im}\lambda_1 \approx \sqrt{a n_0}\, k \left(1 - \nu\tau k^2\right). \tag{3.84}$$

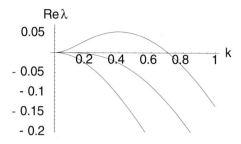

Fig. 3.13. Dependences of the increment of growth $\mathrm{Re}\lambda$ on the wavenumber k for $b = 0.5$ (the *lower* curve), $b = 1.0$ (the *middle* curve) and $b = 1.5$ (the *upper* curve); the other parameters are $\tau = a = \nu = n_0 = 1$

Thus, the instability takes place when $b > \sqrt{a/n_0}$ or, in the original notation, if

$$\left| \frac{d\bar{v}(n_0)}{dn} \right| > \sqrt{\frac{a}{n_0}}. \tag{3.85}$$

This instability condition does not contain the "viscosity" coefficient ν and can therefore be derived by neglecting the respective term in (3.76). It was found by Whitham [100] and later also obtained in [102]. We see that the uniform flow is unstable if the flow velocity $\bar{v}$ is highly sensitive to the density variations, i.e. if the slope of the dependence $\bar{v}(n)$ is sufficiently large at $n = n_0$. The viscous terms, containing the coefficient ν, play an important role by suppressing the instability for short wavelengths. Indeed, as follows from (3.83), only the modes with wavenumbers in the interval $|k| < (2a\nu\tau)^{-1/2}(b^2 n_0 - a)^{1/2}$ will grow above the threshold.

The unstable modes are accompanied by periodic density compression and represent therefore an analog of sound waves. In the long-wave limit the frequency $\omega = \mathrm{Im}\lambda_1$ of these waves is given by $\omega = sk$, where $s_0 = \sqrt{an_0}$ is the speed of sound. The velocity $c(n)$ in (3.60), determining the nonlinear evolution of the density profile, is $c(n) = \bar{v}(n) + n(d\bar{v}/dn)$. If the profile represents only a slight modulation of the uniform state and $n \approx n_0$, this velocity is approximately $c_0 = v_0 - bn_0$. Therefore, we have $b = (v_0 - c_0)/n_0$. Substituting this expression for b into the inequality (3.85), it can be seen that the instability condition is

$$v_0 - c_0 > s_0. \tag{3.86}$$

When the instability threshold is exceeded, all perturbation modes with wavenumbers satisfying the condition

$$2\nu\tau k^2 < \frac{1}{s_0^2}(v_0 - c_0)^2 - 1 \tag{3.87}$$

grow with time. Thus, the instability is related to the growth of sound waves, associated with the spatial modulation of the flow density, and only the modes with sufficiently long wavelengths are unstable.

The nonlinear evolution of the system has been numerically investigated [102]. These studies reveal that the instability of the uniform flow always leads to the development of traveling density clusters. As an example, Fig. 3.14 shows the profiles of density and of local flow velocity in such a cluster. The traveling density clusters are the traffic jams. It is found that the shape and the velocity of these clusters do not depend on the initial conditions and are completely determined by the parameters of the flow. Therefore, they represent self-organized patterns of the traffic flow, similar to traveling pulses in excitable media. When the viscosity is small, the methods of singular perturbation theory can be used to construct approximate analytical solutions for steadily traveling density clusters and to investigate their properties [103].

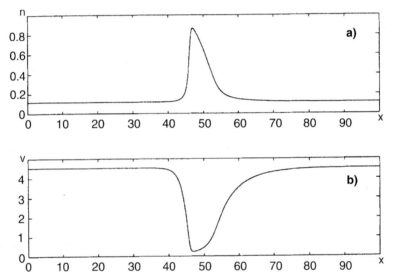

Fig. 3.14. Profiles of density (**a**) and of local flow velocity (**b**) in a traffic cluster (Redrawn from [102])

3.5 Further Reading

The phenomenon of active motion is a ubiquitous property of living beings. With the exception of plants and a few other organisms, living beings that propel themselves can be found at all levels of biological organization [104]. We have already mentioned the bacterium *E. coli*, which is one of the best-studied biological organisms. Such bacteria use propulsion to explore their environment and move towards the nutrient source. Their motion consists of long intervals of straight swimming alternating with sudden tumbles. The tumbles are induced by reversals of the flagellar rotation and lead to random changes of the swimming direction. In the presence of a spatial nutrient gradient, bacteria tend to increase the time of their straight swimming when they move up the gradient, but keep it constant when they find themselves moving down the gradient [105]. Stochastic models of this process were constructed and analyzed by Alt [106] and by Farell et al. [107]. It was shown that the microscopic mechanism of propulsion in *E. coli* involves a molecular motor driven by a proton gradient across the plasma membrane [108–110]. Later on, evidence accumulated that there is also another mechanism involving a sodium-driven rotary motor. Several qualitative proposals explaining the transduction mechanisms can be found in the papers by Junge et al. [111] and Vik and Antonio [112]. Models for sodium- and proton-driven motors were developed by Elston and Oster [113], Elston et al. [114] and Dimroth et al. [115]. It has become clear by now that both motors operate on the same physical principles and can be found in eucaryotic mitochondria as well as in many bacteria [116, 117]. For recent experimental work on how electrochemical energy can be converted into a rotary torque see the papers by Dimroth et al. [118] and by Kaim and Dimroth [119–121], as well as the review by Oster and Wang [122].

Despite its complex organization, the motion of *E. coli* in nutrient gradients can be understood as an effective biased random walk. General descriptions of biological motion based on diffusion models have been constructed [123]. Bell and Tobin have presented an extensive discussion of chemotaxis in various biological systems [124].

Some microorganisms release chemical signals depending on their internal states and intend to establish chemical communication between the cells. When the gradients of the concentration fields of the signal substance are used by microorganisms to direct their active motion, this results in a complex collective dynamics. The most impressive and extensively studied example of such phenomena is provided by the behavior of the slime mold *Dictyostelium discoideum*. The processes of spatiotemporal pattern formation in these systems are described by the chemotaxis–diffusion models. We have already discussed them in Chap. 2.

For an organism to orient according to environmental conditions, it needs sensory devices and information processing capacities, and it should be able to translate this information into useful directional and/or motor control. Berg and Purcell pointed out in their analysis of chemoreception in *E. coli* that the conditions in which the bacteria have to orient at microscopic scales pose severe design requirements on the sensory apparatus. At these scales, transport is effected by diffusion rather than by bulk flow, movement is restricted by viscosity not by inertia, and thermal fluctuations are strong enough to perturb the motion of the bacteria [125]. One of their key results is that the chemotactic sensitivity of *E. coli* approaches that of a cell of optimal design.

An organism may use active motion to fulfill the purpose of getting to a goal, as in the case of *E. coli*. A different type of active motion is found in the case when an organism has not yet identified its goal; this situation corresponds to search behavior. Various aspects of search behavior of cells and arthropods [126] and of insects and larger animals [127, 128] have been studied.

While prokaryotes, such as bacteria, have organelles, like flagellar rotors, as motor systems, various types of propulsion systems are known in eucaryotic cells. For example, many kinds of cells in most animal species, in many protozoa and also in some lower plants are covered by tiny hairlike appendages, about 0.25 μm in diameter, called cilia. Their function is either to move fluid over the cell surface or to propel individual cells in a liquid medium. Propulsion is generated by fields of cilia that show collective whiplike bending motions. Flagellae such as found in sperm or many protozoa are similar to cilia in that they are tiny hairlike structures. They differ in that their bending motion is sinusoidal. In both cases the bending motion is based on the activity of intracellular filament bundles or networks located beneath the cell membrane involving sliding of microtubuli along each other driven by the motor protein dynein. The collective motion of the cilia and flagellae is induced by wavelike excitation of these microtubuli structures [104, 129]. Mathematical models describing these interactions were devised by Brokaw [130, 131], Machemer [132] and Mogami et al. [133]. A recent paper by Camalet et al. [134] shows how wavelike propagating shapes, which propel the filament, can be induced by a self-organized mechanism via a dynamic instability.

Yet another type of active motion can be found in many eucaryotic cells such as amoebae, leukocytes (blood cells) and keratinocytes (tissue cells) floating in suspension or adhering to a substrate. It is based on different types of membrane protrusions and retractions. As these processes induce continuous change of cell shapes, this type of motion is also known as amoeboid motion. It involves the transient formation of actin filaments (F-actin) in parts of the cell, where they are needed until they are eventually disassembled after completion of the task. The contractile filament system consisting of F-actin and the protein myosin interacts with other proteins in the cytosol, cytoskeleton or the plasma membrane, generating propulsion. Despite this common motile mechanism in different cell types, various theories exist ranging from active shear models and mechanisms based on hydraulic or osmotic pressure to those involving surface tension and cytogel expansion (see [135]). Alt and Kaiser [136] describe a model of individual keratinocytes based on continuum equations taking into account various contributions, such as drag-, viscous-, curvature- and other forces acting on the F-actin filament system. Keratinocyte movement plays a role in wound healing, i.e. in covering free space to establish effective cell–cell contacts and to form connected cell tissues under high tension. A good overview of recent work on the various types of individual cell propulsion can be found in the book edited by Alt,

Deutsch and Dunn [137]. Various aspects of modeling wound healing can be found in the papers by Maini et al. [138, 139]. Durand et al. [140] have analyzed adhesion-induced vesicle propulsion in inanimate systems.

Remarkably, the motile machinery of some blood cells (granulocytes) can be isolated. Keller and Bessis [141] have shown that the application of heat causes the leading part of such a cell, containing the motor, to move forward more rapidly. As a result, the cellular motor of a human granulocyte moves out of the cell in a small unit, called the cytokineplast. Initially, this unit is still connected with the cell body by a thin stalk of cytoplasm. Later, the stalk breaks and a separate particle is formed. This biological particle cannot reproduce; it contains only the biochemical subsystem necessary for locomotion which is enclosed inside a membrane. Thus formed cytokineplasts are capable of active motion, including adherence, spreading, random locomotion and directed migration [142]. Gruler has suggested that such self-moving particles can be employed as biological "automobiles" to transport material over a substrate surface [143].

The first experimental examples of self-propulsion in simple physico-chemical systems were published by Dupeyrat and Nakache [144, 145].

Mathematical modeling of collective active motion follows several different directions. One approach is based on the notion of discrete stochastic automata, representing individual self-propelled particles [146]. An example of an automata description for a fish school was discussed in Chap. 2. Vicsek et al. have formulated simple automata models of interacting self-propelled particles [147, 148]. Such models have been subsequently used to investigate the statistical properties of the transition to coherent collective motion and fluctuation behavior in actively moving populations [149, 150]. This research has been reviewed by Czirók and Vicsek [151].

Alternatively, one can specify dynamical equations of motion for all individual particles that explicitly include the interactions between them and/or the action of external fields. This theoretical approach has been applied by Gruler in connection with his experiments on cellular self-motion [152, 153]. Edelstein-Keshet et al. have studied the phenomenon of locust swarms consisting of millions of individuals travelling over large distances from the viewpoint of travelling patterns [154] and have investigated animal swarm formation on the basis of nonlocal interactions [155]. Various aspects of animal groupings have also been studied by Grünbaum [128, 156, 157], Flierl et al. [158] and Chao and Levin [159]. A general analysis of the active motion of particles with "energy depots" has been performed by Ebeling et al. [160, 161] (see also [162]).

Investigations of traffic flow are a practically important and well-developed research field. Already in 1958 Montroll et al. [163] and Komentani and Sasaki [164] suggested that uniform traffic flows on motorways might be unstable with respect to density variations, leading to traffic congestion. In addition to the literature references already given, we should mention here the classic book by Prigogine and Herman [165] and several monographs [166–168]. The transition from free traffic flow to congestion was studied by means of cellular automata by Nagel and Schreckenberg [169] (see also [170] for recent developments). Starting from the Enskog-like kinetic equations, Helbing derived fluid dynamics equations for the dynamics of congested vehicular traffic, accounting for the additional terms due to anisotropic vehicle interactions [171]. This paper is also a good source for recent publications of the various modeling approaches used in the field of traffic dynamics. Models of traffic flow and transport play a role in relation to the planning and development of towns and regions. For models in that broader context see, for example, [172] and [173].

4. Ridden by the Noise ...

The ability to proceed towards a chosen goal or destination is characteristic of many living systems. To drive themselves through physical space, various motors are employed by biological microorganisms, animals and human beings. Such forced self-motion leads to a large degree of autonomy because it allows them to overcome spatial constraints. However, this form of motion is far from economical. An actively propelled particle or organism must act against its own environment and therefore expend considerable energy. An alternative is provided by the mechanisms which exploit intrinsic environmental fluctuations.

The simplest device rectifying fluctuations is the ratchet. It represents a toothed wheel with a catch (pawl) that prevents the wheel from slipping back. When placed into a fluctuating nonequilibrium environment, the ratchet would preferentially move in one direction. A different mechanism is employed by "intelligent" particles that navigate in fluctuating fields. Their active motion is gained through the control of friction or resistance, rather than through forced propulsion. In contrast to passive ratchets, the medium may now be in thermal equilibrium, but the active particles – operated by *demons* – need an external energy supply.

When social systems are considered, the accumulation of wealth and professional success may represent important goals. To reach them, social maneuvering based on a controlled response to fluctuating situations would often be an efficient strategy. In ecological systems, exploiting the environmental fluctuations provides significant evolutionary advantages and can prevent the extinction of a weak species.

4.1 Demons and Ratchets

The foundations of thermodynamics and statistical physics were laid in the 19th century by Boltzmann and Maxwell. The second law of thermodynamics tells that a closed system will tend to approach the state of thermal equilibrium with the maximal possible entropy. Since entropy is a measure of disorder, this law can also be interpreted as implying that the system must always evolve to its least-ordered state. Working on the kinetic theory of gases, Maxwell considered a simple mental experiment that might appear to violate the second law.

Let us take a volume, separated into two boxes by a thin membrane, and assume that it has a tiny hole, so that the boxes are connected. If we fill the volume with gas and wait for a while, thermal equilibrium is eventually reached. In this state, the densities, pressures and temperatures of the gas in both subvolumes must be equal. Imagine now a little *demon* that sits near the hole and watches the molecules. He is so small that he can react, by quickly closing the hole, to individual arriving molecules.

If the demon allows the molecules to pass in one direction and shuts the hole whenever a molecule is coming from the opposite direction, after some time all the molecules will apparently be concentrated in only one of the two boxes. Alternatively, the demon may allow to pass in a certain direction only those molecules that are fast enough. The rapidly moving molecules would then be preferentially located in one of the boxes and the temperature of gas inside it would be higher.

Both outcomes contradict the second law of thermodynamics. Indeed, any breaking of symmetry means an increase in the order of the system. At equilibrium, the statistical states of the gas in the two boxes must therefore be identical. Hence, such a Maxwell demon is not actually possible!

While prohibiting this effect, the second law of thermodynamics does not, of course, tell us exactly why a particular design of a "demon", that may also be a small machine, would not work. At the time of Maxwell, the entire above argument was purely speculative. The molecules, whose existence was hypothesized in the kinetic theory, were never directly observed and even the laws of quantum mechanics, responsible for their formation, were not yet known. Therefore, at that time one would probably just say that the machines, able to manipulate with single molecules, could not be constructed.

Today we know that this is not true. The advances in nanotechnology already allow us to operate with individual molecules. Hence, the question remains: what would physically prevent a nanoscale machine from playing the role of a Maxwell demon? Examining again the above argument, we can see that, to apply the second law of thermodynamics, the whole system, *including the demon*, must be closed, i.e. should not receive energy from outside and dissipate it. This means that our microscopic machine will itself be subject to thermal fluctuations and will evolve to a state of thermal equilibrium. The tiny door, shutting the hole, will swing at random and allow the molecules to pass even when they should have been stopped.

The laws of statistical physics say that, no matter what is the concrete design of our micromachine, at equilibrium the fluctuations inside it will always exactly counterbalance its assumed function and prevent the separation of molecules. This fine universal balance is quite remarkable and shows how fundamental the second law of thermodynamics is.

The converse side of the same argument is that, if the micromachine – a Maxwell demon – is maintained far from thermal equilibrium by supplying to it energy and removing the produced heat, it may function perfectly if it is appropriately designed. Note that the energy would then be used not to move the molecules through the hole, but only to operate the door, evading the effects of thermal fluctuations.

Another implication of the second law of thermodynamics is that heat cannot be converted into mechanical energy in closed equilibrium systems. The patent offices in all countries would therefore automatically reject any proposals for devices in which the thermal energy of an equilibrium system is used to produce work. Again, while prohibiting such machines, statistical physics does not generally tell us what is wrong in their design, i.e. why a particular machine would not function. This can be found only by the careful examination of a concrete proposal.

From 1961 to 1963 Richard Feynman, the great physicist of the 20th century, was giving introductory lectures in physics at the California Institute of Technology. Unlike many introductory courses, he wanted to address them to the most intelligent in the class and make sure, as he said, that "even the most intelligent student was unable to completely encompass everything that was in the lectures – by putting in suggestions of applications of the ideas and concepts in various directions outside the main line of attack" [174]. One of such brilliant sidewalks was his analysis of *ratchets*.

A ratchet is a toothed wheel provided with a catch (pawl) that prevents the wheel from slipping back and allows it to move in only one direction. Figure 4.1 shows a ratchet in the left-hand (transparent) box. Because this device can rotate only clockwise, it is tempting to use it for the rectification of thermal fluctuations and conversion of their energy into mechanical work. Let us take a propeller and place it in the right-hand box filled with the gas. Since the propeller blades are frequently bombarded by the gas molecules, it will perform rotational Brownian motion. Of course, this motion will still be purely random and, averaged over a long time, no steady rotation will be observed. Next, we take a thin rigid rod (an axle) and connect the propeller to the ratchet in the left-hand box. Now, each time when the propeller wants to rotate in the clockwise direction, it can easily do so. On the other hand, if the propeller is pushed in the counterclockwise direction, the ratchet will not allow it to move and hence this motion will be suppressed. As a result, one may expect that this gadget would rectify the rotational Brownian motion of the propeller induced by thermal fluctuations. Perhaps, we can even tie a flea to a thread, attached to the axle, and lift it as the axle rotates in the clockwise direction!

If such a gadget were able to function, this would mean that mechanical energy could be extracted from the equilibrium thermal energy of gas molecules, in contradiction to the second law of thermodynamics. Hence, something must be wrong in our proposal.

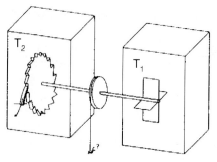

Fig. 4.1. The ratchet (from [174])

We can note that there should be a spring in the ratchet. Indeed, the pawl must return after coming off the tooth, so the spring is necessary. Moreover, energy should be dissipated during the ratchet operation – otherwise the pawl would continue to bounce after hitting the wheel and thus fail to prevent it from rotating in the wrong direction. When energy is dissipated, it is converted into heat. If the ratchet is thermally isolated, its temperature would therefore rise until either the wheel or the pawl melts and the device is destroyed.

To improve the design, we can connect the ratchet to a thermal bath, keeping its temperature at a constant level. But then not only the propeller but also the ratchet would perform Brownian motion. The pawl would jump and occasionally go so high that the wheel might move back. When the pawl rises, the spring is stretched. Suppose that ϵ is the difference in the elastic energies of the spring in the "down" and "up" positions. If the ratchet is in thermal equilibrium at temperature T_1, the probability that it gains this energy through a fluctuation is $\exp(-\epsilon/k_B T_1)$. To move the wheel by one step, the propeller in the right-hand box must perform mechanical work equal to ϵ. This energy can be acquired by the propeller only through a thermal fluctuation. If the gas temperature in the right-hand box is T_2, the probability of such a fluctuation is $\exp(-\epsilon/k_B T_2)$. We see that, if the temperatures are equal ($T_1 = T_2$), these two probabilities coincide. Hence, the number of times that the propeller would accumulate enough energy to turn the wheel is exactly equal to the number of times that the wheel spontaneously rotates in the opposite direction. The wheel would move forth and back, but no persistent motion would be observed.

Thus, the patent office would indeed be perfectly right to reject such a proposal. The psychological difficulty, which many of us might at first have in understanding what is wrong in the project, lies in the fact that the machine must be extremely miniature. Indeed, its dynamics is dominated by the Brownian motion of its parts, which becomes significant only on microscales. We are not familiar with thermal Brownian motions in our everyday life and therefore do not normally possess the intuition needed to understand this phenomenon.

The second law of thermodynamics plays a fundamental role by forbidding certain kinds of process. However, it also indicates a way by which these processes may actually be realized – by creating the conditions under which this law is no longer applicable. For instance, the spontaneous organization and creation of order in equilibrium systems are forbidden by this law. But if a system is open, the second law is not applicable and various self-organization phenomena already become possible.

In his lecture, Feynman went on to show that, if the temperatures in the two boxes are different, the device *would* operate. When $T_1 > T_2$, the wheel in Fig. 4.1 indeed rotates mainly in the clock direction. It is interesting that in the opposite case, when $T_1 < T_2$ and thus the ratchet is more hot than the propeller, the ratchet goes around in the opposite direction. The persistent motion is now possible because the total system is not in a state of thermal equilibrium. Note that, to maintain the temperature difference despite the ratchet operation, heat must be permanently supplied to one of the boxes and removed from the other box.

Of course, this is only one of the possible ways of moving the system out of thermal equilibrium and thus lifting the limitations imposed by the second law. Actually, almost any modification of the ratchet set-up that makes our system open would suffice to achieve the desired function. For example, we may apply to the pawl a weak time-dependent force that would not itself be sufficient to move the pawl so high that it goes over the tooth. Assisted by thermal fluctuations, this weak force would, however, lead to persistent rotation. Alternatively, we can make the fluctuations nonthermal. This can be done, for instance, by pumping external (e.g. acoustic) noise into the right-hand box, containing the propeller. Again, this would make the system nonequilibrium and allow the ratchet operation.

The ratchet might seem to be just a special mechanical gadget. However, we have encountered here a phenomenon that is very general and frequently found in living systems. To see this, let us further examine the ratchet design and extract its principal properties.

The rotational Brownian motion of the ratchet is approximately described by the stochastic differential equation

$$I\frac{d^2\theta}{dt^2} = -\gamma\frac{d\theta}{dt} - \frac{\partial U(\theta)}{\partial \theta} + f(t), \tag{4.1}$$

where θ is the angle variable, which can be viewed as a coordinate, and I is the inertia moment of the ratchet.

The motion takes place in the periodic asymmetric potential $U(\theta)$, a typical example of which is shown in Fig. 4.2. Each period of this potential corresponds to a rotation by one step. It consists of two parts. The slow rise of the potential corresponds to stretching of the spring, increasing its elastic energy. The subsequent sharp potential drop corresponds to falling off the tooth. To rotate the wheel in the positive (clockwise) direction, a force acting in the positive direction and overcoming the weak elastic force of the spring should be applied. If we want, however, to rotate the wheel in the other direction, we must overcome a very strong force that acts in a narrow angle interval where the potential U has a negative slope (physically, to turn back the ratchet we must deform the wheel and act against its elastic energy).

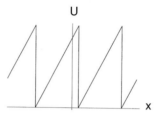

Fig. 4.2. Periodic asymmetric potential

The first term on the right-hand side of (4.1) describes the effect of viscous friction, which accompanies any motion inside a gas. The last term $f(t)$ is the random force, caused by collisions with the gas molecules. The laws of statistical mechanics imply (e.g. see [175]) that at thermal equilibrium this random force must represent a Gaussian delta-correlated noise with $\langle f(t) \rangle = 0$ and

$$\langle f(t_1)f(t_2) \rangle = 2S\delta(t_1 - t_2). \tag{4.2}$$

Moreover, its intensity S is not arbitrary. It is related to the viscous friction coefficient γ by the equation

$$S = \gamma k_{\mathrm{B}} T, \tag{4.3}$$

where T is the temperature and k_{B} is the Boltzmann constant.

If the viscous friction is strong, the inertial term in (4.1) can be neglected, so that this equation takes the form

$$\gamma \frac{d\theta}{dt} = -\frac{\partial U(\theta)}{\partial \theta} + f(t). \tag{4.4}$$

It describes random motion along the coordinate θ in the potential $U(\theta)$ under the action of the random force $f(t)$. The probability density $p(\theta, t)$ of this random process obeys the following Fokker–Planck equation [175, 176]:

$$\frac{\partial p}{\partial t} = \frac{\partial}{\partial \theta}\left[\mu \frac{\partial U(\theta)}{\partial \theta} p \right] + D \frac{\partial^2 p}{\partial \theta^2}, \tag{4.5}$$

where $\mu = 1/\gamma$ is the mobility and $D = S/\gamma^2$ is the (rotational) diffusion constant. As follows from (4.3), these two properties are connected through the Einstein relationship

$$D = \mu k_{\mathrm{B}} T. \tag{4.6}$$

The mean angular velocity is given by

$$\Omega = \left\langle \frac{d\theta}{dt} \right\rangle = -\mu \left\langle \frac{\partial U(\theta)}{\partial \theta} \right\rangle = -\mu \int_0^{2\pi} \frac{\partial U(\theta)}{\partial \theta} p(\theta, t) d\theta, \qquad (4.7)$$

where we have used (4.4) and the statistical property $\langle f(t) \rangle = 0$ of the random force.

The stationary solution of the Fokker–Planck equation (4.5), corresponding to thermal equilibrium, is

$$\bar{p} = Z^{-1} \exp\left(-\frac{U(\theta)}{k_B T} \right), \qquad (4.8)$$

where Z is the normalization factor (the partition function)

$$Z = \int_0^{2\pi} \exp\left(-\frac{U(\theta)}{k_B T} \right) d\theta. \qquad (4.9)$$

The mean angular velocity at thermal equilibrium can be found by substituting (4.8) into (4.7). This yields

$$\Omega = -\mu \int_0^{2\pi} \frac{\partial U(\theta)}{\partial \theta} \bar{p}(\theta) d\theta = -\mu Z^{-1} \int_0^{2\pi} \frac{\partial U(\theta)}{\partial \theta} \exp\left(-\frac{U(\theta)}{k_B T} \right) d\theta = 0 \quad (4.10)$$

for any periodic potential $U(0) = U(2\pi)$. The mean angular velocity is zero and the persistent rotation of the ratchet is never possible at thermal equilibrium. This provides a formal proof of the statement that has already been made above in this section.

The simplest way to bring the ratchet out of thermal equilibrium is to introduce some temporal modulation of its potential, so that $U(\theta)$ is replaced by $U(\theta, t)$ in the stochastic differential equation (4.4). For example, we may take

$$U(\theta, t) = U_0(\theta) + \varphi(t) U_1(\theta), \qquad (4.11)$$

where both $U_0(\theta)$ and $U_1(\theta)$ are periodic functions of the angle θ. Steady rotation can be found even when $\varphi(t)$ represents a simple periodic function of time, e.g. if $\varphi(t) = \sin(\omega t)$. However, the same effect may also be achieved by assuming stochastic modulation, so that $\varphi(t)$ is a random process (which must however be different from the Gaussian delta-correlated noise). Non-stationary ratchets have been investigated by Magnasco [177], Ajdari et al. [178] and Bartussek et al. [179]. The ratchets with non-Gaussian or colored (i.e. not delta-correlated) noise have been studied by Astumian and Bier [180], Doering et al. [181] and Luczka et al. [182]. In all these cases an exact solution of the Fokker–Planck equation was not possible and various approximations had to be used.

Though we have been talking so far only about a ratchet that represents the mechanical device described by Feynman, the results are actually more general. Examining (4.4), one can note that it may also be interpreted as a stochastic equation of motion of a particle in a periodic spatial potential U, if the variable θ is treated as the spatial coordinate of this particle. The particle performs Brownian motion in this potential, induced by the random force $f(t)$. At thermal equilibrium, i.e. if the potential is stationary and $f(t)$ represents a Gaussian delta-correlated noise, there is no preferred direction of the particle's motion and its mean velocity is zero. However, if the periodic potential is temporally modulated, a directed drift of particles can be produced. The same effect can be obtained by introducing an additional "nonequilibrium" noise, which is either non-Gaussian or colored. Thus, the particles may drift in a certain direction despite the fact that no persistent force acting in this direction is applied (i.e. the potential is periodic is space). Since the drift velocity depends in this case on the mobility that is determined by the radius of a particle, the effect can be, for instance, used to separate particles with different sizes.

In biological systems, rectification of thermal fluctuations was first considered by Huxley in his theory of force generation by muscle fibers [183, 184]. The noise-induced transport of particles has been used to explain action of molecular motors in biological cells where nonequilibrium chemical reactions may provide a source of nonthermal fluctuations [185–187, 219].

The ratchets represent simple passive machines that can function only because the medium where they are contained is far from thermal equilibrium and this results in the appearance of nonthermal noises and/or temporal modulation of the medium properties. In contrast to this, Maxwell's demon is an example of an *active* machine in an equilibrium medium. Its persistent functioning becomes possible when such a machine, similar to a living organism, receives energy from an external source and dissipates it into a thermal bath. Maxwell called his hypothetical device a *demon*, implying a certain degree of animation that must be inherent in its actions. The molecular machines in biological cells can actively operate and thus may indeed play the role of demons. An example of active self-motion in fluctuating hydrodynamical fields inside a single cell is considered in the next section.

4.2 Navigation in Fluctuating Fields

The interior of a biological cell is composed of molecules with greatly varying molecular weights, from water and basic monomers to proteins and complex polymers. It also includes larger particles, such as vesicles and other organelles. While low-molecular-weight components representing the bulk of the cytoplasm are still in the state of thermal equilibrium (or close to it), the subsystem of macromolecules and particles is far from thermal equilibrium. It receives energy that is then redistributed inside it, going from one element to another, and is finally dissipated into the liquid forming the bulk of the cytoplasm. The flow of energy passing through this subsystem allows its population to maintain autonomous activity and to produce complex patterns of coherent functional behavior. An impressive example of such behavior is provided by the organization of the intracellular traffic.

Protein molecules are commonly transported in the cells inside lipid vesicles. When a new protein molecule is produced, it is loaded into a budding vesicle, which then moves through the cell until it reaches its target at another location. After the arrival at their targets, vesicles become chemically bound, open up and deliver their contents. The intracellular traffic of vesicles should be fast enough to yield physiologically plausible delivery times in the range of a few seconds. Random Brownian motion of vesicles may suffice for this purpose in small cells, but in larger biological cells special mechanisms should apparently be employed to speed up the delivery.

Adam and Delbrück [188] proposed a mechanism of "dimensionality reduction". According to this mechanism, vesicles first arrive at the cellular membrane and then continue random Brownian motion on it until they meet their targets located on the membrane. Since random motion is thus confined to only two dimensions, the finding of a target is facilitated and occurs within a shorter time. Another possibility of dimensionality reduction is that a vesicle becomes first attached to a one-dimensional cellular structure, such as a polymer microtubule, that leads to the target and then performs random Brownian motion along it. The motion along a microtubule can be further enhanced and made directional by using "molecular motors" that act like noise-driven ratchets. These motors are proteins that can undergo a series of conformational changes, leading to movement along the linear structure. Under normal conditions, the protein would randomly diffuse in either direction. The conformational changes can be made irreversible by binding ATP, for example, which arrives randomly due to Brownian motion, hydrolyzing it to ADP plus organic phosphate and releasing eventually the ADP, before a new cycle begins. These ratchets are called Brownian ratchets because they operate under isothermal conditions transforming chemical energy into directed motion. Kinesin and Dynein are such motor proteins, each of which performs motion in one specific direction along a linear structure. A vesicle connected to such molecules can thus be actively moved. This effect has indeed been observed in an experiment [189] where the vesicles were replaced, for convenience, by tiny rubber spheres. Further examples of dimensionality reduction will be discussed at the end of this chapter.

Though the dimensionality reduction mechanism and the active transport of vesicles along microtubules are realistic, some questions still remain open. For instance, transferring almost all intracellular traffic to the cell membrane may significantly interfere with other important functions of this membrane and it might even be that the membrane has not enough area to accommodate all the required traffic. When the motion of vesicles to their targets is assumed to be guided by the "rails" representing polymer microstructures, each target must have a rail leading to it. In extremely long cells, such as nervous fibers, all vesicles must be basically transported over a large distance in the same direction, and laying a railroad would indeed represent a good solution. In fact, the microtubule array is permanently turned over. The half-life of an individual microtubule is 10 min. Moreover, microtubule structures can be stabilized by capping the outward-bound end to other molecules. It appears that the rail system can be laid down and disassembled very rapidly. Whether such a sophisticated rail system, leading to thousands of different destinations in a normal living cell, is sufficient to ensure all directed transport or whether there are other mechanisms that have to be invoked is still an open question.

The above arguments were essentially based on the assumption that vesicles are passive objects, which either perform random Brownian motion or are moved by molecular motors along the polymer structures. In reality, the vesicles represent structures with rich internal dynamics. The experiments with simple artificially produced vesicles reveal that they may undergo phase transitions, changing their shapes [190]. The biological vesicles used to transport proteins inside the cells are much more complicated. Besides the transported proteins, they include a large number of other proteins incorporated into their membrane and some of the vesicles are even "coated" by such macromolecules. The functions of the membrane proteins are largely unknown. Some of them may be involved in recognizing the target and docking to it.

It is highly plausible that some vesicles may receive chemical energy with the ATP molecules and use it for their active operation. In principle, they can use this energy to propel themselves through the cytoplasm, similar to the active motion of entire biological cells. However, as shown below, this kind of active motion would be very non-economical at the micrometer and submicrometer scales characteristic of the interior of biological cells. At these scales, thermal hydrodynamical fluctuations in the cytoplasm are so strong that they may already effectively drive a particle. The vesicles must only *navigate* in these fluctuating hydrodynamical fields, steering their way to a chosen destination [191].

To explain what is meant by navigation, we first consider a simple example. Under skillful steering, a sailing boat can steadily progress to its destination even when the wind direction changes randomly with time. This is achieved by maximally utilizing the wind velocity when the wind blows in the desired direction and by minimizing the wind action when its direction is reversed. Usually this is done by operating the sail, but the same effect can result from manipulating a floating anchor (resembling a large umbrella): it is taken out of water or folded, when the wind is dragging the boat in the right direction, and is dropped back into water or opened, once the wind's direction has changed.

From a general point of view, three conditions should be present to realize the effect of active navigation. First, there must be some random flows in the considered medium. Second, the medium should include a highly inertial component (water in the above example) which is able to dampen the fluctuations. Third, the navigating agent must be able to detect changes in its motion and actively respond to them.

All these conditions may be satisfied for vesicles inside living cells. Indeed, thermal fluctuations create sufficiently strong random flows in the cytoplasm (the intensity of these flows is estimated below). The inertial component inside a cell is provided by the cytoskeleton which spans the cell's interior. The mesh size of this intracellular polymer structure is often not significantly larger than the diameter of vesicles. Friction is produced when they are dragged by fluid flows through the polymer network. The intensity of such friction is highly sensitive to the size and shape of vesicles. We have already noted that biological vesicles may represent complex molecular machines, capable of active operation. It seems plausible that they can detect the direction of motion and respond appropriately to this information. An internal reference frame is, for instance, provided by the electrostatic fields inside a cell. The detectors of the direction of motion, sensitive to electrostatic fields, can have molecular size. They can trigger conformational changes in the coat proteins, leading to phase transitions in vesicles and the modification of their shapes. Thus, the friction with respect to the polymer mesh can be effectively controlled.

When a vesicle is dragged by a hydrodynamic flow through the polymer mesh, it experiences two forces. The first of them is the Stokes force of viscous friction for a body in a moving fluid,

$$\mathbf{F}_1 = -(\mathbf{u} - \mathbf{v})/\mu, \tag{4.12}$$

where $\mathbf{u}$ is the vesicle velocity, $\mathbf{v}$ is the velocity of the flow and μ is the mobility of the vesicle. Since thermal fluctuating flows are considered, their local velocity is a random function of time, $\mathbf{v} = \mathbf{v}(t)$.

The second force corresponds to the friction produced by the immobile polymer mesh when the vesicle is dragged through it. In the simplest approximation, this force is proportional to the velocity of the vesicle, i.e.

$$\mathbf{F}_2 = -\varkappa\mathbf{u}, \tag{4.13}$$

where $\varkappa$ is the friction coefficient. We assume that the vesicles can detect the direction of their motion and respond to this by changing their friction with respect to the polymer mesh. Therefore, their friction coefficient is a certain function of the velocity, $\varkappa = \varkappa(\mathbf{u})$, reaching a maximum when the flow drags the vesicle away from its target. For example, one can take

$$\varkappa = \varkappa_1 \text{ for } u_x > 0, \text{ and } \varkappa = \varkappa_2, \text{ for } u_x < 0, \tag{4.14}$$

where u_x is the velocity component in the preferred direction of motion and $\varkappa_1 < \varkappa_2$.

The stochastic equation of motion of a vesicle is therefore [191]

$$m\frac{d\mathbf{u}}{dt} = -\varkappa(\mathbf{u})\mathbf{u} - (\mathbf{u} - \mathbf{v}(t))/\mu, \qquad (4.15)$$

where m is the mass of the vesicle. Comparing it with the equation (4.4) of a noise-driven ratchet, we see that the periodic fluctuating potential is now absent, but the moving particle is "intelligent" and can control its friction with respect to the inertial reference frame.

On microscales, i.e. for particles of micrometer and submicrometer size (the typical radius of a biological vesicle is $R \sim 0.1\,\mu\text{m}$), the equation of motion can be further simplified. According to (4.15), the characteristic time required for a vesicle to reach its steady velocity when the velocity $\mathbf{v}$ of the dragging flow is constant is

$$\tau = \frac{m\mu}{1 + \varkappa\mu}. \qquad (4.16)$$

Its upper estimate can be obtained by putting $\varkappa = 0$ in this equation. The mobility μ of a vesicle can be roughly estimated (see [191]) assuming that it has a spherical shape and moves through the fluid with viscosity η. This yields $\mu = (6\pi\eta R)^{-1}$. On the other hand, the mass of a vesicle is roughly given by $m = (4\pi/3)\rho_0 R^3$, where ρ_0 is the vesicle density, which is close to the density of water. Hence, we obtain $\tau \sim \rho_0 R^2/\eta$. Taking the cytoplasm viscosity equal to that of pure water ($\eta = 10^{-2}$ g cm^{-1}s^{-1}), we find that this relaxation time is about 10^{-8} s. This is by many orders of magnitude shorter than all other characteristic timescales of the problem and therefore one can safely assume that the velocity $\mathbf{u}$ of the vesicle adiabatically follows the local fluctuating velocity $\mathbf{v}(t)$ of the flow, i.e. it is given by the algebraic equation

$$[1 + \mu\varkappa(\mathbf{u})]\,\mathbf{u} = \mathbf{v}(t). \qquad (4.17)$$

The absence of inertia in the motion of microparticles is a general property of such objects. In effect, they obey not the Newton law of motion, but the law of Aristotle, according to which the velocity is proportional to the driving force. This was first noted by Purcell in his analysis of bacterial motion [192].

If the velocity dependence of the coefficient of friction is given by (4.14), then (4.17) can easily be solved and we obtain

$$\mathbf{u}(t) = \begin{cases} (1 + \mu\varkappa_1)^{-1}\mathbf{v}(t), & \text{for } v_x(t) > 0 \\ (1 + \mu\varkappa_2)^{-1}\mathbf{v}(t), & \text{for } v_x(t) < 0 \end{cases}. \qquad (4.18)$$

To find the mean drift velocity $\langle\mathbf{u}\rangle$ of the vesicle, the statistical properties of the fluctuating flow velocity $\mathbf{v}(t)$ must be known.

The intensity of the fluctuating fluid flows can be estimated using the theory of hydrodynamical fluctuations that was formulated in 1957 by Landau and Lifshitz. They have shown [193] that thermal fluctuations can be taken into account by introducing random forces into the equations of fluid mechanics:

$$\rho \frac{\partial \mathbf{v}}{\partial t} + \rho \left(\mathbf{v} \cdot \nabla \right) \mathbf{v} = -\nabla p + \eta \nabla^2 \mathbf{v} + \left(\varsigma + \frac{\eta}{3} \right) \nabla \left(\operatorname{div} \mathbf{v} \right) + \mathbf{f}(\mathbf{r}, t), \quad (4.19)$$

$$\frac{\partial \rho}{\partial t} + \operatorname{div}(\rho \mathbf{v}) = 0. \quad (4.20)$$

The random forces $\mathbf{f}(\mathbf{r}, t)$ in the Navier–Stokes equation (4.19) are given by

$$f_i(\mathbf{r}, t) = \sum_{i=1}^{3} \frac{\partial s_{ik}}{\partial x_i}, \quad (4.21)$$

where $s_{ik}(\mathbf{r}, t)$ are random fields representing the components of the fluctuating stress tensor. These random fields are Gaussian and delta-correlated in space and in time, so that $\langle s_{ik}(\mathbf{r}, t) \rangle = 0$ and

$$\langle s_{ik}(\mathbf{r}, t) s_{lm}(\mathbf{r}', t') \rangle = \quad (4.22)$$

$$2 k_\mathrm{B} T \left[\eta \left(\delta_{il} \delta_{km} + \delta_{im} \delta_{kl} \right) + \left(\varsigma - \frac{2\eta}{3} \right) \delta_{ik} \delta_{lm} \right] \delta \left(\mathbf{r} - \mathbf{r}' \right) \delta \left(t - t' \right). \quad (4.23)$$

Here we use the notation $\delta_{ij} = 1$ if $i = j$ and $\delta_{ij} = 0$ if $i \neq j$.

The solution of the stochastic partial differential equations (4.19) and (4.20) is a fluctuating velocity field $\mathbf{v}(\mathbf{r}, t)$. We want to find the power spectrum $S_{ij}(\mathbf{k}, \omega)$ of this random field, defined as[1]

$$S_{ij}(\mathbf{k}, \omega) = \frac{1}{(2\pi)^2} \int \int \langle v_i(\mathbf{r}, t) v_j(\mathbf{r} + \rho, t + \tau) \rangle \exp\left[-i \left(\mathbf{k}\rho - \omega\tau \right) \right] d\rho d\tau.$$

$$(4.24)$$

If the fluid is at rest, its velocity is zero ($\mathbf{v} = 0$) and its density is constant ($\rho = \rho_0$) in absence of thermal fluctuations. The random forces $\mathbf{f}(\mathbf{r}, t)$ induce fluctuating flows. Assuming that these flows are sufficiently weak, we can apply the perturbation approximation to find their statistical properties. By putting $\rho = \rho_0 + \delta\rho$ and $\mathbf{v} = \delta\mathbf{v}$ and neglecting nonlinear terms, we obtain

$$\rho_0 \frac{\partial \mathbf{v}}{\partial t} = -\nabla \delta p + \eta \nabla^2 \mathbf{v} + \left(\varsigma + \frac{\eta}{3} \right) \nabla \left(\operatorname{div} \mathbf{v} \right) + \mathbf{f}(\mathbf{r}, t), \quad (4.25)$$

$$\frac{\partial \delta\rho}{\partial t} + \operatorname{div}(\rho_0 \mathbf{v}) = 0. \quad (4.26)$$

The fluctuating velocity field $\mathbf{v}(\mathbf{r}, t)$ can be decomposed into a sum of the potential and vortex ("transverse") components as $\mathbf{v} = \mathbf{v}^{(\mathrm{l})} + \mathbf{v}^{(\mathrm{t})}$ where $\operatorname{div} \mathbf{v}^{(\mathrm{t})} = 0$ and $\operatorname{rot} \mathbf{v}^{(\mathrm{l})} = 0$. Performing this separation in (4.25) and (4.26), we obtain

---

[1] The readers who are less interested in the calculation details may skip the following derivation and proceed to (4.36).

$$\rho_0 \frac{\partial \mathbf{v}^{(t)}}{\partial t} = \eta \nabla^2 \mathbf{v}^{(t)} + \mathbf{f}^{(t)}(\mathbf{r}, t), \tag{4.27}$$

$$\rho_0 \frac{\partial \mathbf{v}^{(l)}}{\partial t} = -\nabla \delta p + \eta \nabla^2 \mathbf{v}^{(l)} + \left(\varsigma + \frac{\eta}{3}\right) \nabla \left(\operatorname{div} \mathbf{v}^{(l)}\right) + \mathbf{f}^{(l)}(\mathbf{r}, t), \tag{4.28}$$

$$\frac{\partial \delta \rho}{\partial t} + \operatorname{div}(\rho_0 \mathbf{v}^{(l)}) = 0, \tag{4.29}$$

where $\mathbf{f}^{(l)}(\mathbf{r}, t)$ and $\mathbf{f}^{(t)}(\mathbf{r}, t)$ are the potential and the transverse components of the random force field $\mathbf{f}(\mathbf{r}, t)$, such that div $\mathbf{f}^{(t)}(\mathbf{r}, t) = 0$ and rot $\mathbf{f}^{(l)}(\mathbf{r}, t) = 0$.

We see that the potential component $\mathbf{v}^{(l)}$ of the velocity field and the density fluctuations $\delta \rho$ do not enter into the equation (4.27) that determines the fluctuating transverse component $\mathbf{v}^{(t)}$ of the velocity. Therefore, these hydrodynamical fluctuations are independent in the linear approximation.

The liquids are almost incompressible. Therefore (see [193]), the intensity of potential fluctuating velocity flows that are coupled to the density fluctuations is small, and we shall neglect such fluctuations below, focusing our analysis on the stronger transverse velocity fluctuations that are described by (4.27).

It is convenient to perform the Fourier transformation of the fluctuating field $\mathbf{v}^{(t)}(\mathbf{r}, t)$ and define

$$\mathbf{v}^{(t)}(\mathbf{k}, \omega) = \frac{1}{2\pi^2} \int \int \mathbf{v}^{(t)}(\mathbf{r}, t) \exp\left[-i\left(\mathbf{k}\mathbf{r} - \omega t\right)\right] d\mathbf{r} dt. \tag{4.30}$$

Applying this transformation to (4.27), we find

$$-i\omega \rho_0 \mathbf{v}^{(t)}(\mathbf{k}, \omega) = -\eta k^2 \mathbf{v}^{(t)}(\mathbf{k}, \omega) + \mathbf{f}^{(t)}(\mathbf{k}, \omega), \tag{4.31}$$

where $\mathbf{f}^{(t)}(\mathbf{k}, \omega)$ is the Fourier transform of $\mathbf{f}^{(t)}(\mathbf{r}, t)$. The solution of this equation is

$$\mathbf{v}^{(t)}(\mathbf{k}, \omega) = -\frac{\mathbf{f}^{(t)}(\mathbf{k}, \omega)}{i\omega \rho_0 + \eta k^2}. \tag{4.32}$$

Note that both $\mathbf{v}^{(t)}(\mathbf{k}, \omega)$ and $\mathbf{f}^{(t)}(\mathbf{k}, \omega)$ are random here.

The random force field $\mathbf{f}^{(t)}(\mathbf{r}, t)$ must satisfy the equation div $\mathbf{f}^{(t)} = 0$ and therefore its Fourier components $\mathbf{f}^{(t)}(\mathbf{k}, \omega)$ should obey the condition $\mathbf{k} \cdot \mathbf{f}^{(t)}(\mathbf{k}, \omega) = 0$. We can use this condition and write

$$f_i^{(t)}(\mathbf{k}, \omega) = f_i(\mathbf{k}, \omega) - \frac{k_i}{k^2} \left(\mathbf{k} \cdot \mathbf{f}(\mathbf{k}, \omega)\right), \tag{4.33}$$

where $\mathbf{f}(\mathbf{k}, \omega)$ is the Fourier transform of the full fluctuating force field $\mathbf{f}(\mathbf{r}, t)$. The Fourier transformation of (4.21) further yields

$$f_i(\mathbf{k}, \omega) = i \sum_{j=1}^{3} k_j s_{ij}(\mathbf{k}, \omega), \tag{4.34}$$

where $s_{ij}(\mathbf{k}, \omega)$ is the Fourier transform of the fluctuating stress tensor $s_{ij}(\mathbf{r}, t)$. Combining (4.32), (4.33) and (4.34), we obtain

$$v_i^{(t)}(\mathbf{k}, \omega) = -\frac{i}{i\omega\rho_0 + \eta k^2} \sum_{j,l=1}^{3} \left(\delta_{il} - \frac{k_i k_l}{k^2} \right) k_j s_{lj}(\mathbf{k}, \omega). \qquad (4.35)$$

Thus, the Fourier components of the fluctuating velocity field are now expressed in terms of the fluctuating stress tensor $s_{ij}(\mathbf{r}, t)$.

The next step is to use this result and calculate the power spectrum $S_{ij}(\mathbf{k}, \omega)$. This is done by applying the inverse Fourier transformation to (4.35) and finding the velocity–velocity correlation function $\langle v_i(\mathbf{r}, t) v_j(\mathbf{r} + \rho, t + \tau) \rangle$, which can be further substituted into equation (4.24). The final result is [193]

$$S_{ij}(\mathbf{k}, \omega) = \frac{2k_B T}{\rho_0} \left(\delta_{ij} - \frac{k_i k_j}{k^2} \right) \frac{\nu k^2}{\omega^2 + \nu^2 k^4}, \qquad (4.36)$$

where the kinematic viscosity coefficient $\nu = \eta/\rho_0$ is introduced.

As follows from (4.36), the fluctuating velocity field includes pulsations that are characterized by different length scales and timescales. When active navigation of "intelligent" particles in such fluctuating fields is considered, it would be reasonable to assume that these particles cannot respond to very rapid fluctuations, whose timescales exceed a certain characteristic time τ. Moreover, fluctuations with length scales shorter than a certain characteristic length l (determined by the size of a particle) also cannot effect its motion. This means that such particles are able to respond only to the smooth fluctuating velocity field $\widetilde{\mathbf{v}}(\mathbf{r}, t)$, which is obtained by averaging over time intervals τ and volume elements ΔV of linear size l. This fluctuating field will therefore contain only Fourier modes with $|\mathbf{k}| < 2\pi/l$ and $|\omega| < 2\pi/\tau$.

The mean intensity of such velocity fluctuations is given by

$$\langle \widetilde{\mathbf{v}}^2 \rangle_{l\tau} = \sum_{i,j=1}^{3} \int_{|\omega| < 2\pi/\tau} d\omega \int_{|\mathbf{k}| < 2\pi/l} d\mathbf{k} \, S_{ij}(\mathbf{k}, \omega) \qquad (4.37)$$

$$= \frac{4k_B T}{\rho_0} \int_{|\omega| < 2\pi/\tau} d\omega \int_{|\mathbf{k}| < 2\pi/l} d\mathbf{k} \, \frac{\nu k^2}{\omega^2 + \nu^2 k^4}. \qquad (4.38)$$

If the condition $\tau \gg l^2/\nu$ is satisfied, this integral yields an estimate

$$\langle \widetilde{\mathbf{v}}^2 \rangle_{l\tau} \sim \frac{k_B T}{\eta \tau l}. \qquad (4.39)$$

For water at room temperature ($T = 300$ K, $k_B T = 4.14 \times 10^{-14}$ g cm^2 s^{-2}) the viscosity is $\eta = 10^{-2}$ g cm^{-1} s^{-1} and the density is $\rho_0 = 1$ g cm^{-3}. Therefore, we obtain, for example, for thermal pulsations with $l = 10^{-4}$ cm and $\tau = 10^{-2}$ s that $\langle \tilde{\mathbf{v}}^2 \rangle_{l\tau} \sim 10^{-6}$ cm^2 s^{-2}. Hence, the mean velocity of such pulsations is about 10^{-3} cm s^{-1}.

Note that the considered velocity fluctuations represent a Gaussian random field. Indeed, they are given by solutions of the stochastic linear differential equations (4.25) and (4.26) with the Gaussian random forces $\mathbf{f}(\mathbf{r}, t)$. Thus, the probability distribution for the velocities of fluctuating flows, averaged over length l and time τ, is

$$P(\mathbf{v}) = \frac{1}{(2\pi\sigma)^{3/2}} \exp\left(-\frac{\mathbf{v}^2}{2\sigma}\right), \tag{4.40}$$

where $\sigma = \langle \tilde{\mathbf{v}}^2 \rangle_{l\tau}$.

Since the intensity of thermal velocity fluctuations has been estimated, we can now return to the analysis of navigation in these fluctuating flows described by (4.15). Because the vesicles cannot feel fluctuations with the characteristic lengths shorter than their size l and because of the inertiality of their response mechanism, they cannot react to velocity fluctuations that occur within time intervals shorter than a certain characteristic time τ. This means that the fluctuating velocity $\mathbf{v}(t)$ in the equation of motion (4.15) is characterized by the probability distribution (4.40) with a dispersion $\langle \tilde{\mathbf{v}}^2 \rangle_{l\tau}$ given by equation (4.39). The solution of the stochastic differential equation (4.15) yields the fluctuating vesicle velocity $\mathbf{u}(t)$, whose average $\langle \mathbf{u} \rangle$ represents the drift velocity of vesicles.

The calculation of the drift velocity is easily performed if we take into account that, as noted above, the velocity of the microparticles almost instantaneously adjusts to the acting force. According to (4.17), the velocity $\mathbf{u}(t)$ of a vesicle is determined by the flow velocity $\mathbf{v}(t)$ at the same moment of time. Moreover, if the coefficient of friction $\varkappa$ depends only on the direction of the vesicle's motion along the x-axis, i.e. $\varkappa = \varkappa_1$ for $u_x > 0$ and $\varkappa = \varkappa_2$ for $u_x < 0$, we can use (4.18) to find the drift velocity of vesicles:

$$\langle u \rangle = \frac{1}{\sqrt{2\pi\sigma}} \int\limits_0^\infty \frac{v_x \exp\left(-v_x^2/2\sigma\right)}{1 + \varkappa_1\mu} dv_x + \frac{1}{\sqrt{2\pi\sigma}} \int\limits_{-\infty}^0 \frac{v_x \exp\left(-v_x^2/2\sigma\right)}{1 + \varkappa_2\mu} dv_x \tag{4.41}$$

$$= \sqrt{\frac{\sigma}{2\pi}} \frac{(\varkappa_2 - \varkappa_1)\mu}{(1 + \varkappa_1\mu)(1 + \varkappa_2\mu)}. \tag{4.42}$$

Thus, when the friction coefficient is only weakly dependent on the direction of motion, i.e. if $\varkappa_1 = \varkappa_0 - \delta\varkappa$ and $\varkappa_2 = \varkappa_0 + \delta\varkappa$ with $\delta\varkappa << \varkappa_0$, the drift velocity is approximately

$$\langle u \rangle = \sqrt{\frac{2\sigma}{\pi}} \frac{\mu\delta\varkappa}{(1 + \varkappa_0\mu)^2}. \tag{4.43}$$

In the opposite case, where $\varkappa_1 \ll 1/\mu \ll \varkappa_2$, we have

$$\langle u \rangle = \sqrt{\frac{\sigma}{2\pi}}. \tag{4.44}$$

We see that the maximum drift velocity, which may be achieved by active navigation of "intelligent" particles in the fluctuating velocity fields, is determined by the mean velocity of random fluid flows, i.e. $\langle u \rangle \sim \sqrt{\langle \tilde{\mathbf{v}}^2 \rangle}_{l\tau}$. It depends on the characteristic length (l) and time (τ) scales that can be sensed and resolved by the "intelligent" particle. For navigation in thermal fluctuating velocity flows, our estimates give

$$\langle u \rangle \sim \left(\frac{k_B T}{\eta \tau l} \right)^{1/2}. \tag{4.45}$$

As an example, let us assume that an intelligent particle can respond to flow fluctuations with a characteristic time $\tau = 10^{-2}$ s and with the length scale $l = 10^{-5}$ cm. If the viscosity of the cytoplasm is taken to be the same as that of water ($\eta = 10^{-2}$ g cm^{-1} s^{-1}), the maximum drift velocity that might be achieved by active navigation at room temperature ($T = 300$ K, $k_B = 1.38 \times 10^{-16}$ g cm^2 s^{-2}) is estimated according to (4.45) as $\langle u \rangle \approx 10^{-3}$ cm s^{-1}. A vesicle drifting at such a velocity would cross a cell of diameter 10 μm in about a second.

Our analysis of active navigation in thermal fluctuating hydrodynamical fields shows that it can be employed to organize the traffic of particles, such as vesicles, inside biological cells. The conditions needed for active navigation are not stringent and therefore this mechanism of intracellular traffic is plausible, insofar as the physical aspects are concerned. Whether it is actually used by some biological cells is an open question. On the other hand, even if the principles of active navigation are not naturally employed in biology, they may well be utilized in the construction of artificial, man-made systems. Active intelligent particles, navigating in fluctuating thermal fields, can be good transportation vehicles in the molecular-machine devices designed to operate at microscales.

4.3 Long Tails in Probability Distributions

In the section above we have mainly discussed the effects of noise and nonequilibrium fluctuations in microscopic systems. However, if we look at the other extreme and consider large-scale social and economic systems, it can be seen as well that the presence of noise represents an intrinsic aspect of their dynamics. Moreover, some of their statistical properties can only be understood if the selective response of such systems to fluctuations is taken into account.

At the end of the 19th century the social economist Pareto collected statistics on the income and wealth of individuals in many countries at various times in history. In his book he came to the conclusion that "in all places and at all times the distribution of income in a stable economy, when the origin of measurement is at a sufficiently high income level, will be given approximately by the empirical formula $y = ax^{-\nu}$, where y is the number of people having an income x or greater and ν is approximately 1.5"[194]. This observation was confirmed by subsequent studies. For example, Montroll and Shlesinger have analyzed [195] the statistics of data in the full range of incomes in the USA in the 1935/1936 fiscal year, as reported by the Internal Revenue Service, and constructed the distribution shown in Fig. 4.3.

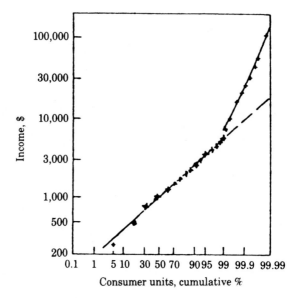

Fig. 4.3. Distribution of families and single individuals by income levels in 1935/1936 in the USA. (From [195])

A convenient way of displaying such statistical data is to use a log-log plot. Suppose that x is the income and $p(x)$ is the statistical distribution of incomes in the economy under consideration. Then the fraction y of the total population having an income x or greater is given by the integral

$$y = \int_x^\infty p(x')dx'. \tag{4.46}$$

If the distribution obeys a power law $y = ax^{-\nu}$, we have $\ln y = \ln a - \nu \ln x$ and therefore the log-log plot of this distribution would represent a straight line with slope $-\nu$.

Examining Fig. 4.3, we see that for very high incomes the distribution indeed follows a power law with exponent $\nu = 1.5$, in excellent agreement with the Pareto rule. This rule holds for the last 1% of the population. Remarkably, Montroll and Shlesinger also found that the main part of the income distribution, in the range between 5 and 97 percent of the total population, can also be well approximated by a power law, but with a different exponent $\nu = 1$.

According to (4.46), the probability distribution $p(x)$ is given by $p(x) = -dy/dx$. Therefore, it follows the power law $p(x) \sim x^{-(\nu+1)}$. In the middle range of incomes the probability distribution has thus the form $p(x) \sim x^{-2}$, whereas for very high incomes the exponent is changed and one gets $p(x) \sim x^{-2.5}$.

The existence of such universal power-law distributions in the economic data is an intriguing problem. But it turns out that similar distributions are also found in other systems of greatly different nature.

Lotka published in 1926 the article "The frequency distribution of scientific productivity" [196]. The article starts with the comment that "it would be of interest to determine, if possible, the part which men of different calibre contribute to the progress of science". A count was made of the number of authors (whose names started with A or B) with different numbers of publications in the decennial index of Chemical Abstracts 1907–1916. A similar procedure was applied to the entire name index of Auerbach's *Geschichtstafeln der Physik* (J. A. Barth, Leipzig, 1910), which covered the entire range of history up to the year 1900 and included only outstanding contributions in physics.

Figure 4.4 shows the log-log plot of the percentage of authors having a given number of publications in Chemical Abstracts (crosses) and in Auerbach's tables (circles). In both cases the points are closely scattered about a straight line with the slope $\nu = 2$. Determined by least squares, the slope of the curve for Auerbach's data is found to be 2.021 ± 0.017, whereas the data from Chemical Abstracts yields a slightly smaller exponent of 1.888 ± 0.007.

Lotka himself became famous for his paper [197], in which he first proposed a model of periodic chemical reaction, and his book *Elements of Physical Biology* [198]. Concerning his discovery of a power-law distribution of scientific publications, he comments that the frequency distributions of this type have a wide range of applicability to a variety of phenomena and "the mere form of such a distribution throws little or no light on the underlying physical relations" [196].

It should be noted that power-law distributions are also known in physics. The classical example is the energy distribution in developed hydrodynamical turbulence derived in 1941 by Kolmogorov [199]. Moreover, the power-law distributions are also characteristic of equilibrium physical systems with second-order phase transitions (e.g., see [200]). However, in contrast to turbulence the power-law fluctuations in these systems are observed only in a narrow

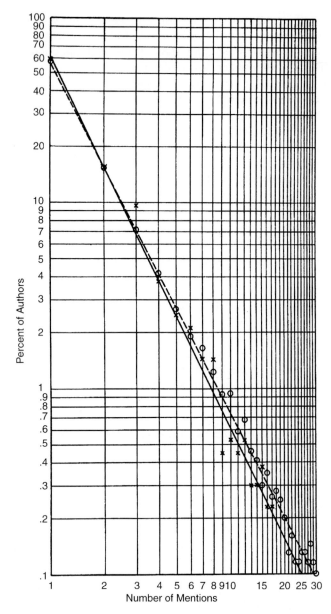

Fig. 4.4. Logarithmic frequency diagram showing the number of authors mentioned once, twice, etc. in Auerbach's tables (points indicated by crosses) and in Chemical Abstracts, letters A and B (points indicated by circles). The fully drawn line indicates points given by the inverse square law, exponent = 2; the dashed line corresponds to exponent = 1.89. (From [196])

interval of the control parameter, i.e. in the proximity of a critical point. This problem has attracted much attention, and sophisticated methods based on the renormalization group were developed for its solution.

The analogies with the statistical physics of phase transitions have led to a conjecture that, if a system shows a persistent power-law distribution, it must contain some intrinsic mechanisms that would bring it into the vicinity of a critical point and keep it there. This concept of *self-organized criticality* has been formulated by Bak et al. [201], who illustrated it by considering an example of an evolving sand pile. It has been subsequently applied to other systems.

However, self-organized criticality is only one of the possible mechanisms that can generate power-law distributions. Below in this section we show that such distributions naturally arise in problems that involve *multiplicative noises*.

Let us first consider a simple system described by the stochastic differential equation

$$\dot{x} = f(t)x, \tag{4.47}$$

where $f(t)$ is the Gaussian delta-correlated (white) noise with $< f(t) >= 0$ and

$$< f(t)f(t') >= 2\sigma\delta(t - t'). \tag{4.48}$$

To understand the model, note that if we put $f = $ const in (4.47), it would correspond to a system having a fixed point $x = 0$. If $f > 0$, this point is unstable and small deviations from it would exponentially increase with time. On the other hand, if $f < 0$, the deviations would exponentially decrease with time and the system would tend to return to the fixed point. Thus, if the coefficient $f(t)$ is not constant and randomly varies around a zero average value, this equation effectively describes a system that alternates between explosion-like growth (when $f(t) > 0$) and exponential decay (when $f(t) < 0$).

We want to show that this simple system already leads to the development of a power-law distribution. First, however, a comment concerning the stochastic differential equation (4.48) should be made. If the noise $f(t)$ in this equation is delta-correlated in time, its right-hand side does not represent a continuous function of time and therefore this equation cannot generally be treated just as a differential equation of a dynamical system with some time-dependent parameters. One can, however, assume that in reality the noise has a small correlation time, much shorter than the characteristic timescales on which we want to consider the dynamics of this system. This assumption is known as the *Stratonovich interpretation* of stochastic differential equations (e.g., see [176] for further discussion). When it is made, both the noise and all the trajectories of the stochastic process $x(t)$ are (almost) continuous on very short scales and the standard procedures of differential calculus can be used.

Introducing the new variable $z = \ln x$, (4.48) can be transformed to the equation

$$\dot{z} = f(t), \tag{4.49}$$

which describes the one-dimensional Brownian motion of a particle with coordinate z. The probability distribution for the coordinate z at time t is therefore

$$p(z,t) = \frac{1}{\sqrt{2\pi\sigma t}} \exp\left(-\frac{z^2}{2\sigma t}\right). \tag{4.50}$$

To determine the corresponding probability density $p(x,t)$ for the variable x, we note that

$$p(x,t) = p(z(x),t)\frac{dz}{dx}, \tag{4.51}$$

and thus obtain the *log-normal distribution*

$$p(x,t) = \frac{1}{x\sqrt{2\pi\sigma t}} \exp\left[-\frac{(\ln x)^2}{2\sigma t}\right]. \tag{4.52}$$

Though the nonstationary distribution (4.52) does not have the exact form of a power law, it is well approximated by the dependence $p(x) \sim 1/x$ in a wide interval of values of the variable x, that grows with time [195]. To demonstrate this, let us consider

$$\ln p(x,t) = -\ln x - \frac{1}{2}\ln(2\pi\sigma t) - \frac{1}{2\pi\sigma t}(\ln x)^2. \tag{4.53}$$

The last term on the right-hand side in this equation can be viewed as the correction that modifies the first term, yielding the power law $p(x) \sim 1/x$. Let us fix a precision parameter $\varepsilon \ll 1$ and find the interval of x where the last term remains small and the modification is negligible. This interval is defined by the inequality

$$\frac{1}{2\pi\sigma t}(\ln x)^2 < \varepsilon\,|\ln x|, \tag{4.54}$$

which is equivalent to

$$\exp(-2\pi\varepsilon\sigma t) < x < \exp(2\pi\varepsilon\sigma t). \tag{4.55}$$

Hence, in the long-time limit, when $t > (\varepsilon\sigma)^{-1}$, the interval of the power-law behavior is exponentially large.

We see that the power-law distribution with the exponent $\nu = 1$ can easily be generated by a system (4.47) with multiplicative noise. But the effects of multiplicative noise can also explain the emergence of power laws with other characteristic exponents. To show this, let us now consider a system described by the stochastic differential equation

$$\dot{x} = (-a + f(t))x + b, \tag{4.56}$$

which includes a small positive constant term b. The principal role of this term is that it prevents the system from coming too close to the point $x = 0$, while not affecting its behavior at large x. Note that in the absence of noise this system has a stable state with $x = x_1 = b/a$.

The probability distribution $p(x, t)$ for the stochastic process, defined by (4.56) in the Stratonovich interpretation, obeys the Fokker–Planck equation (see [176])

$$\frac{\partial p}{\partial t} = -\frac{\partial}{\partial x}\left[(-ax + \sigma x + b)\,p\right] + \sigma\frac{\partial^2}{\partial x^2}\left(x^2 p\right). \tag{4.57}$$

Its stationary solution satisfies the equation

$$-\frac{\partial}{\partial x}\left[(-ax + \sigma x + b)\,p\right] + \sigma\frac{\partial^2}{\partial x^2}\left(x^2 p\right) = 0, \tag{4.58}$$

which can be integrated once, yielding

$$-(-ax + \sigma x + b)\,p + \sigma\frac{d}{dx}\left(x^2 p\right) = J, \tag{4.59}$$

where J is a constant of integration. This constant must be zero, as follows from the following arguments:

Though the trajectories of the stochastic process $x(t)$ defined by (4.56) may come very close to the point $x = 0$, they can never cross this point. This means that the point $x = 0$ represents the so-called *natural boundary* for the considered stochastic process. Since this boundary cannot be reached, the probability distribution $p(x)$ must vanish in the limit $x \to 0$. According to (4.59), this is only possible if $J = 0$ in this equation.

The linear equation (4.59) with $J = 0$ is easily integrated if we introduce $Q(x) = x^2 p(x)$ and write it as

$$\sigma\frac{d\ln Q}{dx} + (a - \sigma)\frac{1}{x} - b\frac{1}{x^2} = 0. \tag{4.60}$$

Using additionally the normalization condition

$$\int_0^\infty p(x)dx = 1 \tag{4.61}$$

we finally obtain

$$p(x) = \frac{1}{\Gamma\left(a/\sigma\right)}\left(\frac{b}{\sigma}\right)^{a/\sigma}\frac{1}{x^{1+a/\sigma}}\exp\left(-\frac{b}{\sigma x}\right), \tag{4.62}$$

where $\Gamma\left(a/\sigma\right)$ is the gamma-function of the argument a/σ.

The exponential term in the distribution (4.62) approaches unity at $x \gg b/\sigma$, and hence in the limit of $x \to \infty$ this distribution follows a power law

$$p(x) \propto \frac{1}{x^{1+a/\sigma}}. \tag{4.63}$$

Its exponent $\nu = 1 + a/\sigma$ can take various values, depending on the parameters a and σ.

Using the obtained stationary probability distribution, we can also calculate the statistical mean value $< x >$ of the variable x in this steady state. The result is

$$< x >= \frac{b}{a - \sigma}. \qquad (4.64)$$

The calculated mean value $< x >$ is thus *different* from the steady state $x = b/a$ that is found in the absence of noise. Moreover, when the noise intensity σ is increased, the mean value $< x >$ grows and becomes infinite at $\sigma = a$. It is interesting to note that at the boundary of this instability, i.e. for $\sigma = a$, the power law (4.63) is characterized by the exponent $\nu = 2$.

The above results were obtained, assuming the Stratonovich interpretation of the stochastic differential equation (4.56). If the Ito interpretation is used instead, the only modification is that the parameter a should be replaced by $a + \sigma$ in the probability distributions (4.62) and (4.63) and in (4.64). The existence of a stationary power-law distribution for multiplicative stochastic processes with injection has been demonstrated in 1973 by Kesten [202] for a model that represented a discrete version of the stochastic differential equation (4.56).

To conclude this section, we present a different example of the probability theory where a power-law distribution also appears [203]. Suppose that we can generate independent positive random numbers h with some probability density $p(h)$. Let us choose at random one such number h and take it as a reference level. Then we go on and generate a sequence of other random numbers $h_1, h_2, \ldots, h_n$ which is terminated when a random number is found that exceeds the set reference level h, so that $h_1, h_2, \ldots, h_{n-1} < h$ and $h_n > h$. This procedure is repeated with new random choices of the reference levels h, yielding a certain distribution of lengths n of generated sequences. To determine this distribution, we first note that the probability $P(h, n)$ to generate such a sequence of n independent random numbers for a fixed reference level h is given by

$$P(h, n) = [q(h)]^{n-1} [1 - q(h)], \qquad (4.65)$$

where $q(h)$ is the probability that a random number does not exceed h, i.e.

$$q(h) = \int_0^h p(h')dh', \qquad (4.66)$$

and $1 - q(h)$ is the probability that a random number exceeds the level h. Next, we take into account that the reference level h in (4.65) is itself a random number obeying the same probability distribution $p(h)$. Therefore, the probability to find a sequence of length n with a random choice of the reference level is given by the integral

$$P(n) = \int_0^\infty P(h,n)p(h)dh. \tag{4.67}$$

At first glance, it seems that the form of the distribution $P(n)$ would depend on the choice of the probability density $p(h)$. This is not true, however, as can be shown by a simple calculation. We substitute $P(h,n)$ given by (4.65) into (4.67) and use the fact that, according to (4.66), the relationship $dq(h) = p(h)dh$ holds. This yields

$$P(n) = \int_0^\infty [q(h)]^{n-1}\,[1 - q\,(h)]\,p(h)dh = \int_0^1 [q(h)]^{n-1}\,[1 - q\,(h)]\,dq(h), \tag{4.68}$$

$$= \int_0^1 q^{n-1}(1-q)dq = \frac{1}{n(n+1)}. \tag{4.69}$$

Hence, the distribution over sequence lengths is actually *universal*, and for large n it has the power-law form

$$P(n) \approx \frac{1}{n^2}. \tag{4.70}$$

The results of our analysis can be applied to studies of social systems. Let us consider a population of agents (individuals or companies) that produce the same kind of goods and sell them on the market. Suppose that the competition in this market is so strong that only an agent that currently has the best quality can sell its product. While this top agent goes on selling, other agents try to create a product of a higher quality and thus lead the race. The innovation process can roughly be modelled as the random generation of new products with varying quality h, characterized by a certain statistical distribution $p(h)$. The number n of attempts made by other competing agents until they finally override the leader will then have, according to our analysis, a power-law distribution (4.70) with the exponent $\nu = 2$. Furthermore, this number n can also be viewed as determining the time T, during which a given agent is leading the race. Assuming constant production rate and sale rates, the total income x of the leading agent is proportional to T and, hence, to n. Therefore, the distribution of the incomes of the competing agents will also follow the power law

$$P(x) \propto \frac{1}{x^2}. \tag{4.71}$$

Similar mechanisms may also operate in science and determine the publication statistics of individual authors and research laboratories.

Of course, the above arguments should not be taken too seriously. Both economics and science are complex fields of human endeavour and it would be naive to assume that such simple models may suffice for their characterization.

4.4 Noise as a Resource

Earlier in this chapter we have seen that intelligent microparticles can employ thermal fluctuations to power their directed motion towards targets. Turning now to the macroscopic level of ecological and social systems, we may ask whether biological species and humans can also benefit from noise and use it as a resource that improves their relative performance in the competition process.

We consider in this section a simple ecological system with two species that consume the same nutrient. The dynamics of this system is described by the following equations for the concentrations X_1 and X_2 of both species and the nutrient concentration Y:

$$
\begin{aligned}
\dot{X}_1 &= (B_1 Y - A_1) X_1 \\
\dot{X}_2 &= (B_2 Y - A_2) X_2 \\
\dot{Y} &= Q - (C_1 X_1 + C_2 X_2) Y
\end{aligned}
\qquad (4.72)
$$

The reproduction rates of both species are proportional to the nutrient concentration. The nutrient is supplied at a constant rate Q. Its disappearance rate is proportional to concentrations of both feeding species.

In contrast to the classical Lotka–Volterra system, which is conservative and shows oscillations [197, 204], the dynamics of the model (4.72) is dissipative. Since both species compete for the same resource, their coexistence is not possible in the system under consideration. It can easily be checked that it has only one attractive stationary state:

$$
Y^0 = \frac{A_1}{B_1}, \; X_1^0 = \frac{Q B_1}{C_1 A_1}, \; X_2^0 = 0, \qquad (4.73)
$$

when the condition

$$
\frac{A_1}{B_1} < \frac{A_2}{B_2} \qquad (4.74)
$$

is satisfied. This state corresponds to the extinction of the weak species X_2.

The coexistence of both species is not possible even if the rate of the nutrient supply fluctuates in time, so that $Q = Q(t)$. To show this, we write the first two equations as

$$
\frac{d}{dt} (\ln X_1) = B_1 Y - A_1 \qquad (4.75)
$$

$$\frac{d}{dt}(\ln X_2) = B_2 Y - A_2. \tag{4.76}$$

Multiplying them by B_2 and B_1, respectively, and subtracting the second equation from the first equation, we obtain

$$B_2 \frac{d}{dt}(\ln X_1) - B_1 \frac{d}{dt}(\ln X_2) = -A_1 B_2 + A_2 B_1. \tag{4.77}$$

Its integration yields

$$B_2 \ln X_1 - B_1 \ln X_2 = (-A_1 B_2 + A_2 B_1) t + \text{ const}, \tag{4.78}$$

so that the relationship

$$X_2(t) = \text{const } [X_1(t)]^{B_2/B_1} \exp\left[-\left(A_2 - \frac{B_2}{B_1}A_1\right)t\right] \tag{4.79}$$

holds. Therefore, if the condition (4.74) is satisfied and the concentration X_1 of the first species does not increase indefinitely with time, the concentration X_2 of the second species goes to zero with time. Hence, the second species undergoes extinction in its competition with the stronger first species. Note that the relationship (4.79) does not include Q, and the fact that the rate Q is constant was not used in its derivation.

Next, we modify the model and consider the situation where the considered ecological system is distributed in space, i.e. described by local densities $X_1(\mathbf{r}, t), X_2(\mathbf{r}, t)$ and $Y(\mathbf{r}, t)$. It will be assumed that the rate Q of the nutrient supply fluctuates in space and time, i.e.

$$Q = Q_0 + f(\mathbf{r}, t), \tag{4.80}$$

where $f(\mathbf{r}, t)$ is a Gaussian noise function with the correlation function

$$\langle f(\mathbf{r}, t) f(\mathbf{r}', t') \rangle = S(\mathbf{r} - \mathbf{r}', t - t'). \tag{4.81}$$

As an example, we will use the noise described by

$$S(\mathbf{r} - \mathbf{r}', t - t') = s_0 \delta(t - t') \exp\left[-\frac{(\mathbf{r} - \mathbf{r}')^2}{r_0^2}\right]. \tag{4.82}$$

Its intensity is s_0 and the correlation radius is r_0; this noise is delta-correlated in time.

Finally, it will be assumed that the weak species X_2 is mobile and diffuses through the medium with a diffusion constant D. Hence, the model is described by the stochastic differential equations:

$$\begin{aligned}
\dot{X}_1 &= (B_1 Y - A_1) X_1, \\
\dot{X}_2 &= (B_2 Y - A_2) X_2 + D\nabla^2 X_2, \\
\dot{Y} &= Q_0 - (C_1 X_1 + C_2 X_2) Y + f(\mathbf{r}, t).
\end{aligned} \tag{4.83}$$

The analysis of this model reveals [205–207] that its behavior is qualitatively different from that described by (4.72). Beginning from a certain critical noise intensity, extinction of the second weak species does not take place and both species can coexist in this ecological system. Hence, the system undergoes *a noise-induced transition.*

Though this result is general, the analysis is simplified if a number of additional assumptions is made. The transition occurs at relatively low noise intensities, where fluctuations are still small, if in the absence of noise the second species is only slightly less efficient than the strong first one, i.e. if the small parameter

$$\varepsilon = \frac{A_2}{B_2} - \frac{A_1}{B_1} \tag{4.84}$$

is present in the model ($\varepsilon << 1$).

Introducing deviations $\delta X_1(\mathbf{r}, t)$ and $\delta Y(\mathbf{r}, t)$ from the uniform stationary state (4.73), the equations (4.83) can be written in the form

$$
\begin{aligned}
\delta \dot{X}_1 &= B_1 X_1^0 \delta Y + B_1 \delta X_1 \delta Y, \\
\dot{X}_2 &= -(A_2 - B_2 Y^0) X_2 + B_2 X_2 \delta Y + D \nabla^2 X_2, \\
\delta \dot{Y} &= -C_1 X_1^0 \delta Y - C_1 Y^0 \delta X_1 - C_2 Y^0 X_2 - C_1 \delta X_1 \delta Y - C_2 X_2 \delta Y + f(\mathbf{r}, t).
\end{aligned}
\tag{4.85}
$$

If the noise intensity is sufficiently low, the nonlinear terms in the first and the third equations are small and can be neglected. The nonlinear term $B_2 X_2 \delta Y$ in the second equation is also small, but we should not neglect it because the first linear term in this equation is proportional to the small parameter ε. This partial linearization yields

$$
\begin{aligned}
\delta \dot{X}_1 &= B_1 X_1^0 \delta Y, \\
\dot{X}_2 &= -\varepsilon B_2 X_2 + B_2 X_2 \delta Y + D \nabla^2 X_2, \\
\delta \dot{Y} &= -C_1 X_1^0 \delta Y - C_1 Y^0 \delta X_1 - C_2 Y^0 X_2 + f(\mathbf{r}, t).
\end{aligned}
\tag{4.86}
$$

It is convenient to introduce new variables

$$u = \frac{B_2}{B_1} \delta X_1, \quad x = \frac{C_2}{B_2} Y^0 X_2, \quad v = \delta Y. \tag{4.87}$$

Moreover, we shall measure time in units of $1/B_2$ and the coordinates in units of the correlation radius r_0 of the noise $f(\mathbf{r}, t)$. In these notations equations (4.86) become

$$
\begin{aligned}
\dot{u} &= v, \\
\dot{x} &= -\varepsilon x + xv + \chi \nabla^2 x, \\
\dot{v} &= -\gamma v + \Omega^2 u - x + \xi(\mathbf{r}, t),
\end{aligned}
\tag{4.88}
$$

where

$$\gamma = \frac{C_1}{B_2} X_1^0, \quad \Omega^2 = \frac{B_1 C_1}{B_2^2} X_1^0 Y^0, \quad \chi = \frac{D}{B_2 r_0^2}, \tag{4.89}$$

and $\xi(\mathbf{r}, t)$ is the new noise function with the correlation function

$$\langle \xi(\mathbf{r}, t)\xi(\mathbf{r}', t')\rangle = \sigma\left(\mathbf{r} - \mathbf{r}', t - t'\right). \tag{4.90}$$

In the example (4.82) it is given by

$$\sigma\left(\mathbf{r}, t\right) = \sigma_0 \delta(t) \exp\left(-r^2\right), \tag{4.91}$$

where the noise intensity is $\sigma_0 = S_0/B_2$.

Taking into account that $v = \dot{u}$, we thus obtain

$$\dot{x} = -\varepsilon x + x\,\dot{u} + \chi \nabla^2 x, \tag{4.92}$$

$$\ddot{u} + \gamma\,\dot{u} + \Omega^2 u = -x + \xi(\mathbf{r}, t). \tag{4.93}$$

This system has a clear structure. Equation (4.92) describes the evolution of the rescaled population density $x(\mathbf{r}, t)$ of the weak species under the action of the noise $\psi(\mathbf{r}, t) = \frac{\partial}{\partial t} u(\mathbf{r}, t)$. The population grows in those spatial regions where at a given moment the condition $\psi(\mathbf{r}, t) > \varepsilon$ holds, and decays in the rest of the medium. These spatial regions can be viewed as reproduction areas of the weak species. The principal role of diffusion in (4.92) is that it spreads the population out of the reproduction areas and maintains some nonvanishing population level anywhere in the medium. Equation (4.93) describes the evolution of the rescaled population density of the strong species. The deviations $u(\mathbf{r}, t)$ of this population density from its steady level have damped oscillatory dynamics with an oscillation frequency Ω and a damping constant γ. The random force $\xi(\mathbf{r}, t)$ generates fluctuations of the field $u(\mathbf{r}, t)$.

The analysis of the stochastic system (4.92) and (4.93) is greatly simplified if the damping constant is much smaller than the oscillation frequency ($\gamma \ll \Omega$). According to (4.89) and (4.73), this implies that the mean rate Q_0 of the nutrient supply should be small:

$$Q_0 \ll \frac{A_1^2}{B_1}. \tag{4.94}$$

Below, we assume that this condition is satisfied and that furthermore $\varepsilon \ll \gamma$.

We first consider the situation at the onset of coexistence, when the population density of the weak species is still extremely small and the term x can therefore be neglected in (4.93). Then, this linear stochastic differential equation can be solved by the Fourier transform:

$$u_{\omega \mathbf{k}} = \frac{1}{(2\pi)^3} \int \int u\left(\mathbf{r}, t\right) \exp\left[-i\left(\omega t - \mathbf{kr}\right)\right] dt d\mathbf{r}, \tag{4.95}$$

and

$$\xi_{\omega \mathbf{k}} = \frac{1}{(2\pi)^3} \int \int \xi\left(\mathbf{r}, t\right) \exp\left[-i\left(\omega t - \mathbf{kr}\right)\right] dt d\mathbf{r} \tag{4.96}$$

for the two-dimensional medium. The solution is

$$u_{\omega \mathbf{k}} = \frac{\xi_{\omega \mathbf{k}}}{-\omega^2 + \Omega^2 + i\gamma\omega}. \tag{4.97}$$

The resonance at $\omega = \pm\Omega$ leads to a great enhancement of the fluctuation modes with this frequency.

Averaging (4.92) yields the dynamical equation for the mean rescaled population density $\bar{x}$ of the weak mobile species[2]:

$$\frac{d\bar{x}}{dt} = -\varepsilon\bar{x} + \langle \delta x \, \dot{u} \rangle, \tag{4.98}$$

where the deviations $\delta x(\mathbf{r}, t) = x(\mathbf{r}, t) - \bar{x}(t)$ are introduced and we have taken into account that $\langle u \rangle = 0$. Subtracting (4.98) from (4.92), we also obtain the dynamical equation for the density fluctuations:

$$\dot{\delta x} = -\varepsilon\delta x + \bar{x} \, \dot{u} + \delta x \, \dot{u} + \chi\nabla^2\delta x. \tag{4.99}$$

The evolution of the mean population density $\bar{x}$ is therefore characterized by the timescale ε^{-1} and, because $\varepsilon \ll \gamma \ll \Omega$, it is slow. On the other hand, the fluctuations δx have rapid dynamics with the characteristic timescale Ω^{-1}. This means that $\bar{x}$ can be considered as a constant parameter in (4.99) and the term $\varepsilon\delta x$ can be neglected there. Moreover, the term $\delta x \, \dot{u}$ in this equation is much smaller than the term $\bar{x} \, \dot{u}$ and can also be neglected if $\langle \delta x^2 \rangle \ll \bar{x}^2$. We shall assume that this condition is valid and later verify it. Therefore, (4.99) simplifies to

$$\dot{\delta x} = \bar{x} \, \dot{u} + \chi\nabla^2\delta x, \tag{4.100}$$

and its solution can be found by the Fourier transform,

$$\delta x_{\omega \mathbf{k}} = \frac{i\omega\bar{x}u_{\omega \mathbf{k}}}{i\omega + \chi k^2}. \tag{4.101}$$

The correlator $\langle \delta x \, \dot{u} \rangle$ can now be determined:

$$\langle \delta x \, \dot{u} \rangle = \int \int \frac{\omega^2\bar{x}\sigma(\omega, k)d\omega d\mathbf{k}}{(i\omega + \chi k^2)\left[(-\omega^2 + \Omega^2)^2 + \gamma^2\omega^2\right]}, \tag{4.102}$$

where $\sigma(\omega, k)$ is the noise spectrum,

$$\sigma(\omega, k) = \frac{1}{(2\pi)^3} \int \int \sigma(\mathbf{r}, t) \exp\left[-i(\omega t - \mathbf{k}\mathbf{r})\right] dt d\mathbf{r}. \tag{4.103}$$

The integration over ω in (4.102) can be performed approximately, if we take into account that the main contribution to this integral comes from the

---

[2] Those readers who are not interested in the technical details of the derivation may proceed to equation (4.108).

regions $\omega \approx \pm\Omega$. The result is

$$\langle \delta x \, \dot{u} \rangle = \frac{\pi \bar{x}}{\gamma} \int \frac{\chi k^2 \sigma(\Omega, k)}{\Omega^2 + \chi^2 k^4} \, d\mathbf{k}. \tag{4.104}$$

So far our analysis has been valid for noise with any correlation function $\sigma(\mathbf{r}, t)$. Below we assume that it is given by (4.91) and therefore has the spectrum

$$\sigma(\omega, \mathbf{k}) = \frac{\sigma_0}{8\pi^2} \exp\left(-\frac{1}{4} k^2\right). \tag{4.105}$$

In the case of the two-dimensional medium under consideration we thus have

$$\langle \delta x \, \dot{u} \rangle = \frac{\sigma_0 \bar{x}}{\Omega 20 \gamma} F\left(\frac{\Omega}{\chi}\right), \tag{4.106}$$

where the function $F(z)$ is defined as

$$F(z) = \frac{z}{4} \int_0^\infty \frac{k^3}{z^2 + k^4} \exp\left(-\frac{1}{4} k^2\right) dk. \tag{4.107}$$

Substituting (4.106) into (4.98) yields

$$\frac{d\bar{x}}{dt} = \left[\frac{\sigma_0}{\Omega \gamma} F\left(\frac{\Omega}{\chi}\right) - \varepsilon\right] \bar{x}. \tag{4.108}$$

Hence, reproduction of the weak species begins when the noise intensity σ_0 exceeds the threshold σ_c given by the equation

$$\sigma_c = \frac{\varepsilon \gamma \Omega}{F(\Omega/\chi)}. \tag{4.109}$$

Figure 4.5 shows the dependence of the critical noise intensity σ_c on the dimensionless diffusion constant $\chi = D(B_2 r_0^2)^{-1}$, computed using (4.107).

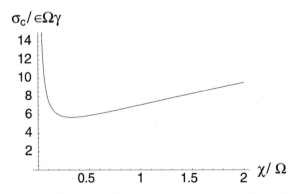

Fig. 4.5. Dependence of the critical noise intensity σ_c on the dimensionless diffusion constant $\chi = D(B_2 r_0^2)^{-1}$, computed using (4.107)

We see that σ_c reaches its minimum value at intermediate diffusion constants $\chi \sim \Omega$. For slow diffusion (i.e. when $\chi \ll \Omega$), the critical noise intensity can be estimated as

$$\sigma_c \approx \frac{\varepsilon \gamma \Omega^2}{2\chi}. \tag{4.110}$$

If diffusion is fast ($\chi \gg \Omega$), the approximate estimate is

$$\sigma_c \approx \frac{8\varepsilon\gamma\chi}{|\ln(\Omega/\chi)|}. \tag{4.111}$$

The above results were obtained under the condition $\langle \delta x^2 \rangle \ll \bar{x}^2$, whose validity can now be verified. Using (4.101), we find

$$\langle \delta x^2 \rangle = \bar{x}^2 \int \int \frac{\omega^2 \sigma(\omega,k) d\omega d\mathbf{k}}{(\omega^2 + \chi^2 k^4)\left[(-\omega^2 + \Omega^2)^2 + \gamma^2 \omega^2\right]} \tag{4.112}$$

$$\approx \frac{\pi\bar{x}^2}{\gamma} \int \frac{\sigma(\Omega,k)d\mathbf{k}}{\Omega^2 + \chi^2 k^4} = \frac{\bar{x}^2 \sigma_0}{4\gamma\chi^2} \int_0^\infty \frac{\exp(-k^2/4)\,kdk}{(\Omega/\chi)^2 + k^4}. \tag{4.113}$$

The noise intensity σ_0 in this equation can be taken at the critical point, where $\sigma_0 = \sigma_c$. For slow diffusion ($\chi \ll \Omega$), a simple estimate can be obtained:

$$\langle \delta x^2 \rangle \approx \frac{\varepsilon}{\chi}\bar{x}^2. \tag{4.114}$$

In the opposite case of fast diffusion ($\chi \gg \Omega$), we find

$$\langle \delta x^2 \rangle \approx \frac{\pi\varepsilon}{2\Omega|\ln(\Omega/\chi)|}\bar{x}^2. \tag{4.115}$$

Thus, the validity of this condition can always be ensured by choosing a sufficiently small value of the parameter ε.

The evolution equation (4.108) for the mean population density $\bar{x}$ was derived by neglecting the influence of the reproducing weak species on the fluctuations in the subsystem, "strong species – nutrient", i.e. by neglecting the term $\bar{x}$ in (4.93). If we retain this term, still assuming that $\bar{x}$ is sufficiently small, the evolution equation (4.108) is modified and takes the form

$$\frac{d\bar{x}}{dt} = \left[\frac{\sigma_0}{\Omega\gamma}F\left(\frac{\Omega}{\chi}\right) - \varepsilon\right]\bar{x} - \varkappa\bar{x}^2 \tag{4.116}$$

with a positive coefficient $\varkappa$. This coefficient is given by the integrals

$$\varkappa = 4 \int \int \frac{\chi k^2 \omega^4 \sigma(\omega,k) d\omega d\mathbf{k}}{(\omega^2 + \chi^2 k^4)^2 \left[(\Omega^2 - \omega^2)^2 + \gamma^2 \omega^2\right]} \tag{4.117}$$

$$\approx \frac{4\pi\Omega^2}{\gamma} \int \frac{\chi k^2 \sigma(\Omega,k)d\mathbf{k}}{(\Omega^2 + \chi^2 k^4)^2} = \frac{\Omega^2 \sigma_0}{\gamma} \int_0^\infty \frac{\chi k^3 \exp(-k^2/4)}{(\Omega^2 + \chi^2 k^4)^2}dk. \tag{4.118}$$

The noise intensity σ_0 in these integrals can be taken at the critical point, so that $\sigma_0 = \sigma_c$. In the limit of slow diffusion ($\chi \ll \Omega$) we get $\varkappa \approx 4\varepsilon$, whereas in the opposite limit of strong diffusion $\varkappa \approx 2\varepsilon \left| \ln \left(\Omega/\chi \right) \right|^{-1}$.

The evolution equation (4.116) describes the logistic growth of the weak species that sets in after the transition. The (rescaled) stationary mean population density of this species for $\sigma_0 > \sigma_c$ is

$$\bar{x} = \frac{1}{\Omega \gamma \varkappa} F \left(\frac{\Omega}{\chi} \right) (\sigma_0 - \sigma_c). \tag{4.119}$$

Note that, according to (4.87), the actual mean population density $\overline{X}_2$ of the weak species is related to $\bar{x}$ as $\overline{X}_2 = \left(B_2/C_2 Y^0 \right) \bar{x}$. Moreover, using (4.83), one can derive the following exact relationship linking the mean population densities $\overline{X}_1$ and $\overline{X}_2$ of both species:

$$\overline{X}_1 = X_1^0 - \frac{A_2 B_1 C_2}{A_1 B_2 C_1} \overline{X}_2. \tag{4.120}$$

Thus, in a system where the nutrient concentration is not uniform and fluctuates in space and in time, the weak mobile species can *coexist* with the strong species. The coexistence begins at a certain critical intensity of fluctuations in the local rate of the nutrient supply, which depends on the diffusion constant of the weak species. This result was first derived in 1978 by Mikhailov for three-dimensional media and large diffusion constants [205] (see also [206, 207]).

The coexistence of species in ecological systems is usually possible only when they do not entirely rely upon the same resource (i.e. type of nutrient). This is also true for the considered ecological system – in absence of fluctuations. The environmental variability apparently represents an additional "resource" that can be utilized by the mobile species.

The physical mechanism underlying this effect is similar to the rectification of fluctuations discussed at the beginning of this chapter. The response of the weak species to the nutrient fluctuations is strongly asymmetric. Indeed, if the concentration of the nutrient is locally increased by a significant amount, this leads to strong reproduction and a rapid increase in the local population density. If, on the other hand, the local nutrient concentration is significantly decreased for the same interval, this may only lead to the loss of the local population present in the considered region before the fluctuation has taken place. This loss will, however, soon be replenished through diffusion from the surrounding regions of the medium.

Noise produces reproduction islands with an enhanced nutrient concentration. It excites damped oscillations with characteristic frequency ΩB_2 in the subsystem formed by the nutrient and the strong species. Hence, the typical lifetime of a reproduction island is determined by the period of such oscillations and can be estimated as $\tau \sim (\Omega B_2)^{-1}$. Within this time the diffusing weak species would spread over the area of radius $(D\tau)^{1/2} \sim (D/\Omega B_2)^{1/2}$.

This radius should be compared with the characteristic spatial size of an island given by the noise correlation radius r_0.

When $r_0 \gg (D\tau)^{1/2}$, diffusion is so weak that it cannot provide efficient transport of the multiplied population from the reproduction island. In the opposite case, when $r_0 \ll (D\tau)^{1/2}$, diffusion is so strong that it immediately removes any newly produced particle from the reproduction island, preventing its further multiplication. Hence, a significant reproduction rate enhancement on fluctuations can be expected only for intermediate diffusion constants, when the condition $r_0 \sim (D\tau)^{1/2}$ is satisfied. According to (4.89), this condition is equivalent to $\chi \sim \Omega$. Looking at Fig. 4.5, we see that the critical noise intensity needed for coexistence is indeed strongly reduced in this region.

It is interesting to note that diffusional mixing in the system under consideration at the coexistence threshold is still so strong that the fluctuations in the population density of the weak mobile species are small compared with the mean population level (i.e. $\langle \delta x^2 \rangle \ll \bar{x}^2$). Moreover, the fluctuations will be approximately Gaussian when Gaussian noise is acting. Hence, the system does not show in this case the long probability tails, characterized by power laws, that we discussed for systems with multiplicative noise in the previous section.

Though our analysis was performed for a particular example of an ecological system, it has general implications. There are profound analogies between competition between species in ecological systems and social market phenomena, where competing individual agents and their groups are the main actors. The lesson one can learn from the analysis of this simple model is that flexibility can lead to evolutionary advantages in fluctuating social environments.

4.5 Further Reading

Scaling and power laws are not only important in economics and social sciences, but also in linguistics and information theory. Zipf´s law states that the frequency of words in natural languages is inversely proportional to their length. Remarkably, this property is independent of the language under consideration. The historical context of Zipf´s law, its derivation and relation to lexicographic trees are discussed in [208]. The statistical analysis of language texts has started in the 1950s with Shannon's famous paper [209] on the statistical properties of English texts. Various tools and methods, used not only to study linguistic texts but also other symbol sequences such as music or DNA, are reviewed in the book by Ebeling et al. [210]. Zipf´s law is also known in social geography. The fraction of cities with population n follows a power law with an exponent of -2. Manrubia and Zanette proposed a reaction–diffusion model for the growth of urban agglomerations and reproduced this power-law behavior [211]. Mathematical aspects of systems with multiplicative noise are discussed in the book by Horsthemke and Lefever [212] and in review articles [213, 214]. Sornette has recently investigated stochastic processes with multiplicative noise that produce intermittency characterized by a power-law distribution [215].

Various ratchet phenomena are involved in the motility of biological cells. While it was generally known that polymerization can provide enough energy [216], only recently have the actual mechanisms been found which explain how chemical energy is transformed into mechanical forces. The experiments have shown that polymerization can generate forces capable of overcoming the bending energy of a lipid bilayer without the aid of molecular motors. In one of these experiments Janmey et al.[217] loaded actin monomers into spherical liposomes (lipid bilayers) and then triggered polymerization. Subsequently, it was found that spikes, resembling filopodia, extruded from the liposomes. Myamoto and Totani [218] described a similar phenomenon using tubulin. Following these experimental observations, Peskin et al. [219] constructed a model applicable to different types of cellular protrusions, including filopodia, lamellipodia and acrosomal extensions.

Such polymerization ratchets also play a role in generating the force driving chromosome movements inside a cell. Here microtubules, which are linear polymers composed of asymmetric globular subunits and which can be found in all eukaryotic cells, participate in force generation [220, 221]. Mitchison [222] and Inoue and Salmon [223] have shown that the forces can reach magnitudes of hundreds of piconewtons and that both polymerization and depolymerization are involved in their generation. Dogterom and Yurke [224] performed experiments in which they grew microtubule polymers against a wall and estimated the force–velocity relation. This work has shown that the polymerization ratchet model [219] gives the correct order of magnitude of the growth rate at a given force load, but it quantitatively overestimates the experimental data. Mogilner and Oster extended the original model and obtained a reasonable agreement between experiment and theory [225–227].

Most proteins of the mitochondrial matrix are synthesized as cytosolic precursors, which are imported through a proteinaceous pore that spans the two mitochondrial membranes [228]. Protein import across the inner membrane is ATP-driven [229, 230] and involves a special mitochondrial enzyme [231, 232]. A Brownian ratchet mechanism has been proposed [233, 234] according to which this enzyme binds from the inside to the precursor chain in the membrane, thereby inhibiting reverse diffusion of the latter and causing a net unidirectional movement of the chain into the mitochondrium [219, 234]. Such a mechanism can indeed account for qualitative features. As more experimental details were revealed, an alternative mechanism was however proposed that also involves rectification of noise. This latter "power-stroke" mechanism assumes that the same special mitochondrial enzyme binds to the protein precursor and induces a conformational change which actively pulls the precursor chain through the pore [235, 236]. Chauwin et al. have compared the two mechanisms and shown that the power-stroke mechanism is more efficient and therefore more likely to be involved in mitochondrial protein import [237].

In Sect. 4.2 we have already mentioned motor proteins, such as kinesin or dynein, which can move along linear tracks. A series of conformational changes based on binding and hydrolization of ATP, leading to different affinities of the motor protein to the linear track, converts Brownian motion (ATP being supplied through diffusion) into directed motion. Somewhat similar is the mechanism of a correlation ratchet [238] where asymmetries in the binding affinities of the protein to the linear track are used for propulsion. The energy provided by ATP hydrolysis induces alternation of weak and strong binding states between the track and the motor and hence generates motion along the track.

Rectification of noise explains force generation in RNA polymerase during the process of DNA transcription. In this process, an enzyme molecule of RNA polymerase (RNAP) moves along a DNA strand and transcribes the DNA base pair sequence into a RNA sequence [239–241]. It further polymerizes this sequence into a RNA strand and executes several operations on its surface. Yin et al. have shown in laser trap experiments, where RNAP was fixed to a substratum and a DNA strand was attached to a small polymer bead kept in position by a laser beam, that RNAP indeed exerts a mechanical force on DNA [242]. By measuring the force needed to move the bead out of the laser trap, they have experimentally determined a load–velocity dependence. The experiment has shown that the action of the enzyme can be seen as rectification of noise, in which hydrolysis of nucleotides is transformed into mechanical motion. Wang et al. formulated a model for the observed processes, demonstrated that the nucleotide triphosphoshate binding free energy can drive the observed processes and calculated a load–velocity curve in agreement with the experimental data [243].

5. Dynamics with Delays and Expectations

All living beings are on a road that starts with their birth and ends with their death. The coordinate along this road is the age of a biological organism. A population typically consists of individuals of various ages that collaborate or compete each with another. The characteristic property of such interactions is that they are nonlocal and that individuals of largely different ages can interact. The distribution over ages, representing the demographic structure of a population, evolves with time and may show complex temporal behavior.

An important age-dependent property is the reproduction capacity. The birth rate in a population is determined by a number of adult individuals inside it. Moreover, only mature individuals would usually play a significant role in exploiting the environment and interactions with other species. Because of a time gap between birth and the reproductive age, the response of a population is generally delayed. The delays can lead to dynamic instabilities and the emergence of demographic waves.

On the other hand, the actions of individuals in a human society are typically based on their evaluation of the current political, social and economic situation and forecasting of the future trends. In its most clear form, this behavior is seen in financial markets, where various stocks are purchased or sold depending on their expected performance. Since any forecasting can only represent a future projection, based on the information collected over a period of time, it effectively leads to the introduction of delays into the economic and social dynamics. Consequently, the collective behavior of a self-forecasting population is subject to intrinsic instabilities and crises.

5.1 The Age Dimension

The age τ is simply the time passed after the birth of a given individual. Therefore, this variable satisfies the trivial dynamical equation

$$\frac{d\tau}{dt} = 1. \tag{5.1}$$

A population consists of individuals with different ages and its demographic distribution is described by the density $n(\tau, t)$, defined in such a way that

$n(\tau, t)d\tau$ yields the number of individuals with ages in the interval from τ to $\tau + d\tau$ at time t. The evolution equation for this distribution can easily be constructed.

Two processes contribute towards the change of the population inside a given age interval. Firstly, the population is aging and therefore younger individuals would arrive into the considered interval, while older individuals would leave it. Moreover, some individuals may die. Hence, the evolution equation is

$$\frac{\partial n}{\partial t} = -\frac{\partial n}{\partial \tau} - \gamma n, \tag{5.2}$$

where γ is the death rate coefficient.

This equation should be complemented by a boundary condition at $\tau = 0$. It follows from the definition of the density $n(\tau, t)$, that the quantity $n(\tau = 0, t)$ represents the rate of appearance of new individuals in the system and thus should be equal to the birth rate u. Therefore, the boundary condition is

$$n(\tau = 0, t) = u. \tag{5.3}$$

The death rate coefficient γ is a function of age τ. Additionally, it can depend on the amount of the available food. The latter is determined by the rate q of food consumption by the entire population,

$$q(t) = \int_0^\infty \alpha(\tau)n(\tau, t)d\tau, \tag{5.4}$$

where $\alpha(\tau)$ is the food consumption rate for a single individual of age τ. The food consumption by adult organisms is significantly higher. Therefore, the dependence $\alpha(\tau)$ has a maximum at a certain age, corresponding to mature individuals. When food consumption is increased, the amount of available food diminishes and this leads to an increase in the death rate. Hence, we have $\gamma = \gamma(\tau, q)$, where $\gamma(\tau, q)$ is an increasing function of q.

The birth rate u is proportional to the number of individuals in the reproductive age. If $\theta(\tau)$ is the probability that an individual of age τ would reproduce per unit time, this rate is

$$u(t) = \int_0^\infty \theta(\tau)n(\tau, t)d\tau. \tag{5.5}$$

The reproduction probability rate θ can also depend on the amount of food and, hence, on the rate of food consumption (5.4). It should be expected that reproduction becomes less probable when less food is available. Hence, $\theta = \theta(\tau, q)$ is a decreasing function of the variable q. Equations (5.2) to (5.5) constitute a model for the age dynamics. It is interesting to compare it with the models of the traffic flows discussed in

Chap. 4. If we interpret age τ as a physical coordinate, equation (5.2) describes the motion of particles in one direction at a constant velocity $v = 1$. New particles start their motion at $\tau = 0$. The particles are removed from the flow at a rate γ that increases with τ.

An important property of the evolution equation (5.2) is its nonlocality with respect to the coordinate τ which enters the model through the integral dependences (5.4) and (5.5). The individuals may interact and influence the behavior of each other, no matter what is the difference in their ages. Indeed, the individuals characterized by different "internal coordinates" τ may well find themselves in close proximity in real space. This leads to new effects which are absent for the traffic flows.

Let us consider a population where the death rate coefficient γ is constant for all ages and the reproduction rate has the form $\theta(\tau, q) = f(q)\chi(\tau)$ with some functions $f(q)$ and $\chi(\tau)$. Using (5.4) and (5.5), we find that the birth rate is then given by

$$u(t) = f\left[\int_0^\infty \alpha(\tau)n(\tau, t)d\tau\right]\int_0^\infty \chi(\tau)n(\tau, t)d\tau. \qquad (5.6)$$

On the other hand, integration of (5.2) yields

$$n(\tau, t) = u(t - \tau)e^{-\gamma\tau}. \qquad (5.7)$$

This result is obvious: individuals of age τ at time t are those ones that were earlier born at time $t - \tau$ and have survived till moment t. Substituting this into (5.6), we derive a closed evolution equation for the birth rate:

$$u(t) = f\left[\int_0^\infty \alpha(\tau)e^{-\gamma\tau}u(t - \tau)d\tau\right]\int_0^\infty \chi(\tau)e^{-\gamma\tau}u(t - \tau)d\tau. \qquad (5.8)$$

Note that because of the factors $e^{-\gamma\tau}$ the integrations are here limited to the ages $\tau \lesssim 1/\gamma$. This evolution equation is integral: the birth rate at a given time is determined by its past values.

The description is further simplified if individuals consume food and reproduce only at a fixed age τ_0, i.e. that $\alpha(\tau) = \alpha_0\delta(\tau - \tau_0)$ and $\chi(\tau) = \chi_0\delta(\tau - \tau_0)$. In this case (5.8) is reduced to

$$u(t) = f\left[\alpha_0 e^{-\gamma\tau_0}u(t - \tau_0)\right]\chi_0 e^{-\gamma\tau_0}u(t - \tau_0). \qquad (5.9)$$

When a solution for the birth rate $u(t)$ is found, the age structure of the population at time t can be determined as $n(\tau, t) = u(t - \tau)e^{-\gamma\tau}$. Thus, the complete solution to the original evolution equation (5.2) will be constructed.

By performing a time shift $t \to t + \tau_0$ and introducing a new function $F(z) = e^{-\gamma \tau_0} f(\alpha_0 e^{-\gamma \tau_0} z)$, equation (5.9) can be written in the form of a *map*:

$$u(t + \tau_0) = F[u(t)]. \tag{5.10}$$

If the birth rates u for initial moments t_0 in the interval $0 \le t_0 < \tau_0$ are given, the birth rates at any subsequent moment can be determined by repeatedly applying this map.

Above, we have shown several typical models of age-structured populations. By making different assumptions about reproduction, death and food consumption, the general evolution equation (5.2) for the age distribution can lead to a variety of simplified dynamical descriptions. The common feature of all such models is that they include *delays*. Their principal origin is clear: the individuals, which are presently born, would be able to significantly contribute to food consumption and to reproduce only after some time has passed and they are mature. In the next section we consider dynamical instabilities and waves that are induced by such delays.

5.2 Demographic Waves

Population statistics often shows significant variations of birth rates. "Baby booms" in human societies periodically give rise to propagating waves, first filling to the top limit all kindergartens, then schools, and finally bringing problems with pension payments to a suddenly increased number of elderly citizens. Of course, some increases in the birth rate are clearly explained by concrete economical and sociopsychological processes in particular societies. For example, the end of the Second World War triggered a baby boom in Europe in the middle of the 20th century. However, the population dynamics may also be intrinsically unstable and lead to the *spontaneous* development of demographic waves. Such instabilities play an important role for animal populations in ecological systems.

First, we consider a population whose evolution is described by the map (5.10). To be specific, we shall take $f(q) = \theta_0 \left[1 + (q/q_0)^\nu\right]^{-1}$, so that the mapping function is $F(x) = ax \left[1 + (x/x_0)^\nu\right]^{-1}$ with the coefficients $a = \theta_0 e^{-\gamma \tau_0}$ and $x_0 = (q_0/\alpha_0) e^{\gamma \tau_0}$. Introducing a new rescaled dynamical variable $x(t) = (\alpha_0/q_0) u(t)$ and measuring time in units of τ_0, the map (5.10) is then written as

$$x(t + 1) = G[x(t)], \tag{5.11}$$

where

$$G(x) = \frac{ax}{1 + x^\nu}. \tag{5.12}$$

Starting with the initial condition $x_0 = x(t_0)$ at some time t_0 and repeatedly applying this map, a sequence $x_1 = x(t_0 + 1), x_2 = x(t_0 + 2), \ldots,$ $x_k(t_0 + k), \ldots$ is generated. Generally, the terms in this sequence will be different. However, for an initial condition, corresponding to a *fixed point* of the map, all of them will be equal. Fixed points x^* of the map (5.11) satisfy the equation $x^* = G(x^*)$. It can easily be seen that for $a > 1$ it has two solutions: $x_1^* = 0$ and $x_2^* = (a-1)^{1/\nu}$. Below, we consider only the nontrivial fixed point x_2^*, which will be denoted as x^*.

The evolution of small initial deviations δx from the fixed point is described by the linearized map

$$\delta x(t+1) = -A\delta x(t), \tag{5.13}$$

where $A = -dG/dx$ at $x = x^*$. The general solution of (5.13) is $\delta x(t_0 + k) = (-1)^k A^k \delta x(t_0)$. Therefore, the deviations grow with time and the fixed point is unstable, if $|A| \geq 1$. For this map we find $A = a^{-1}[(\nu-1)(a-1)-1]$. Hence, the instability takes place if $a > a_c$ where $a_c = \nu(\nu-2)^{-1}$ (note that the exponent ν must be sufficiently large, i.e. $\nu > 2$).

The first instability leads to the growth of oscillations with period $T_0 = 2$. By running iterations of the map (5.11), it can easily be checked that such oscillations indeed develop and are stable within a certain interval of the parameter a. When they take place, the variable x is repeated in every second generation, alternating between two different values $x^{(1)}$ and $x^{(2)}$. The phase of the oscillations is determined by the initial condition.

The map (5.11) is discrete, but time is continuous for the considered ecological system. To obtain the solution at any time t, we should specify initial values of variable x inside the entire interval $0 \leq t_0 < 1$. Starting from any particular point t_0 inside this interval and repeatedly applying the map (5.11), we determine the sequence values of x at moments $t_0, t_0 + 1, \ldots, t_0 + k, \ldots$ which are separated by unit time intervals. This sequence represents the evolution history originating from moment t_0. The solution at *all* times t is the union of solutions starting from all possible points t_0 inside the initial unit interval. According to the map (5.11), the evolution histories starting from two different moments are completely independent, even when these two moments are close to one another.

An example of the evolution with continuous time described by the map (5.11) is shown in Fig. 5.1. In this simulation the time is divided into small steps $\Delta t = 0.02$ and initial values $x(t_0)$ are chosen independently at random for each of the 50 moments inside the interval from 0 to 1. To visualize the evolution, a special diagram is here constructed. Imagine that we take a thin strip of paper, and choose the coordinate along the strip to represent the time and plot on this long strip the temporal evolution of the variable $x(t)$, using a grayscale code with darker regions corresponding to smaller values of x. We cut the strip into equal segments of length $T = 1$, each containing 50 subsequent states of the system and starting from $t = 0$. These strip segments are then placed vertically, as columns, one after another. In this way, a diagram as displayed in Fig. 5.1 is constructed. Its advantage is that the states of the system, separated by the delay time $T = 1$, are neighbors along the horizontal direction and therefore their oscillations can easily be recognized. Effectively, each horizontal band shows an evolution history that starts from a certain initial moment t_0. Examining Fig. 5.1, we indeed notice that, after a short transient, oscillations with period $T = 2$ are observed for each evolution history. However, the evolution histories starting from different initial moments are independent even when these moments are separated by just one time step Δt.

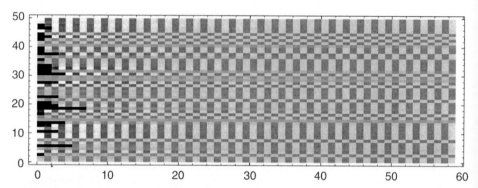

Fig. 5.1. Evolution diagram for the map (5.11) with the nonlinear function (5.12) for $\nu = 5$ and $a = 1.68$. Time is divided into small discrete steps $\Delta t = 0.02$, so that the delay is equal to 50 time steps. Each next column shows the states of the system within the following 50 steps. Random initial conditions

The discontinuity of the obtained solution $x(t)$ is the direct consequence of the lack of interactions between the evolution histories starting from different initial moments. It is not realistic and represents a deficiency of the simple model considered. To remove it, we note that, in reality, the individuals forming a population can never reproduce (and consume food) exactly at age τ_0, but only within a certain interval of ages near τ_0. Hence, we must analyze the effects of dispersion in the reproduction ages.

Let us slightly modify the map (5.11) and write it as

$$x(t+1) = G[\bar{x}(t)], \qquad (5.14)$$

where $\bar{x}(t)$ represents a weighted average of $x(t)$ inside a narrow interval centered at time t, i.e.

$$\bar{x}(t) = \int_{-\Delta\tau}^{\Delta\tau} \sigma(s)x(t+s)ds. \qquad (5.15)$$

Here $\sigma(s)$ is a function that has a sharp maximum at $s = 0$ and is identically zero outside of a narrow interval of width $\Delta\tau$ around this point.

It is convenient to assume that $\sigma(s)$ satisfies the normalization condition

$$\int_{-\Delta\tau}^{\Delta\tau} \sigma(s)ds = 1. \tag{5.16}$$

Clearly, the evolution equation (5.14) has the same fixed point x^* as the original map (5.11). The stability of the stationary solution $x(t) = x^*$ can again be analyzed by considering the dynamics of small perturbations $\delta x(t)$. Linearizing (5.14) around x^*, we find

$$\delta x(t+1) = -A \int_{-\Delta\tau}^{\Delta\tau} \sigma(s)\delta x(t+s)ds. \tag{5.17}$$

The solution of this linear evolution equation has the form

$$\delta x(t) = \sum_{l} C_l e^{\lambda_l t}, \tag{5.18}$$

where λ_l are the roots of the characteristic equation

$$e^{\lambda} = -A \int_{-\Delta\tau}^{\Delta\tau} \sigma(s)e^{\lambda s}ds. \tag{5.19}$$

The characteristic equation (5.19) can be solved for any particular function $\sigma(s)$, thus yielding the values of λ.

Considering only such relatively small λ that $\lambda\Delta\tau \ll 1$, an approximate general analysis is possible. In this case the exponent $e^{\lambda s}$ only slowly changes inside the interval from $-\Delta\tau$ to $\Delta\tau$ and we can use the quadratic approximation $e^{\lambda s} \approx 1 + \lambda s + \frac{1}{2}\lambda^2 s^2$. Substituting this into (5.19) yields

$$e^{\lambda} = -A\left(1 + \epsilon\varsigma_1\lambda + \epsilon^2\varsigma_2\lambda^2\right), \tag{5.20}$$

where $\epsilon = \Delta\tau$ is a small parameter and the coefficients ς_1 and ς_2 are given by the integrals

$$\varsigma_1 = \frac{1}{\Delta\tau} \int_{-\Delta\tau}^{\Delta\tau} \sigma(s)s\,ds, \tag{5.21}$$

$$\varsigma_2 = \frac{1}{2\Delta\tau^2} \int_{-\Delta\tau}^{\Delta\tau} \sigma(s)s^2\,ds. \tag{5.22}$$

Note that the coefficient ς_2 is always positive, whereas ς_1 may be either positive or negative.

We seek the roots λ of equation (5.20) as a series in powers of the small parameter ϵ, i.e. write $\lambda = \lambda^{(0)} + \epsilon\lambda^{(1)} + \epsilon^2\lambda^{(2)} + \ldots$. Substituting this into (5.20) and equating terms that have the same orders in ϵ, we obtain

$$\lambda_l \approx \lambda_l^{(0)} + \epsilon\varsigma_1\lambda_l^{(0)} + \epsilon^2\varsigma_2\left[\lambda_l^{(0)}\right]^2, \tag{5.23}$$

where $\lambda_l^{(0)}$ are (complex) roots of the equation $e^{\lambda\tau_0} = -A$ given by

$$\lambda_l^{(0)} = \ln A + i\,(2l+1)\,\pi, \quad l = 0, 1, 2, 3, \ldots \tag{5.24}$$

The complex exponents λ_l can be written as $\lambda_l = \mu_l + i\omega_l$. In the vicinity of the instability point we have $\ln A \ll 1$. Keeping only the leading terms in (5.23), we thus obtain

$$\mu_l = \ln A - \epsilon^2\left(\frac{1}{2}\varsigma_1^2 + \varsigma_2\right)(2l+1)^2\,\pi^2, \tag{5.25}$$

$$\omega_l = (1 + \epsilon\varsigma_1)\,(2l+1)\,\pi. \tag{5.26}$$

The modes $l = 0, 1, 2, 3, \ldots$ correspond to oscillations with different periods $T_l = 2\pi/\omega_l$ which are close to $\frac{2}{2l+1}$. These oscillations grow with time if real parts of the respective exponents are positive, i.e. if $\mu_l > 0$.

As follows from (5.25), the largest real part μ_l always corresponds to the mode with $l = 0$ and period $T_0 \approx 2$. As the control parameter A is increased, this mode begins to increase at the instability point $A = A_0$ with

$$A_0 = 1 + \epsilon^2\left(\frac{1}{2}\varsigma_1^2 + \varsigma_2\right)\pi^2. \tag{5.27}$$

Note that this critical value is thus a little higher than in absence of dispersion (i.e. for $\epsilon = 0$). In a narrow neighborhood of the instability point, all other modes with $l \geq 1$, and thus shorter temporal periods, remain damped. The period of the first unstable mode is approximately $T_0 = 2(1 - \epsilon\varsigma_1)$. Thus, it is not any longer exactly equal to the doubled delay time. Depending on the sign of the coefficient ς_1, the oscillation period T_0 can be larger or smaller than 2.

The linear stability analysis does not allow us to tell what dynamic behavior of the system will be established as the result of the instability. This question should be answered by performing numerical simulations of the model (5.14). Our simulations were performed using the Gaussian function $\sigma(s) = (\pi D)^{-1/2}\exp\left(-s^2/D\right)$ with $D = 0.0016$. Note that under this choice we have $\varsigma_1 = 0$ and thus a dispersion-induced change in the oscillation period is not expected here. The time was divided into small discrete steps $\Delta t = 0.02$. As the initial condition, independent random values from the interval $[0, 1]$ were assigned to all points in the interval $1 > t \geq 0$. The mapping function (5.12) with $\nu = 5$ was employed. In absence of dispersion the map (5.11) has in this case an instability leading to periodic oscillations at $a > a_c$ where $a_c = \nu(\nu - 2)^{-1} \approx 1.66\ldots$.

Figure 5.2 shows the nonlinear evolution described by equation (5.14) for the same control parameter $a = 1.68$ as in Fig. 5.1 for the map (5.11). We see that the system evolves in this case to periodic stable oscillations of period 2. Though we have again started with random initial conditions, the solution $x(t)$ is now perfectly smooth. Thus, the deficiency of the model (5.11) is indeed removed.

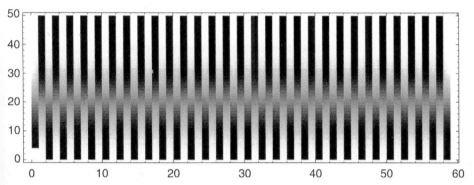

Fig. 5.2. Evolution diagram for the modified (5.14) with $\nu = 5$, $a = 1.68$, and $D = 0.0016$. Random initial conditions. Each column displays the system evolution within the next interval $T = 1$

Examining the profile of the solution $x(t)$ which is displayed in Fig. 5.3 and corresponds to the space–time diagram shown in Fig. 5.2, we find that the system shows harmonical oscillations. These oscillations represent *demographic waves*.

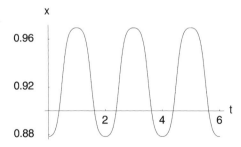

Fig. 5.3. Time dependence of the birth rate $x(t)$ yielded by equation (5.14). The same parameter values as in Fig. 5.2

To see this, we note that the age distribution $n(\tau, t)$ of the population at time t is related to the birth rate $x(t)$ by $n(\tau, t) = e^{-\gamma\tau} x(t - \tau)$. Thus, when the function $x(t)$ is known, the temporal evolution of the age distribution can easily be reconstructed. Figure 5.4 displays the temporal evolution of thus calculated age distribution. Time runs in this diagram from left to right and age increases upwards along the vertical axis. The local density is shown on a gray scale, with brighter levels corresponding to its higher values. Periodical increases of the birth rate $x(t)$ lead to repeated "baby booms" that correspond to maxima of the age distribution. As the generation of a "baby boom" gets older, these maxima drift towards greater ages. Thus, waves of the population density are periodically generated and propagate up along the age dimension.

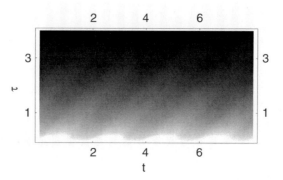

Fig. 5.4. Demographic waves. The evolving age distribution $n(\tau, t)$ is shown as a density plot, with darker regions corresponding to lower densities. Time runs from left to right (the total time interval is $T = 4$). Age increases along the vertical axis from $\tau = 0$ to $\tau = 2$. The death rate coefficient is $\gamma = 1$; other parameters are the same as in Fig. 5.2

It should be emphasized that the development of demographic waves in the considered model is the result of an intrinsic internal instability of this system. Close to the instability onset, the waves are almost harmonic. As the control parameter a is increased, the dynamical behavior of the system becomes more complex and further instabilities appear. Depending on the parameter a, this system may show stable oscillations with high periods and can also generate random oscillations.

5.3 A Model of the Market Crash

In contrast to primitive animals, the decision making of humans is essentially based on prognoses. Very often human behavior does not represent a direct response to the current situation, but rather reflects our various expectations. Thinking of the future spans all spheres of social activity, from the individual and family life to politics and economics. In market economies, where the driving stimulus is profit gained through the selling of goods and services, the correct forecasting of price trends brings significant benefits in the competition with other agents.

Thus, the build-up of expectations is an intrinsic element of social dynamics that must be incorporated into the mathematical models constructed for its description. This should lead to a large class of models of cooperative behavior where the actions of individual agents are strongly influenced by prognoses made by the agents about their future collective evolution. Today, theoretical investigations of such models are still at an early stage.

One general property of dynamics with expectations is however obvious. The trends are found by examining the history of a process and any prognosis can only represent an *extrapolation* of the past dynamical behavior of the same system. Hence, when extrapolation procedures are explicitly included into a description, this results in the dynamical models where the next state of a system is determined not only by the present state of the system, but also by its states at earlier, or *delayed*, moments. This means that, from a mathematical viewpoint, dynamical systems with expectations belong to the category of differential delay models, some examples of which have been discussed above in this chapter.

While forecasting can strongly improve the performance of an individual agent, it may also induce instabilities in the collective dynamics of an ensemble. Indeed, we have already noticed in Sect. 5.2 that delays can entail instabilities and lead to complex oscillations. The most efficient forecasting strategy may well turn out to be the most dangerous, since it generates crises that might undermine the very existence of a system.

Below in this section we consider a simple model of a system whose dynamics is influenced by extrapolation of its own previous trends. We formulate this model as a toy description of dynamical processes in the stock exchange market. The authors are not financial experts and we certainly do not claim that this model and its concrete conclusions apply to real stock exchanges. The aim of our study will be to illustrate the origin of intrinsic instabilities in self-forecasting dynamical ensembles.

The prices of stocks fluctuate. Short-term fluctuations result not only from objective developments, but also reflect variations in the selling and buying strategies of investors. While the majority of investors pursue long-term institutional interests and do not respond to short-scale price changes, financial speculators try to earn money through quick purchases and sales based on their subjective estimates of current price trends. We shall consider only the effects of such speculatory transactions.

Let $p(t)$ be the price of a given stock at time t and suppose that $u(t)$ is the speculatory investment in this stock. The decision to increase or decrease the investment is made by comparing the forecast price $p^*(t + \tau_0)$, which is expected at a future time $t+\tau_0$, with the current stock price $p(t)$. A speculator will buy additional stock if a price rise is forecast and sell it otherwise. The scale of buying or selling will depend on the magnitude of the expected price difference $\Theta(t) = p^*(t+\tau_0) - p(t)$. Moreover, it will generally depend on how much is already invested in this stock, so that the additional purchase or sale will represent a fraction of the existing investment. In a rough model linear proportionality can be assumed. This would lead to a dynamical equation of the form

$$\frac{du}{dt} = \Gamma\Theta(t)u. \tag{5.28}$$

Such an equation is, however, deficient, because according to it the investment u may indefinitely increase. In reality, the investment is limited by the available capital and a speculator would not be willing to make too high investments in a single stock even if its performance appears to be extremely good. This leads to the saturation of growth, which can be phenomenologically taken into account by adding a nonlinear term, so that (5.28) becomes

$$\frac{du}{dt} = \Gamma\Theta(t)u - \mu u^\nu, \tag{5.29}$$

where the exponent ν is sufficiently high.

The price dynamics is controlled by various processes. We consider a market characterized by steady growth so that all prices slowly increase. Furthermore, the price of a given stock increases if it is currently preferentially purchased and decreases if it is mainly sold. Thus, as a simple dynamical equation that describes the price variation we can choose

$$\frac{dp}{dt} = \alpha + \beta\frac{du}{dt}, \tag{5.30}$$

where the parameter α determines the growth rate and the coefficient β specifies the response of the price to speculation transactions. The integration of (5.30) yields

$$p(t) = p_0 + \alpha t + \beta u(t), \tag{5.31}$$

where p_0 is a constant.

To forecast the future price $p^*(t+\tau_0)$, different strategies can be employed. In our toy model we shall assume that the forecasting is done by simple linear extrapolation of the existing trend. This means that $p^*(t+\tau_0)$ is estimated as $p^*(t+\tau_0) = p(t) + \Lambda(t)\tau_0$, where $\Lambda(t)$ is the current rate of the price variation. This rate is in turn determined by comparing the prices at time t and at the corresponding *delayed* time moment $t-\tau$, so that $\Lambda(t) = (p(t) - p(t-\tau))/\tau$. Hence, the forecast price is

$$p^*(t+\tau_0) = p(t) + \frac{\tau_0}{\tau}\left(p(t) - p(t-\tau)\right), \tag{5.32}$$

and the expected price difference is given by $\Theta(t) = (\tau_0/\tau)\left(p(t) - p(t-\tau)\right)$. Using (5.31), it can be written as

$$\Theta(t) = \alpha\tau_0 + \frac{\beta\tau_0}{\tau}\left(u(t) - u(t-\tau)\right). \tag{5.33}$$

Substituting (5.33) into (5.29), we obtain a closed equation that describes the investment dynamics in our toy model,

$$\frac{du}{dt} = \Gamma \alpha \tau_0 u + \frac{\Gamma \beta \tau_0}{\tau} \left[u(t) - u(t - \tau) \right] u - \mu u^\nu. \tag{5.34}$$

As we show below, this equation describes instabilities that can be interpreted as market crashes.

To simplify its analysis, we shall measure time in units of the delay τ and use the rescaled investment variable $x(t) = (\mu\tau)^{1/(\nu-1)} u(t)$, so that this equation has the form

$$\frac{dx}{dt} = ax + b\left[x(t) - x(t - 1) \right] x - x^\nu, \tag{5.35}$$

where the parameters are

$$a = \alpha\tau\tau_0\Gamma, \quad b = \frac{\beta\tau_0\Gamma}{(\mu\tau)^{1/(\nu-1)}}. \tag{5.36}$$

Note that these two parameters have a simple interpretation, i.e. the parameter a represents the growth rate whereas the coefficient b characterizes the vulnerability of prices with respect to speculation.

The differential equation (5.35) has the fixed point $x_0 = a^{1/(\nu-1)}$, which corresponds to a steady market. Let us analyze the stability of this steady state with respect to small perturbations. Introducing $x(t) = x_0 + \delta x(t)$, substituting this into (5.35) and linearizing, we find

$$\frac{d\delta x}{dt} = x_0 \left[b\delta x(t) - b\delta x(t - 1) - (\nu - 1)x_0^{\nu-1}\delta x(t) \right]. \tag{5.37}$$

The solution of this linear equation can be sought in the form $\delta x(t) = \delta x(0)e^{\lambda t}$. The exponent λ satisfies the algebraic equation,

$$\lambda = bx_0 - bx_0 e^{-\lambda} - (\nu - 1)x_0^{\nu-1}. \tag{5.38}$$

Generally, its roots are complex, so that $\lambda = \gamma + i\omega$. By separately taking the real and the imaginary parts of (5.38), we find

$$\gamma = bx_0(1 - e^{-\gamma}\cos\omega) - (\nu - 1)x_0^{\nu-1}, \\ \omega = bx_0 e^{-\gamma}\sin\omega. \tag{5.39}$$

The steady state becomes unstable when the real part of the exponent λ changes its sign and becomes positive. Hence, the instability boundary is determined by the condition $\gamma = 0$. Putting $\gamma = 0$ in (5.39) yields the two equations,

$$bx_0(1 - \cos\omega) = (\nu - 1)x_0^{\nu-1}, \\ \omega = bx_0 \sin\omega. \tag{5.40}$$

Because we have $x_0 = a^{1/(\nu-1)}$, these two equations define a curve in the parameter plane (a, b) which represents the instability boundary of the steady state.

Note that equations (5.40) can be transformed and written as

$$a = \frac{\omega(1 - \cos\omega)}{(\nu - 1)\sin\omega},$$
(5.41)

$$b = \frac{\omega}{a^{1/(\nu-1)}\sin\omega}.$$
(5.42)

Thus, they define in the parametric form the instability boundary where $\gamma = 0$.

The instability boundary computed for $\nu = 3$ according to (5.41) and (5.42) is shown in Fig. 5.5. The steady state is unstable with respect to the onset of oscillations above this curve. We see that, if the parameter b specifying the relative effect of speculations on the stock price, is small, the market remains stable at any growth rate a. When this parameter is however increased, oscillations develop in an interval of the growth rate.

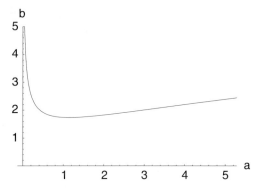

Fig. 5.5. The market instability boundary ($\nu = 3$). Oscillations take place above this curve

Suppose that initially the growth rate was small and then it started to slowly increase, i.e. $a(t) = a_0 + rt$. Integrating (5.35), we can find the response of the market to this evolution. If the parameter b is only slightly larger than its minimum value, at which instability is possible, the market develops small-amplitude fluctuations within a relatively narrow interval of time, while transversing the instability window (Fig. 5.6a). At larger values of b, strong oscillations develop when the instability boundary is crossed (Fig. 5.6b). In these oscillations the investment x drops down almost to zero.

The development of large-amplitude oscillations can be interpreted as a market crash. Actually, once such great oscillations have suddenly appeared, the behavior of trading agents cannot be expected to remain rational. Panic decisions and governmental interventions may follow. Clearly, the

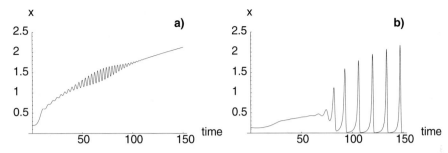

Fig. 5.6. Time dependence of the investment variable x under a steady increase of the growth rate a for **(a)** $b = 1.8$, $\nu = 3$, $a = 0.03t$ and **(b)** $b = 3.0$, $\nu = 3$, $a = 0.003t$

simple model given by equation (5.35) would no longer be applicable in this special situation. It can only describe the onset of a market crash.

The market instabilities in the considered simple model originate from the perfectly rational behavior of the traders. Each agent analyzes the price trend and tries to optimize its gain by appropriately adjusting its actions. This works nicely if the predicted process is independent and cannot be influenced by the undertaken decisions. Indeed, that would be the case if only a few minor traders are involved and their actions do not perturb the prices. The situation is different if all trading agents follow the same strategy. Then their own decisions can strongly influence the market dynamics and make the actual output greatly different from the forecast. Collectively, the agents behave like a swarm that forms its own environment.

5.4 Further Reading

Maps are interesting mathematical objects that arise in various applications. Complex dynamics in ecological systems described by simple maps was first analyzed by May [244]. The general dynamical properties of maps are discussed, for instance, in the book [245].

Mackey, Glass and An der Heiden have considered several physiological systems described by a class of differential delay equations [246–248]. Let $z(t)$ be some controlled physiological variable. Then its net rate of change dz/dt will be given by the difference between its production rate P and its rate of destruction D, that is

$$\frac{dz}{dt} = P(z) - D(z). \tag{5.43}$$

The attractors of this differential equation can only be fixed points, corresponding to stable steady states of the system. The system cannot oscillate and will monotonously approach a steady state. A completely different behavior is observed if either P or D (additionally) depend on the variable z

taken at a delayed time moment $t - \tau$. In this case $z(t)$ may display not only sustained and regular oscillations, but also show chaotic dynamics. Note that equation (5.35) also belongs to this class of differential delay equations, though the mechanisms leading to delays are different in this model.

The instabilities caused by delays in physiological systems have been used to explain the origin of so-called dynamical diseases characterized by qualitative changes in the dynamics of physiological control systems [246–248]. Time delays in the physiological control system were recognized to be involved in the Cheyne–Stokes respiration that accompanies a congestive heart failure and is found in obesive individuals or after neural brainstem lesions. This type of respiration is characterized by breathing patterns with regular waxing and waning of ventilation. The time delay in the equations of the ventilation control system arises because there is a blood transit time from the brainstem, where ventilation is determined by chemoreceptors, to the lungs, where CO_2 elimination takes place.

A different dynamic disease in which delays play a role is chronic myelogenous leukemia, which is a neoplastic disorder of the hemotopoietic system characterized by a massive increase in circulating neutrophils. Neutrophils are permanently supplied and removed from the blood. Since the committed neutrophil precursor cells require a certain time to mature, the actual influx of neutrophils is a function of the white blood cell population some time ago. Hence, the effective time delay in this system is the cell maturation time. The model for neutrophil production was also the first example of associating intrinsic chaos in a continuous time deterministic system with a pathological process.

The time delays in the recurrent inhibition due to conduction and synaptic delays in the hippocampus seem to play a role in epileptic seizures. Review papers which also partly contain the examples given above have been published by Glass and Mackey [247] and by Mackey and Milton [249]. Further interesting examples can be found in publications [250–252].

Systems with time delays are also known in physics and chemistry. The presence of delays in the differential equations that describe the system's dynamics has been used to explain some effects of optical bistability [253, 254]. Khrustova et al. have shown that delays may explain the origin of chaotic oscillations in catalytic surface reactions [255, 256]. In the latter case the system represents a population of adsorbate islands which are globally coupled through the gas phase. The islands form and grow, until they become reactive at a certain age. The delays result from the fact that only reactive islands which have reached this age can influence the island generation rate. Thus, the situation is similar to that found for ecological systems.

An important feature of economic systems is that the state of an economy at a given time depends in an essential way on its agents' expectations about the future. The actual dynamics followed by the economy is thus determined by the way in which the agents forecast this future (in terms of profits, market development, technological progress, etc.) as a function of their information on the past, while possibly learning about the structure of their environment.

Classical economic theory asserts that in the long run the economy settles down to a steady state or converges towards stable cycles. The real time series of economic data, however, often show deviations from such ideal behavior in the form of irregular patterns, crashes and other more complex types of phenomena. Nonetheless, for a long time economists maintained the hypothesis of entirely rational behavior and perfect foresight. Until the middle of the 1980s the mainstream of economic thinking attributed irregular deviations from time series to exogenous perturbations (for a review see Grandmont and Malgrange [257]). Contrasting with earlier macroeconomic studies, more recent approaches try to model explicitly the intertemporal optimizing behavior of agents. This makes it possible to analyze how expectations interact with the internal dynamics of an economy to generate autonomous fluctuations. Expectation-driven business cycles can thus be obtained that are compatible with individual optimization, self-fulfilling expectations and competitive market clearings.

Complex deterministic dynamics, including chaos, was found in the models where agents optimize their behavior over an infinite horizon (i.e. their predictions are not confined to overlapping generations). Deneckere and Pelikan [258] established necessary and sufficient conditions on the return function and the discount parameter, leading to deterministic cycles or chaos. Boldrin and Montrucchio [259] investigated standard capital theory in which the dynamic behavior of an economy is described by a policy function defined as a continuous map, attributing to any given capital stock a certain future capital stock which is optimal according to an intertemporal objective. While regular capital accumulation paths are normally assumed, the authors found that chaotic paths can also represent possible solutions in a neoclassical model of optimal growth. Furthermore, Dana and Montrucchio [260] obtained chaotic deterministic dynamics in an infinite horizon duopoly model. A good source of information on the relationship between the stability of economic cycles and expectations (without having knowledge about the expectation formation process) can be found in Grandmont and Laroque [261] and the papers cited therein.

In previous examples the deterministic chaotic behavior results from expectations of a single agent or from a set of homogeneous expectations of a population of agents. A different methodological approach in the theory of endogenous business cycles is related to the concept of "sun-spot equilibrium" [262, 263]. Such equilibria can arise in an economy which is itself deterministic but where, as economic agents believe, prices and quantities are affected by some random factors (the so called "sun spots"). The "sun spot equilibria" are reached if the developing equilibrium prices and quantitites agree with the agents' expectations. Azariadis et al. [264, 265] established sufficient conditions for the existence of stationary stochastic business cycles, driven by random but nevertheless self-fulfilling expectations. Guesnerie [266] treated the case of sun-spot equilibria for an arbitrary number of commodities.

The notion of perfect rationality also plays an important role in modeling stock market dynamics. The conventional argument is that opportunistic trading by rational investors will eliminate any gains that can be made through predictable patterns. What is left are the perturbations in asset values caused by more or less random news. Since the end of the 1980s there has been an increased interest in the role of psychological forces in erratic market movements and especially in the precipitous drops in equity prices that occasionally follow prolonged advances, such as occurred in the crash of 1929 or more recently on the Black Monday of October 1987. Of particular interest in this context is how alternating periods of generally rising or generally falling prices, so called bull and bear markets, follow from the heterogeneous set of behaviors and expectations about the future development of investors. Day and Huang [267] proposed a deterministic excess demand model of stock market behavior that generates stochastically fluctuating prices and randomly switching bear and bull markets. Smith et al. [268] ran computer experiments to investigate the convergence of stock market traders' expectations, changes in asset dividend value and asset price adjustments for traders with a finite temporal horizon.

Over the last ten years, a number of theories has been proposed where multiple equilibria and path dependence play an important role (see, for example, [269, 270]). The rational expectations theory asks what forecasting model (or expectations) is consistent with an actual time series. This is a valid approach, but it assumes that agents can somehow deduce in advance what model will be the correct one. Moreover, it requires that any agent surely knows that all other agents shall use the same forecasting model.

Arthur posed the question of what happens when the forecasting models are not obvious and must be formed individually by agents that have no knowledge about others' knowledge and forecasting methods. This becomes particularly important in systems that allow multiple equilibria. He proposed a model [271] in which agents guess other agents' behavior and revise their hypothesis based on new information. In contrast to the "sun spot equilibrium" theory, Arthur's model leads to the dynamics characterized by two persistent classes of expectations among agents. The ratio of mean numbers of agents, sharing a particular expectation class, is a fixed point of this dynamical system, although individual agents are allowed to alternate between both types of expectation. Thus, the time series of the number of agents belonging to a particular class fluctuate chaotically around the mean values. Because agents compete to be in a class, the author speaks of an ecology that these beliefs create.

This approach has been further developed in a model where Arthur et al. [272, 273] extended the idea of agents that cannot deduce other agents' behavior but that have to discover it to the situation of a stock market. In this model, agents continuously create and use multiple market hypotheses, i.e. individual, subjective, expectational models of future prices. They get dividends based on their currently most accurate hypothesis within an "artificial stock market" in a computer simulation. The authors found two regimes in which such a computerized market can operate. If the parameters are chosen such that the agents update their hypotheses slowly, the diversity of expectations collapses into homogeneous rational ones, as in conventional finance theory. When the rate of updating the hypotheses is increased, the market undergoes a phase transition into a complex dynamical regime resembling the behavior of real markets.

An important question is the emergence of cooperation. Public transportation only works if there are no free riders. Contracts between agents require that both parties respect the rules. In everyday life, one may ask oneself how many times one will invite acquaintances for dinner if they never invite one in return. An executive in an organization does favors for another executive in order to get favors in exchange. Team work relies to a good deal on mutual respect and anticipation of other's behavior. The emergence of cooperation does not only play a role in an economic context. Mastering conflict situations in politics, many examples from behavioral biology, as well as the general sociological question of how cooperation in a world of egoists can emerge at all, fall into the same class of problems. An abstract formulation of the problem in a game-theoretical context is provided by the so-called *Prisoner's Dilemma.*

Imagine two players A and B, who can either cooperate or defect. Each of their encounters may have four possible outcomes: both cooperate, both defect, and the cases when one of them cooperates and the other one defects. Any of these cases is associated with a certain payoff. If both defect, both have the same the payoff. But if they would have both cooperated, they would have been even better off. Yet, none of them knew beforehand how the other one was going to behave. Each player has a certain expectation about how the other might behave and what the payoff would be in each case. Past encounters are also included in the reckoning.

At the end of the 1970s Axelrod, Hamilton and Dion [274–276] organized a computer tournament to investigate the outcome of repeated two-agent Prisoner's Dilemma encounters. Participants were invited to contribute computer programs that would encode agent behavior with various types of expectation methods, and with delays as well as weighting and averaging functions. The authors then let the population of agents, represented by the various contributed computer programs, go through repeated Prisoner's Dilemmas and observed the evolution of the population. Interestingly, neither strategies that always cooperated nor those that always defected were the most successful ones. Likewise, complex strategies involving punishment of former defectors were not very successful. A simple strategy called TIT-FOR-

TAT turned out to be the most successful. It starts with cooperation and then always does what the other player did on the previous move.

An evolutionary dimension was introduced by playing series of tournaments and by bringing only the more successful players into the next generations, while eliminating less scoring ones. In addition, in each generation a strategy was represented proportional to its scoring. Remarkably, TIT-FOR-TAT came out as the winner of such an evolution. Although these tournaments represented highly simplified conflict situations taking into account past behavior and expectations, they have shown how cooperation can emerge in populations with various competing strategies.

Today, the Prisoner's Dilemma is widely used as a metaphor in a large number of problems in economics, political and social sciences, evolutionary and behavioral biology, and psychology. In the economic and political context one is particularly interested in understanding the role of time delays and various expectations. In some other cases, the evolutionary aspect, i.e. the formation of evolutionary stable strategies is of central interest.

An interesting application of the Prisoner's Dilemma has been investigated by Huberman and Glance [277] in the context of the free rider problem. Here a group of agents works together to produce some overall utility to the whole group. If in such a situation the gain to an individual agent for cooperation is less than its cost for participating, the agent may rationally choose not to cooperate and instead to free ride on the efforts of the others. No sustained cooperation ensues if this logic holds for all agents. A dilemma arises if the benefit to be accrued by overall cooperation offsets individual costs, since rationality on the part of each individual leads to failure in achieving a collective good beneficial to all. Huberman and Glance have studied this problem as a repeated n-person Prisoner's Dilemma. The expectations of each agent consist of two components. Firstly, each individual believes that future aggregate collective behavior is directly influenced by the individual's choices in inverse proportion to the group size, and secondly, interaction is of finite duration characterized by a certain time horizon. It has been found that in such a situation individuals cooperate if they perceive the fraction cooperating to be greater than some critical amount and defect otherwise. This strategy is reminiscent of the successful TIT-FOR-TAT strategy in two-player Prisoner's Dilemma. Agent based models of competition and collaboration beyond the Prisoner's Dilemma are described in [78].

Another study [278] was devoted to an investigation of the diversity of expectations among agents coupled to reward mechanisms. The agents that behaved well were profiting here at the expense of the others. The observed overall dynamics was characterized by cycles of almost stable behavior interrupted by sudden crashes. This process was accompanied by an ever-changing diversity in the composition of the system. As time progressed, the crashes became separated by increasingly long intervals, but remained unpredictable.

Further examples of dynamics with expectations can be found in Huberman and Glance [279].

The problems of financial markets are increasingly attracting the attention of theoretical physicists. An interesting and important aspect is the statistical analysis of time series which represent the stock price dynamics. As noticed by Mandelbrot [280], variations of prices are apparently described by probability distributions with long power-law tails, similar to the Pareto distribution which we have discussed in Chap. 4. Statistical studies, confirming this conclusion, have been performed by Mantegna for the Milan stock exchange [281] and by Mantegna and Stanley for the New York stock exchange [282]. Refined statistical methods, based on the theory of random matrices, are now used to distinguish the effects of pure noise from the meaningful information contained in the financial time series (see Bouchaud [283], Laloux et al. [284], and Plerou et al. [285]). For a general discussion, see the recent books by Bouchaud and Potters [286] and by Mantegna and Stanley [287].

Several different phenomenological models of stock exhange markets have been proposed and investigated. Takayasu et al. [288] introduced a model of a market consisting of dealers with deterministic nonlinear interactions and showed that the price fluctuations could be an effect of deterministic chaos. Stochastic models of markets have been constructed by Levy and Solomon [289], Caldarelli et al. [290], and Sato and Tahayasu [291]. These models yield the power-law distributions of price variations and relate their origin to the presence of multiplicative noise in the stochastic evolution equations. Bouchaud and Cont [292] have used a simple model based on a nonlinear stochastic differential equation to describe stock market fluctuations and crashes.

6. Mutual Synchronization

Biological organisms do not only actively move themselves in physical coordinate space. Any such organism has a complex internal dynamics. In some cases this dynamics is cyclic, so that an organism persistently goes through a closed sequence of states and operates as a clock. This form of internal dynamics can be viewed as an active motion along a certain cyclic coordinate. Interaction between individual elements in large ensembles of such oscillators may lead to mutual entrainment of their cycles and thus to the emergence of coherent internal dynamics. The onset of mutual synchronization in oscillator ensembles resembles a phase transition.

The synchronization phenomena depend on the properties of interactions between the oscillators. When interactions cause only slight acceleration or slowing down of individual cyclic motions, they are described by the Kuramoto model of phase dynamics. Different behavior is found when interactions are exchanged in the form of short intensive pulses acting on the oscillators. Both external noise and structural inhomogeneity of the ensemble make the synchronization more difficult and can destroy it.

The emergence and breakdown of coherent collective motion are not a unique property of the internal dynamics. Similar effects are possible when cyclic collective motions of swarms in physical coordinate space are considered.

6.1 Interacting Clocks

Fireflies emit light at regular intervals, typically twice per second. Large populations of such insects may cover the trees and, when the cycles of individual insects are synchronized, these whole trees are seen rapidly flashing – an impressive example of coherent cooperative dynamics [308]. Similar phenomena are known at the cellular level. Oscillations in single neurons or in the cells forming pacemaker nodes in the heart may become synchronous, so that a large-amplitude periodic signal is generated. It was Norbert Wiener [293, 294] who first suggested that physiological rhythms might reflect the mutual synchronization of myriads of individual oscillatory processes.

The actual processes underlying the cyclic behavior of a biological organism are complex and may greatly differ from one organism to another.

However, to describe the response of biological oscillators to external perturbations and consider mutual synchronization in such ensembles, it would often suffice to treat a living organism as an automaton that repeatedly goes through a cycle of its internal states. As noted by A. T. Winfree [295], the cyclic internal motion can generally be characterized by introducing the *phase*. If the number of states inside the cycle is large, the phase variable is continuous. A biological oscillator is like a clock whose every cycle corresponds to a full rotation of its hand over the dial. The phase ϕ can be viewed as the angle coordinate on this dial. Each rotation corresponds to an increase of 2π. The phase values differing by an integer number of rotations correspond to identical oscillator states. The individual cyclic dynamics of a clock can thus be described by a simple evolution equation

$$\dot{\phi} = \omega, \tag{6.1}$$

where ω is its frequency, which represents the velocity of motion over the phase coordinate. The temporal period is $T = 2\pi/\omega$.

The clocks may interact. Under external action, a clock can accelerate or slow down its cyclic motion. Below in this section we investigate the behavior of two coupled clocks with different frequencies ω_1 and ω_2. We shall show that, under certain conditions, the motion of the clocks can spontaneously synchronize so that they would rotate at a common frequency which may be different from both ω_1 and ω_2.

This system is described by a set of two dynamical equations:

$$\begin{aligned} \dot{\phi}_1 &= \omega_1 + U_{12}(\phi_1 - \phi_2) \\ \dot{\phi}_2 &= \omega_2 + U_{21}(\phi_2 - \phi_1) \end{aligned}, \tag{6.2}$$

where the function U specifies interactions between the clocks. We assume that the interactions depend only on the phase difference. Because phases differing by 2π correspond to the same physical state of the system, U_{12} and U_{21} must be periodic functions with period 2π.

A closed equation for the evolution of the phase difference $\theta = \phi_1 - \phi_2$ is obtained by subtracting the second equation in (6.2) from the first one:

$$\dot{\theta} = \Delta\omega + u(\theta). \tag{6.3}$$

Here we have introduced the notations $\Delta\omega = \omega_1 - \omega_2$ and $u(\theta) = U_{12}(\theta) - U_{21}(-\theta)$.

When synchronization has occurred and both clocks rotate at the same frequency, the phase difference θ should remain constant. Hence, synchronous states correspond to fixed points of the differential equation (6.3). These points are given by the roots of the algebraic equation

$$-\Delta\omega = u(\theta). \tag{6.4}$$

As an example, Fig. 6.1 shows the graphical solution of (6.4) for the function $u(\theta) = -\varepsilon \sin\theta$. The roots θ_s and θ_u represent intersection points of the graph of this function with the horizontal line $U = -\Delta\omega$ inside the interval $0 \leq \theta < 2\pi$.

A synchronous state is stable if small perturbations of the phase difference fade away with time. Linearizing equation (6.3) at a fixed point θ_0, we find that the deviation $\delta\theta$ from this point obeys the equation

$$\dot{\delta\theta} = A\delta\theta, \tag{6.5}$$

with $A = dU/d\theta$ at $\theta = \theta_0$. Hence, the deviations decrease with time and therefore the fixed point θ_0 is stable if the derivative A is negative. Looking at Fig. 6.1, we note that stable synchronization corresponds only to the root θ_s. The other root θ_u yields an absolutely unstable synchronous solution.

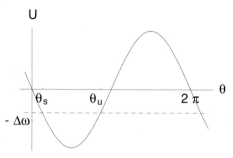

Fig. 6.1. Graphical solution of equation (6.4) for the function $u(\theta) = -\varepsilon \sin\theta$, $\varepsilon = 1$ and $\Delta\omega = 0.4$

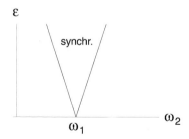

Fig. 6.2. Synchronization region in the parameter plane (ω_2, ε)

Equation (6.4) has solutions only if the frequency difference $\Delta\omega$ lies within the interval of values of the function $U(\theta)$, i.e. if $U_{\min} \leq \Delta\omega \leq U_{\max}$, where $U_{\min}$ and $U_{\max}$ are the minimum and the maximum values of this function. In the example shown in Fig. 6.1 we have $u(\theta) = -\varepsilon \sin\theta$ and therefore the condition simply reads as $|\Delta\omega| \leq \varepsilon$. The synchronization region in the parameter plane (ω_2, ε) is plotted in this case in Fig. 6.2. It grows for larger amplitudes U_0 and thus can be described as a *tongue*.

The frequency Ω of the common synchronous motion of two coupled clocks is given by

$$\Omega = \omega_1 + U_{12}(\theta_0) = \omega_2 + U_{21}(-\theta_0), \qquad (6.6)$$

where θ_0 is the phase difference determined by (6.4). Note that our analysis also includes the case when one of the clocks is freely running and generates a force acting on the other clock. This corresponds to the choice of $U_{21} = 0$ in (6.2). In this case the first clock is entrained by the free-running second one. Then it rotates at frequency $\Omega = \omega_2$.

Outside of the entrainment region the motions of the two oscillators are not synchronous. However, they may still be significantly influenced by the interactions. To simplify the analysis, we shall now consider the case of symmetric interactions: $U_{12}(\theta) = -U_{21}(-\theta) = -\frac{1}{2}\varepsilon\sin\theta$. Equation (6.3) now has the explicit form

$$\dot\theta = \Delta\omega - \varepsilon\sin\theta. \qquad (6.7)$$

Integration of this equation yields

$$\theta(t) = 2\arctan\left[\frac{\varepsilon}{\Delta\omega} + \sqrt{1 - \frac{\varepsilon^2}{\Delta\omega^2}}\tan\left(\frac{1}{2}\sqrt{\Delta\omega^2 - \varepsilon^2}\,t\right)\right]. \qquad (6.8)$$

Figure 6.3 shows the time dependence of the phase difference $\theta = \phi_1 - \phi_2$ described by this solution for two clocks with frequencies that lie close to the synchronization range. We see that the phase difference remains approximately constant inside a wide time interval, as should have been observed under the entrainment conditions. Then, however, the phase difference quickly increases by 2π, indicating that one of the clocks makes a full rotation with respect to the other. After that a relatively long interval of approximate entrainment, followed by another quick rotation, takes place and so on.

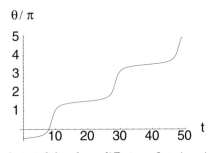

Fig. 6.3. Time dependence of the phase difference $\theta = \phi_1 - \phi_2$ of two clocks outside of the synchronization region; $\varepsilon = 1$ and $\Delta\omega = 1.05$

The phase dynamics of individual clocks can easily be determined in this case. Since $U_{12}(\phi_1 - \phi_2) = -U_{21}(\phi_2 - \phi_1)$, summing the two equations (6.2) yields

$$\frac{d}{dt}\left(\phi_1 + \phi_2\right) = \omega_1 + \omega_2, \tag{6.9}$$

and therefore we have $\phi_1 + \phi_2 = (\omega_1 + \omega_2)\,t + \text{const}$. Thus, we obtain

$$\phi_1(t) = \tfrac{1}{2}\left(\omega_1 + \omega_2\right)t + \tfrac{1}{2}\theta(t) + \phi_1^0$$
$$\phi_2(t) = \tfrac{1}{2}\left(\omega_1 + \omega_2\right)t - \tfrac{1}{2}\theta(t) + \phi_2^0 \tag{6.10}$$

The clocks do not rotate at a constant angular velocity, i.e. the oscillations are not harmonic. The periods T_1 and T_2 of the two clocks can be found from the conditions $\phi_1(t + T_1) = \phi_1(t) + 2\pi$ and $\phi_2(t + T_2) = \phi_2(t) + 2\pi$. The respective frequencies $\Omega_1 = 2\pi/T_1$ and $\Omega_2 = 2\pi/T_2$ are given by

$$\Omega_{1,2} = \frac{1}{2}\left(\omega_1 + \omega_2\right) \pm \frac{1}{2}\sqrt{\left(\omega_1 - \omega_2\right)^2 - \varepsilon^2}, \tag{6.11}$$

for $\varepsilon < |\omega_1 - \omega_2|$. When entrainment takes place ($\varepsilon > |\omega_1 - \omega_2|$), both oscillators have the same frequency $\Omega = \tfrac{1}{2}\left(\omega_1 + \omega_2\right)$. Similar results can be obtained for other choices of the interactions U_{12} and U_{21} between the clocks in (6.2).

6.2 The Synchronization Transition

We have seen above that two interacting clocks can synchronize their motion, provided that the difference of their natural frequencies is small enough. This suggests that a similar behavior may also be found in large ensembles of interacting clocks. Mutual synchronization of many interacting clocks is considered below in this section.

The natural frequencies of individual elements in an ensemble are characterized by statistical dispersion and, generally, one would find both small and large frequency differences in these systems. The clocks with close natural frequencies tend to synchronize, whereas the clocks with greatly different frequencies act against synchronization. The consequence of this competition is that the synchronization begins at a certain minimum interaction intensity, collectively determined by all members of the ensemble. Close to the threshold, mutually entrained clocks form only a small fraction of the entire population. The synchronization transition bears features that make it similar to phase transitions in equilibrium physical systems.

The problem of the synchronization transition in large populations of biological oscillators was formulated by A.T. Winfree [295] who has also provided some quantitative estimates for this phenomenon. Its full solution for a system of interacting clocks has been found by Y. Kuramoto [299]. Our analysis in this section will follow his arguments.

We consider an ensemble consisting of N individual clocks, each characterized by its own frequency ω_i ($i = 1, 2, \ldots, N$). The collective dynamics of this ensemble is described by a set of N equations

$$\dot{\phi}_i = \omega_i + \sum_{j=1}^{N} U_{ij}(\phi_i - \phi_j). \tag{6.12}$$

According to these equations, the interactions are additive, i.e. the influence of all other oscillators in the ensemble on a given oscillator is the sum of the individual influences. Moreover, the interactions U_{ij} depend only on the phase differences $\phi_i - \phi_j$ for the respective pairs of elements.

Since the ensemble is large ($N \gg 1$), its frequency distribution can be characterized by a continuous distribution function $g(\omega)$, such that $Ng(\omega)d\omega$ yields the number of elements with frequencies in the interval from ω to $\omega + d\omega$. Though the following analysis is general, we shall often use as an example the Lorentz distribution,

$$g(\omega) = \frac{\gamma}{\pi} \frac{1}{(\omega - \omega_0)^2 + \gamma^2}. \tag{6.13}$$

This distribution has the form of a narrow symmetric peak of width γ with the center at $\omega = \omega_0$.

The interactions between the clocks will be described by the functions

$$U_{ij}(\phi_i - \phi_j) = -\frac{\varepsilon}{N} \sin(\phi_i - \phi_j). \tag{6.14}$$

Their intensity is therefore specified by the parameter ε. The intensity is the same for any chosen pair of elements.

To characterize the state of the ensemble, we introduce the phase density $n(\phi, t)$. This density is defined in such a way that $Nn(\phi, t)d\phi$ yields the number of elements in the ensemble which have phases in the interval from ϕ to $\phi + d\phi$ at time t. The phase density satisfies the normalization condition

$$\int_{-\pi}^{\pi} n(\phi, t)d\phi = 1. \tag{6.15}$$

This function is periodic, i.e. $n(\phi + 2\pi, t) = n(\phi, t)$ for any ϕ and t.

When all phases are equally probable, the phase distribution is flat, so that $n(\phi) = \text{const}$. Deviations from this flat distribution would indicate the presence of phase correlations in the system. We can characterize the degree of synchronization by the complex order parameter $\Lambda = \sigma e^{i\Psi}$ defined as

$$\sigma e^{i\Psi} = \int_{-\pi}^{\pi} n(\phi, t)e^{i\phi}d\phi. \tag{6.16}$$

Note that $\sigma = 0$ for the uniform phase distribution $n(\phi) = $ const. On the other hand, if all oscillators have the same phase, i.e. $n(\phi) = \delta(\phi - \phi_0)$, we have $\sigma = 1$ and $\Psi = \phi_0$.

In the limit $N \to \infty$ we can write equations (6.12) with interaction functions (6.14) as

$$\dot{\phi}_i = \omega_i - \frac{\varepsilon}{2iN} \sum_{j=1}^{N} \left(e^{i\phi_i - i\phi_j} - e^{-i\phi_i + i\phi_j} \right) \tag{6.17}$$

$$= \omega_i - \frac{\varepsilon}{2i} \int_{-\pi}^{\pi} n(\phi, t) \left(e^{i\phi_i - i\phi} - e^{-i\phi_i + i\phi} \right) d\phi \tag{6.18}$$

$$= \omega_i - \frac{\varepsilon}{2i} \left(e^{i\phi_i} \int_{-\pi}^{\pi} n(\phi, t) e^{-i\phi} d\phi - e^{-i\phi_i} \int_{-\pi}^{\pi} n(\phi, t) e^{i\phi} d\phi \right). \tag{6.19}$$

Recognizing that, according to (6.16), the integrals in the last line in (6.17) are given by $\sigma e^{-i\Psi}$ and $\sigma e^{i\Psi}$, we obtain the phase dynamics equation

$$\dot{\phi}_i = \omega_i - \varepsilon\sigma \sin(\phi_i - \Psi). \tag{6.20}$$

This is an interesting result. We see that interactions between any given oscillator i and all other oscillators in the ensemble can be effectively described as an interaction with a *single* oscillator with phase Ψ, characterized by the interaction intensity $\varepsilon\sigma$.

We limit our subsequent analysis to the steady synchronous regime where $\Psi(t) = \Omega t + \Psi_0$ and $\sigma(t) = $ const. Introducing phase deviations $\theta_i = \phi_i - \Omega t - \Psi_0$, (6.20) is then written as

$$\dot{\theta}_i = \Delta\omega_i - \varepsilon\sigma \sin\theta_i, \tag{6.21}$$

where $\Delta\omega_i = \omega_i - \Omega$. It coincides with (6.7), and therefore we can now use the results of our previous analysis of two coupled clocks in Sect. 6.1.

All oscillators are divided into two groups, depending on their frequency difference $\Delta\omega_i$. The first group is formed by the elements with $|\Delta\omega_i| < \varepsilon\sigma$. These elements are entrained and perform oscillations with a collective frequency Ω. The elements belonging to the second group ($|\Delta\omega_i| > \varepsilon\sigma$) are not entrained, but their phase motion is influenced by the interactions. Thus, the phase distribution $n(\theta)$ with respect to the phase θ in this steady regime can be written as the sum of the phase distributions corresponding to the two groups, i.e. as

$$n(\theta) = n_{\mathrm{I}}(\theta) + n_{\mathrm{II}}(\theta). \tag{6.22}$$

The relative phase θ of an entrained oscillator with the natural frequency ω is determined, according to (6.21), by

$$\omega - \Omega = \varepsilon\sigma \sin\theta, \tag{6.23}$$

where only the root $-\pi/2 < \theta < \pi/2$ that corresponds to stable entrainment (see Sect. 6.1) should be taken.

Hence, the number $n_I(\theta)d\theta$ of elements in the first group with phases in the interval from θ to $\theta + d\theta$ is given by the number of oscillators with frequencies in the interval from ω to $\omega + d\omega$, where ω is given by (6.23), i.e. $n_I(\theta)d\theta = g(\omega)d\omega$. Therefore, we have

$$n_I(\theta) = g\left(\omega\left(\theta\right)\right)\frac{d\omega}{d\theta}. \tag{6.24}$$

Differentiation of (6.23) yields $d\omega/d\theta = \varepsilon\sigma\cos\theta$ and thus the phase distribution of oscillators in the first group is given by

$$n_I(\theta) = g\left(\Omega + \varepsilon\sigma\sin\theta\right)\varepsilon\sigma\cos\theta, \text{ for } -\pi/2 < \theta < \pi/2. \tag{6.25}$$

Note that $n_I(\theta)$ vanishes outside of the interval $-\pi/2 < \theta < \pi/2$.

For an oscillator with frequency ω in the second group, such that $|\omega - \Omega| > \varepsilon\sigma$, the phase θ is not fixed. This element will move at a velocity $d\theta/dt$ that is determined by (6.23) and depends on the current phase θ. The probability $p_\omega(\theta)d\theta$ of finding this element inside the phase interval from θ to $\theta + d\theta$ is simply given by the relative time spent by the moving element inside this interval (remember that phases θ differing by 2π are physically equivalent). Hence, we have

$$p_\omega(\theta)d\theta = Z\left|\frac{d\theta}{dt}\right|^{-1}d\theta, \tag{6.26}$$

where Z is the normalization constant, determined by the condition

$$\int\limits_{-\pi}^{\pi} p_\omega(\theta)d\theta = 1, \tag{6.27}$$

and therefore given by

$$Z = \left[\int\limits_{-\pi}^{\pi}\left|\frac{d\theta}{dt}\right|d\theta\right]^{-1}. \tag{6.28}$$

Taking into account that according to (6.23) $d\theta/dt = \omega - \Omega - \varepsilon\sigma\sin\theta$ and performing the integration in (6.28) we thus find

$$p_\omega(\theta) = \frac{\left[(\omega - \Omega)^2 - \varepsilon^2\sigma^2\right]^{1/2}}{2\pi\left|\omega - \Omega - \varepsilon\sigma\sin\theta\right|}. \tag{6.29}$$

Equation (6.29) determines the probability density that an oscillator with a given frequency ω will have the phase θ. The relative number of oscillators with this frequency is given by the function $g(\omega)$. Hence, the phase distribution for the oscillators in the second (non-entrained) group is

$$n_{\mathrm{II}}(\theta) = \int_{\Omega+\varepsilon\sigma}^{\infty} g(\omega)p_\omega(\theta)d\omega + \int_{-\infty}^{\Omega-\varepsilon\sigma} g(\omega)p_\omega(\theta)d\omega. \qquad (6.30)$$

The phase distributions (6.25) and (6.30) include the quantities σ and Ω, which remain unknown. To find them, the definition (6.16) of the order parameter Δ should be used. In the considered steady regime, equation (6.16) can be written in terms of the relative phase θ as

$$\sigma e^{i(\Omega t + \Psi_0)} = \int_{-\pi}^{\pi} n(\theta) e^{i(\theta + \Omega t + \Psi_0)} d\theta. \qquad (6.31)$$

Dividing both sides by $\exp\left[i\left(\Omega t + \Psi_0\right)\right]$, we obtain

$$\sigma = \int_{-\pi}^{\pi} n(\theta) e^{i\theta} d\theta. \qquad (6.32)$$

This equation must hold separately for its real and imaginary parts. Thus, it is equivalent to two different equations:

$$0 = \int_{-\pi}^{\pi} n(\theta) \sin\theta d\theta, \qquad (6.33)$$

and

$$\sigma = \int_{-\pi}^{\pi} n(\theta) \cos\theta d\theta. \qquad (6.34)$$

Substituting $n(\theta) = n_{\mathrm{I}}(\theta) + n_{\mathrm{II}}(\theta)$ into (6.33) and (6.34) and integrating, we would obtain a set of two algebraic equations that determine the parameters σ and Ω. The explicit form of these equations and their solutions depend on the function $g(\omega)$ that describes the frequency distribution in the ensemble.

When the distribution $g(\omega)$ is centered at a certain frequency ω_0 and has a symmetric shape, i.e. $g(\omega_0 + \delta\omega) = g(\omega_0 - \delta\omega)$, the collective frequency Ω will coincide with ω_0. Indeed, if we choose $\Omega = \omega_0$, the phase distributions $n_{\mathrm{I}}(\theta)$ and $n_{\mathrm{II}}(\theta)$ will be symmetric, so that $n(\theta) = n(-\theta)$. But then the integral in (6.33) is zero and this equation is identically satisfied. Equation (6.34) directly determines in this case the parameter σ.

As an example, we construct the solutions for the Lorentz frequency distribution (6.13). Substituting this function $g(\omega)$ into (6.25) and (6.30) and integrating, we obtain the phase distributions for both oscillator groups:

$$n_{\mathrm{I}}(\theta) = \frac{\gamma\varepsilon\sigma\cos\theta}{\pi\left(\varepsilon^2\sigma^2\sin^2\theta + \gamma^2\right)}, \quad \text{for } -\frac{\pi}{2} < \theta < \frac{\pi}{2}; \tag{6.35}$$

$$n_{\mathrm{II}}(\theta) = \frac{\gamma}{2\pi}\frac{\sqrt{\varepsilon^2\sigma^2 + \gamma^2} - \varepsilon\sigma\,|\cos\theta|}{\varepsilon^2\sigma^2\sin^2\theta + \gamma^2} \tag{6.36}$$

The total phase distribution $n(\theta) = n_{\mathrm{I}}(\theta) + n_{\mathrm{II}}(\theta)$ is

$$n(\theta) = \frac{\gamma}{2\pi}\frac{\sqrt{\varepsilon^2\sigma^2 + \gamma^2} + \varepsilon\sigma\cos\theta}{\varepsilon^2\sigma^2\sin^2\theta + \gamma^2}. \tag{6.37}$$

After substitution of (6.37) into (6.34) and easy integration, we obtain the algebraic equation

$$\sigma = \frac{1}{\varepsilon\sigma}\left(-\gamma + \sqrt{\varepsilon^2\sigma^2 + \gamma^2}\right), \tag{6.38}$$

whose solution is

$$\sigma = \begin{cases} \sqrt{1 - \frac{2\gamma}{\varepsilon}}, & \text{for } \varepsilon \geq 2\gamma \\ 0, & \text{for } \varepsilon < 2\gamma \end{cases}. \tag{6.39}$$

We see that synchronization starts at a definite interaction intensity $\varepsilon_c = 2\gamma$, determined by the statistical dispersion of the oscillator frequencies. Close to the synchronization threshold, the order parameter is small, i.e. $\sigma \sim \sqrt{\varepsilon - \varepsilon_c}$. This behavior is similar to what is known in the mean-field theory for second-order phase transitions in equilibrium physical systems.

Figure 6.4 shows the computed distributions $n_{\mathrm{I}}(\theta)$ and $n_{\mathrm{II}}(\theta)$ given by (6.35) and (6.36). Near the synchronization threshold (Fig. 6.4a) the fraction of entrained elements, corresponding to the distribution $n_{\mathrm{I}}(\theta)$, is small and the phase distribution $n_{\mathrm{II}}(\theta)$ of the non-entrained elements is almost flat. Far from the threshold (Fig. 6.4b) almost all the elements are entrained and form a strong peak at $\theta = 0$.

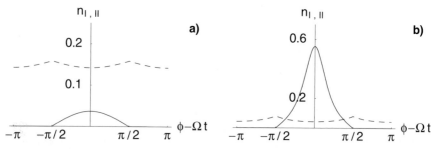

Fig. 6.4. Phase distributions $n_{\mathrm{I}}(\phi - \Omega t)$ (*solid line*, entrained clocks) and $n_{\mathrm{II}}(\phi - \Omega t)$ (*dash line*, non-entrained clocks) for the Lorentz frequency distribution with $\gamma = 0.5$ for two different values of the coupling strength: (**a**) $\varepsilon = 1.005$ and (**b**) $\varepsilon = 1.5$

One should remember that the actual phase is $\phi = \Omega t + \Psi_0 + \theta$. The distributions seen in Fig. 6.4 are therefore plotted in the rotating coordinate frame. The entrained elements form a cluster that rigidly moves (i.e. rotates) at velocity Ω. The non-entrained elements move at different velocities with respect to the cluster. They slow down or accelerate their motion depending on the distance from the rigid traveling cluster.

6.3 The Influence of Noise

So far we have considered ensembles of clocks (i.e. phase oscillators) whose motion was purely deterministic. In the absence of interactions, these clocks rotated at different constant angular velocities. The synchronization essentially represented the formation of a coherently rotating group inside this ensemble. Now we shall analyze the behavior of ensembles formed by identical phase oscillators under the action of noise.

The considered system is described by a set of stochastic differential equations

$$\dot{\phi}_i = \omega_0 - \frac{\varepsilon}{N} \sum_{j=1}^{N} \sin\left(\phi_i - \phi_j\right) + \xi_i(t), \tag{6.40}$$

where $\xi_i(t)$ are independent white noises, such that $\langle \xi_i(t) \rangle = 0$ and

$$\langle \xi_i(t)\xi_k(t') \rangle = 2S\delta_{ik}\delta(t - t'). \tag{6.41}$$

Note that since the rotation frequencies of all the clocks are the same, we can always go into the rotating frame $\phi_i \to \phi_i - \omega_0 t$ and thus eliminate the term with ω_0 in (6.40). Hence, without any loss of generality we can set $\omega_0 = 0$ in these equations.

First, we briefly discuss the behavior of identical coupled clocks in the absence of noise. The synchronization means in this case that the individual phases of all clocks become equal, i.e. $\phi_i = \Psi_0$ for any clock i. To investigate the stability of such a synchronous state, we can assume that the phase of *one* of the oscillators is slightly perturbed, so that $\phi_i = \Psi_0 + \delta\phi_i(t)$. Substituting this into (6.40) and linearizing this equation, we obtain

$$\dot{\delta\phi} = -\frac{\varepsilon}{N} \sin \delta\phi_i \approx -\frac{\varepsilon}{N}\delta\phi_i. \tag{6.42}$$

The perturbation fades away and thus the synchronous state is stable if $\varepsilon > 0$.

When noise is present and therefore the phases fluctuate, we can construct the statistical phase distribution $n(\phi)$ that yields the probability density to find phase ϕ. Using this distribution, the complex order parameter $\Lambda = \sigma e^{i\Psi} = \langle e^{i\phi} \rangle$ can be again defined. In the limit $N \to \infty$ we have

$$\frac{1}{N} \sum_{j=1}^{N} e^{i\phi_j} = \int_{-\pi}^{\pi} n(\phi)e^{i\phi}d\phi \equiv \langle e^{i\phi} \rangle. \tag{6.43}$$

Therefore, in this limit we can transform (6.40) as

$$\dot{\phi}_i = -\frac{\varepsilon}{2iN} \sum_{j=1}^{N} \left[e^{i(\phi_i - \phi_j)} - e^{-i(\phi_i - \phi_j)} \right] + \xi_i(t) \tag{6.44}$$

$$= -\frac{\varepsilon}{2i} \left[e^{i\phi_i} \frac{1}{N} \sum_{j=1}^{N} e^{-i\phi_j} - e^{-i\phi_i} \frac{1}{N} \sum_{j=1}^{N} e^{i\phi_j} \right] + \xi_i(t) \tag{6.45}$$

$$= -\frac{\varepsilon\sigma}{2i} \left(e^{i\phi_i - i\Psi} - e^{-i\phi_i + i\Psi} \right) + \xi_i(t). \tag{6.46}$$

This yields

$$\dot{\phi}_i = -\varepsilon\sigma \sin(\phi_i - \Psi) + \xi_i(t). \tag{6.47}$$

Introducing the relative phase $\theta_i = \phi_i - \Psi$ and dropping the index i, we finally arrive at the equation

$$\dot{\theta} = -\varepsilon\sigma \sin\theta + \xi(t). \tag{6.48}$$

This stochastic partial differential equation corresponds to the following Fokker–Planck equation for the probability density $n(\theta, t)$ (see [306]):

$$\frac{\partial n}{\partial t} = \frac{\partial}{\partial \theta} (\varepsilon\sigma \sin\theta \; n) + S \frac{\partial^2 n}{\partial \theta^2}. \tag{6.49}$$

Its stationary solution is determined by

$$S \frac{dn}{d\theta} = -\varepsilon\sigma \sin\theta \; n. \tag{6.50}$$

Hence, the phase distribution $n(\theta)$ in the statistical steady state of this system has the form

$$n(\theta) = C \exp\left(\frac{\varepsilon\sigma}{S} \cos\theta \right), \tag{6.51}$$

where C is the normalization constant,

$$C = \left[\int_{-\pi}^{\pi} \exp\left(\frac{\varepsilon\sigma}{S} \cos\theta \right) d\theta \right]^{-1} = \left[2\pi I_0 \left(\frac{\varepsilon\sigma}{S} \right) \right]^{-1}, \tag{6.52}$$

and $I_0(z)$ is the modified Bessel function of the first kind.

Using the identity $\sigma e^{i\Psi} = \langle e^{i(\theta + \Psi)} \rangle = \langle e^{i\theta} \rangle e^{i\Psi}$ and the symmetry of the distribution (6.51), we find that the order parameter σ is given by

$$\sigma = \langle \cos\theta \rangle = \int_{-\pi}^{\pi} n(\theta) \cos\theta \; d\theta. \tag{6.53}$$

Substituting (6.51) and integrating, we obtain

$$\sigma = \frac{I_1\left(\frac{\varepsilon\sigma}{S}\right)}{I_0\left(\frac{\varepsilon\sigma}{S}\right)}. \tag{6.54}$$

This is an algebraic equation that determines the order parameter σ in the steady statistical state. To analyze its solutions, it is convenient to write (6.54) in the form

$$\frac{S}{\varepsilon}y = \frac{I_1(y)}{I_0(y)}, \tag{6.55}$$

where $y = \varepsilon\sigma/S$. The function $f(y) = I_1(y)/I_0(y)$ is plotted in Fig. 6.5. For small y it can be approximated as $f(y) \approx (1/2)y - (1/16)y^3$. The roots of (6.55) are given by the intersection points of this curve with the straight line $(S/\varepsilon)y$. The trivial solution $y = 0$, corresponding to $\sigma = 0$, is always possible. Additionally, a root with $y \neq 0$ and $\sigma \neq 0$ exists for $S/\varepsilon \leq 1/2$.

Figure 6.6 shows the dependence of the order parameter σ on the ratio S/ε, obtained by numerical solution of (6.54). Close to the synchronization threshold it can be approximated as

$$\sigma = \begin{cases} \sqrt{2 - 4S/\varepsilon}, \text{ for } S \leq \varepsilon/2 \\ 0, \text{ for } S > \varepsilon/2 \end{cases}. \tag{6.56}$$

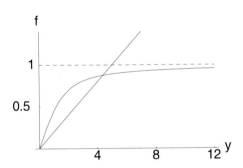

Fig. 6.5. Graphical solution of equation (6.55)

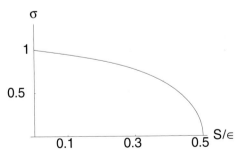

Fig. 6.6. Dependence of the order parameter σ on the ratio S/ε

In absence of noise the phases of all the oscillators are identical and therefore $\sigma = 1$. The introduction of a weak noise results in a decrease of the order parameter, proportional to the noise intensity S. For larger noise intensities, the order parameter σ rapidly falls and reaches zero at the critical point $S_c = \varepsilon/2$. When $S > S_c$, the synchronization is absent ($\sigma = 0$) and the phase distribution is flat (i.e. $n(\phi) = \mathrm{const.}$).

Thus, temporal fluctuations play a role similar to the static heterogeneity studied in Sect. 6.2. Both the dispersion of individual oscillation frequencies and the fluctuations, caused by an external noise, make synchronization more difficult and eventually lead to its disappearance.

6.4 Noise-Induced Breakdown of Coherent Active Motion

Mutual synchronization can also take place for active motion, previously considered in Chap. 3. In contrast to the systems of elements with internal cyclic dynamics, the synchronous state corresponds now to a coherent collective motion of the entire population (i.e. of a swarm) in the physical coordinate space. Below in this section we consider a system of interacting self-moving particles that shows a transition from coherent translational motion to random oscillations [296].

The dynamics of this system is decribed by equations (3.25). For simplicity, we shall study only the one-dimensional model, where each particle i is characterized by a single spatial coordinate x_i. In contrast to Sect. 3.1, where collective motions of swarms with local interactions were analyzed, we discuss now the opposite case of *long-range* attractive interactions between particles. The interaction potential $U(x)$ is chosen to have the parabolic form, $U(x) \sim x^2$. This assumption might seem unrealistic, since the interactions indefinitely grow with the separation x between two particles, i.e. $\nabla U \sim x$. However, in our subsequent analysis we require only that the parabolic approximation holds for separations not exceeding the diameter of the population cloud. At larger distances the potential U may be constant, so that the interactions are absent.

The model is therefore described by a set of N dynamical equations:

$$\ddot{x}_i + \left(\dot{x}_i^2 - 1\right)\dot{x}_i + \frac{a}{N}\sum_{j=1}^{N}(x_i - x_j) = f_i(t), \quad (i = 1, \ldots, N). \tag{6.57}$$

The coeffiicient a characterizes the intensity of the interactions and can be viewed as a parameter that specifies the strength of coupling in the population. The independent random forces $f_i(t)$ are modeled as white noise of intensity S, so that

$$\langle f_i(t)f_j(t')\rangle = 2S\delta_{ij}\delta(t - t'). \tag{6.58}$$

Note that equations (6.57) are invariant with respect to an arbitrary translation in the coordinate space.

Introducing the average coordinate $\bar{x}(t)$ of the swarm,

$$\bar{x}(t) = \frac{1}{N} \sum_{j=1}^{N} x_j(t), \tag{6.59}$$

(6.57) in the absence of noise can be written in the form

$$\ddot{x}_i + \left(\dot{x}_i^2 - 1 \right) \dot{x}_i + a(x_i - \bar{x}) = 0. \tag{6.60}$$

Thus, if the swarm does not move as a whole, i.e. $\bar{x}(t) = $ const., the particles perform oscillations with frequency $a^{1/2}$. In such a state the phases of individual oscillations and the spatial location $\bar{x}$ of an oscillating swarm are arbitrary. On the other hand, the system (6.60) also has two collapsed states where the coordinates of all particles are identical, i.e. $x_i = \bar{x}$ for any i. These states correspond to translational motions $\bar{x}(t) = \bar{x}_0 + ut$ of the entire swarm with velocities $u = \pm 1$ and can therefore be viewed as analogs of the synchronous motion considered above in this chapter. Both kinds of state are stable with respect to small perturbations and thus represent different coexisting attractors of the dynamical system (6.60).

If weak noise is present, the collapsed states transform into traveling clouds of particles that coherently move at a certain mean velocity. The oscillatory state corresponds now to a noisy limit cycle. Figure 6.7 shows three typical states of a system of 100 particles with $a = 10$ subject to noise

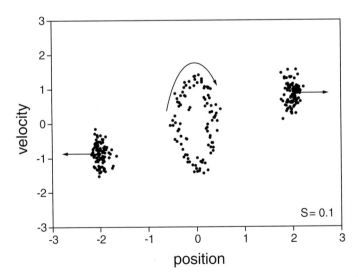

Fig. 6.7. Three typical states of a system of 100 particles with $a = 10$ subject to noise with $S = 0.1$

with $S = 0.1$, obtained by numerical integration of stochastic differential equations (6.57) with different initial conditions.[1] The swarm can form a diffuse cloud traveling either to the left or to the right. Within such clouds, each particle performs random motions which are superimposed on the collective translation. The swarm also can be found in the random oscillatory state where the overall translational motion is absent.

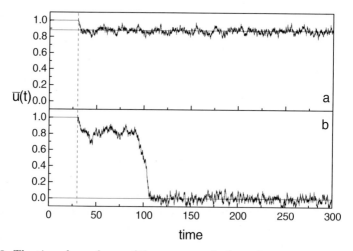

Fig. 6.8. The time dependence of the average velocity u for a swarm with $N = 100$ particles and $a = 10$ at two different noise intensities: (**a**) $S = 0.1$ and (**b**) $S = 0.12$. The noise is turned on at time $t = 30$. Horizontal lines indicate asymptotic mean velocity values

When the noise intensity is increased, the synchronous traveling states disappear and the swarm undergoes an abrupt transition into its random oscillatory state. This breakdown of coherent collective motion is illustrated in Fig. 6.8, where the calculated average swarm velocity

$$\overline{u}(t) = \frac{1}{N} \sum_{j=1}^{N} \dot{x}_j\,(t) \tag{6.61}$$

is plotted as function of time for two different noise intensities. We see that if the noise is relatively weak (Fig. 6.8a), its introduction at moment $t = 30$ only slightly decreases the average velocity of the coherent cloud that fluctuates around a certain finite level. If, however, the noise intensity exceeds a threshold, the effect of the noise introduction is qualitatively different (Fig. 6.8b).

---

[1] In numerical simulations of the finite difference scheme corresponding to the considered stochastic differential equations, noise was introduced by generating at each time step a random number ξ with uniform distribution in the interval $(-\xi_0, \xi_0)$. This choice corresponds to having noise of intensity $S = \xi_0^2/6\Delta t$, where Δt is the time step. This noise is delta-correlated in time, but non-Gaussian.

Within some time interval, the swarm continues to travel at a somewhat decreased strongly fluctuating average velocity. Then, it suddenly starts to decelerate and soon reaches a steady state where the average velocity $\bar{u}(t)$ fluctuates around zero. Inspection of the distribution of particles in the ensemble shows that in this state the system has been attracted by the noisy limit cycle.

Hence, the system undergoes a *noise-induced transition* from the condition of multistability with two kinds of attractors to a situation where only one of them exists. The coherent clouds observed for small noise intensities are no longer possible for $S > S_c$ and the system is always found in the state of noisy oscillations.

Figure 6.9 displays the dependence of the asymptotic mean velocity $\bar{u}$ of the traveling swarm on the noise intensity S for three different values of the coupling coefficient a. The mean velocity monotonously decreases with the noise intensity, until a certain critical noise intensity is reached and coherent swarm motion becomes impossible. The mean velocity at the critical point is still relatively large: $\bar{u} \approx 0.8$. The critical noise intensity S_c becomes lower for smaller values of a. The behavior of the swarm is characterized by a strong hysteresis. If the breakdown of the synchronous motion has occurred, the subsequent decrease of the noise intensity leaves the system in the oscillatory state with zero mean velocity down to $S = 0$.

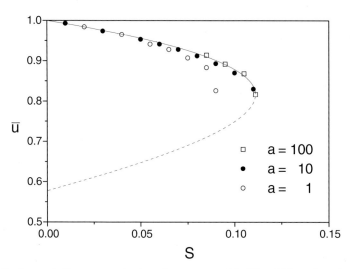

Fig. 6.9. Asymptotic mean velocity of a swarm with $N = 100$ as function of the noise intensity for three different values of the parameter a

The noise-induced transition can be explained by a simple theoretical analysis. Summing all equations (6.57) for different particles i and taking into account that the noises acting on individual particles are not correlated (i.e. $\sum_{i=1}^{N} f_i(t) = 0$), the evolution equation for the average swarm velocity $\bar{u}(t)$ is obtained:

$$\frac{d\bar{u}}{dt} + \frac{1}{N} \sum_{i=1}^{N} \dot{x}_i^3 - \bar{u} = 0. \tag{6.62}$$

Let us introduce for each particle its deviation $y_i = x_i - \bar{x}$ from the average swarm position. Then we can write

$$\frac{1}{N} \sum_{i=1}^{N} \dot{x}_i^3 = \bar{u}^3 + 3\Theta\bar{u} + \frac{1}{N} \sum_{i=1}^{N} \dot{y}_i^3, \tag{6.63}$$

where the average square dispersion of the swarm is defined by $\Theta = N^{-1} \sum_{i=1}^{N} \dot{y}_i^2$.

The last cubic term in this equation can be neglected if the distribution of particles in the traveling cloud is symmetric. The numerical simulations (see [296]) indicate that this is a good approximation for sufficiently large values of the coupling constant a. With this approximation, (6.62) takes the form

$$\frac{d\bar{u}}{dt} + \left(\bar{u}^2 - 1\right)\bar{u} + 3\Theta\bar{u} = 0. \tag{6.64}$$

On the other hand, deviations of particles from the swarm center obey the stochastic differential equation

$$\ddot{y}_i + \left(3\bar{u}^2 - 1\right)\dot{y}_i + ay_i + 3\bar{u}\left(\dot{y}_i^2 - \Theta\right) + \left(\dot{y}_i^3 - \frac{1}{N}\sum_{i=1}^{N}\dot{y}_i^3\right) = f_i(t), \tag{6.65}$$

which is derived by subtracting (6.62) from (6.57). Assuming that the deviations $\dot{y}_i$ are relatively small and linearizing this equation, we obtain

$$\ddot{y}_i + \left(3\bar{u}^2 - 1\right)\dot{y}_i + ay_i = f_i(t). \tag{6.66}$$

In this approximation the deviations for different particles i represent, therefore, statistically independent random processes. This allows us to replace the ensemble average in the dispersion Θ by the statistical mean taken over independent random realizations of such processes, defined by (6.66).

Hence, we have derived a closed set of equations (6.64) and (6.66) that approximately describe the swarm. We want to investigate steady statistical states of this system. The stationary solutions of (6.64) are

$$\bar{u} = \pm\sqrt{1 - 3\Theta}, \tag{6.67}$$

and $\bar{u} = 0$. The latter solution corresponds to the resting swarm. Examining equation (6.66), we notice that it yields damped oscillations only if $\Gamma = 3\bar{u}^2 - 1 > 0$, i.e. only if the mean velocity of the swarm is sufficiently large.

To find the dispersion Θ in the statistical steady state corresponding to a traveling swarm, we introduce Fourier components

$$y(\omega) = \frac{1}{\sqrt{2\pi}} \int_{-\infty}^{\infty} y(t)e^{i\omega t} dt \qquad (6.68)$$

(for simplicity we drop now the index i). Applying the Fourier transform to both sides of (6.66), we obtain

$$-\omega^2 y(\omega) - i\omega \Gamma y(\omega) + a y(\omega) = f(\omega). \qquad (6.69)$$

Here $f(\omega)$ is the Fourier component of the random force $f_i(t)$. As follows from (6.58), we have

$$\langle f(\omega)f(\omega')\rangle = \frac{1}{2\pi} \int_{-\infty}^{\infty} \int_{-\infty}^{\infty} \langle f(t)f(t')\rangle \, e^{i\omega t + i\omega' t'} dt dt' \qquad (6.70)$$

$$= \frac{S}{2\pi} \int_{-\infty}^{\infty} \int_{-\infty}^{\infty} \delta(t - t')e^{i\omega t + i\omega' t'} dt dt' = S\delta(\omega + \omega'). \quad (6.71)$$

Solving (6.69) yields

$$y(\omega) = \frac{f(\omega)}{a - \omega^2 - i\omega\Gamma}, \qquad (6.72)$$

and therefore

$$\langle y(\omega)y(\omega')\rangle = \frac{S\delta(\omega + \omega')}{\left(a - \omega^2\right)^2 + \omega^2 \Gamma^2}. \qquad (6.73)$$

Finally, we note that

$$\Theta = \left\langle \dot{y}_i^2 \right\rangle = -\frac{1}{2\pi} \int_{-\infty}^{\infty} \int_{-\infty}^{\infty} \omega\omega' \, \langle y(\omega)y(\omega')\rangle \, e^{-i(\omega+\omega')t} d\omega d\omega'. \qquad (6.74)$$

Substituting (6.73) into (6.74) and integrating, we obtain

$$\Theta = \frac{S}{3\bar{u}^2 - 1}. \qquad (6.75)$$

Since $\bar{u}$ is determined by Θ (see (6.67)), we have thus derived an algebraic equation for the dispersion Θ. This quadratic equation can easily be solved. Its roots are

$$\Theta_{1,2} = \frac{1}{9}\left(1 \pm \sqrt{1 - 9S}\right). \qquad (6.76)$$

The respective mean velocities of the swarm are given by

$$\bar{u}_{1,2} = \sqrt{\frac{1}{3}\left(2 \pm \sqrt{1 - 9S}\right)}. \qquad (6.77)$$

Note that these velocities do not depend on the parameter a.

We have plotted the obtained solutions as solid and dashed lines in Fig. 6.9. The lower branch is apparently unstable, since it approaches the value $\bar{u} = 1/\sqrt{3}$ at $S = 0$, i.e. in the absence of the noise. The traveling state solutions disappear when the critical noise intensity $S_c = 1/9 = 0.11\ldots$ is reached. At this critical point the mean swarm velocity is $\bar{u} = \sqrt{2/3} \approx 0.82\ldots$ and the mean dispersion of particles in the cloud is $\Theta = 1/9 = 0.11\ldots$.

Comparing the theoretical prediction with the numerically determined values of the mean swarm velocity, also shown in Fig. 6.9, we can see that this approximation provides good estimates of the swarm velocity and the critical noise intensity when the coupling parameter a is relatively high ($a = 10$ and $a = 100$). At small values of a, the deviations from the numerical results become significant near the breakdown threshold.

For a standing swarm ($\bar{u} = 0$), the deviations y_i obey in the limit $N \to \infty$ the nonlinear stochastic differential equation

$$\ddot{y}_i + \left(\dot{y}_i^2 - 1\right)\dot{y}_i + ay_i = f_i(t), \qquad (6.78)$$

which is similar to the classical Van der Pol equation (such nonlinear oscillators were considered by Lord Rayleigh [297]) Therefore, particles in the swarm perform in this state perturbed limit-cycle oscillations with a random distribution of phases. This state exists at any noise intensity S and is approached when the noise-induced breakdown of coherent motion takes place at $S = S_c$.

6.5 Synchronous Chaos

Interacting elements, forming a population, may have much more complex individual dynamics than periodic oscillations. It is now well-known that even relatively simple systems can display chaotic behavior in absence of any applied noise. An interesting question is whether large populations, consisting of such intrinsically chaotic elements, will also undergo a transition leading to synchronous collective chaos. Below in this section we show that coupling between the elements can always be introduced so that synchronization takes place.

Suppose that the internal state of an individual element is described by a set of K dynamical variables $\mathbf{w}(t) = \{w_1(t), w_2(t), \ldots, w_K(t)\}$ whose evolution is governed by the system of ordinary diferential equations

$$\begin{aligned} \dot{w}_1 &= F_1(w_1, w_2, \ldots, w_K) \\ \dot{w}_2 &= F_2(w_1, w_2, \ldots, w_K) \\ &\cdots \\ \dot{w}_K &= F_K(w_1, w_2, \ldots, w_K) \end{aligned} \qquad (6.79)$$

It is convenient to write this system of equations in the short form as

$$\dot{\mathbf{w}} = \mathbf{F}(\mathbf{w}). \qquad (6.80)$$

When the initial condition $\mathbf{w}(0)$ is fixed, equations (6.80) determine an orbit $\mathbf{w}(t)$ in the internal space of the considered element. We shall consider only elements whose orbits remain bounded at all times.

The dynamics of an element is chaotic if its orbits are exponentially unstable with respect to small perturbations of the initial conditions (e.g., see [306] for an extensive discussion). This property is quantified through the introduction of *Lyapunov exponents*.

Let us take an orbit $\mathbf{w}(t)$ of the dynamical system (6.80), originating from the point $\mathbf{w}(0)$, and another orbit $\mathbf{w}'(t) = \mathbf{w}(t) + \boldsymbol{\eta}(t)$ that originates from a nearby initial point $\mathbf{w}'(0) = \mathbf{w}(0) + \boldsymbol{\eta}(0)$. While the difference between these two orbits is still sufficiently small, the deviation $\boldsymbol{\eta}(t)$ satisfies the linear evolution equation

$$\dot{\boldsymbol{\eta}} = \widehat{G}(t)\boldsymbol{\eta}, \tag{6.81}$$

where $\widehat{G}$ is a time-dependent matrix with the components

$$G_{kl}(t) = \frac{\partial F_k(\mathbf{w})}{\partial w_l}, \tag{6.82}$$

taken on the non-perturbed orbit $\mathbf{w}(t)$.

If the matrix $\widehat{G}$ were constant, the solution of the linear differential equation (6.81) would be given by a superposition of exponentials

$$\boldsymbol{\eta}(t) = \sum_n C_n \exp\left(\alpha_n t\right) \mathbf{e}_n, \tag{6.83}$$

where α_n and $\mathbf{e}_n$ are the different eigenvalues and eigenvectors of the matrix $\widehat{G}$ and the constant coefficients C_n are determined by the initial conditions. When the matrix $\widehat{G}$ is time-dependent but does not increase indefinitely with time, the general theory of linear ordinary differential equations predicts [303] that the solution still has the form (6.83), though the coefficients C_n are now functions of time, growing more slowly than any time exponential.

The Lyapunov exponents λ_n are defined as the real parts of the eigenvalues, i.e. as $\lambda_n = \mathrm{Re}\,\alpha_n$. When all Lyapunov exponents are negative, the perturbation $\boldsymbol{\eta}(t)$ decreases with time and the orbits converge. The presence of at least one positive Lyapunov exponent means that the perturbation will generally exponentially grow with time, implying an instability with respect to perturbations of the initial conditions.

Such a strong instability means that the evolution of the considered system cannot be deterministically predicted. Indeed, we can usually know and specify the initial condition only up to a certain precision. But the orbits, initially separated by a distance shorter than this precision, would soon go very far one from another. Hence, the behavior of this system may only be described in probabilistic terms and the solution of the differential equation (6.80) is essentially a stochastic process. Today, many examples of dynamical systems that possess this property are known [304].

What would happen if we couple many identical systems with such chaotic dynamics? At a first glance, it might seem that the behavior of this ensemble must be even more unpredictable and unstable than that of a single element. Yet, this impression is wrong. Under an appropriate choice of coupling between the elements, the ensemble would display a collective behavior which we can describe as *synchronous chaos*. In this regime the states of all elements are always the same, but their collective temporal evolution follows a chaotic orbit.

Suppose that we have N identical dynamical systems (6.80). To distinguish these elements, we shall use the index $i = 1, 2, \ldots, N$, so that $\mathbf{w}_i(t)$ will denote the set of internal variables for the ith element. Let us consider an ensemble of coupled elements whose collective evolution is described by the equations

$$\dot{\mathbf{w}}_i = \mathbf{F}(\mathbf{w}_i) + \varepsilon\,(\overline{\mathbf{w}} - \mathbf{w}_i)\,, \tag{6.84}$$

where $\overline{\mathbf{w}}(t)$ is the global average, defined as

$$\overline{\mathbf{w}}(t) = \frac{1}{N} \sum_{i=1}^{N} \mathbf{w}_i(t), \tag{6.85}$$

and the coefficient ϵ specifies the strength of the coupling between the elements.

According to (6.84), each element is influenced by a force which is proportional to the difference between the current state of the given element and the average state of the elements in the ensemble at the considered moment. If the states of all elements are identical, we have $\mathbf{w}_i = \overline{\mathbf{w}}$ and hence this force vanishes. In this synchronous regime the evolution of a single element may still be chaotic.

Fujisaka and Yamada [305] have shown that for sufficiently strong coupling the synchronous regime always becomes stable for an ensemble described by (6.84). This can be proven by the following simple arguments: Let $\mathbf{w}^s(t)$ be the synchronous orbit of the ensemble, such that $\mathbf{w}_i(t) = \mathbf{w}^s(t) = \overline{\mathbf{w}}(t)$ for any i. We introduce small perturbations of this orbit, $\mathbf{w}_i(t) = \mathbf{w}^s(t) + \delta\mathbf{w}_i(t)$, and study their time evolution. Linearizing equations (6.84) with respect to $\delta\mathbf{w}_i$, we obtain

$$\dot{\delta\mathbf{w}}_i = \widehat{G}(t)\delta\mathbf{w}_i + \varepsilon(\overline{\delta\mathbf{w}} - \delta\mathbf{w}_i), \tag{6.86}$$

where $\widehat{G}(t)$ is the time-dependent matrix defined by (6.82) and taken along the synchronous orbit $\mathbf{w}^s(t)$ and $\overline{\delta\mathbf{w}}$ is given by

$$\overline{\delta\mathbf{w}} = \frac{1}{N} \sum_{i=1}^{N} \delta\mathbf{w}_i. \tag{6.87}$$

Next we note that the perturbation $\delta\mathbf{w}_i$ for each element i can be decomposed into a sum $\delta\mathbf{w}_i = \overline{\delta\mathbf{w}} + \delta\mathbf{w}_i^{\mathrm{T}}$. The first term in this sum is the same for all elements i. Hence, if the second term is absent, the perturbed motion would still be synchronous, though corresponding to a perturbed collective orbit. The second term $\delta \mathbf{w}_i^{\mathrm{T}}$ introduces individual *transverse* deviations from the synchronous collective motion. A perturbation destroys the synchronous collective motion only if such transverse deviations grow with time.

Summing up equations (6.86) for all $i = 1, 2, \ldots, N$ yields

$$\overline{\dot{\delta \mathbf{w}}} = \widehat{G}(t)\overline{\delta \mathbf{w}}. \tag{6.88}$$

Substracting this from (6.86), we obtain the following system of equations for the transverse deviations:

$$\dot{\delta \mathbf{w}}_i^{\mathrm{T}} = \widehat{G}(t)\delta \mathbf{w}_i^{\mathrm{T}} - \varepsilon \delta \mathbf{w}_i^{\mathrm{T}}. \tag{6.89}$$

Introducing new variables $\boldsymbol{\eta}_i(t) = \delta \mathbf{w}_i(t)e^{\varepsilon t}$ and writing equations (6.89) for these variables, we get

$$\dot{\boldsymbol{\eta}}_i = \widehat{G}(t)\boldsymbol{\eta}_i, \tag{6.90}$$

which coincides with (6.81). The general solution of (6.90) is therefore given by (6.83), so that

$$\delta \mathbf{w}_i^{\mathrm{T}}(t) = \sum_n C_{i,n}(t) \exp\left[(\alpha_n - \varepsilon)\, t\right] \mathbf{e}_n, \tag{6.91}$$

where $C_{i,n}(t)$ are time-dependent coefficients, determined by initial conditions and, at most, increasing more slowly than any time exponential.

Thus, all transverse deviations from the synchronous orbit vanish with time, if the condition

$$\operatorname{Re} \alpha_n < \varepsilon \tag{6.92}$$

is satisfied for any n. Recalling that $\lambda_n = \operatorname{Re} \alpha_n$ represent Lyapunov exponents of single-element dynamics, this condition says that *the synchronous motion is (linearly) stable if the coupling strength ε exceeds the maximal positive Lyapunov exponent λ*. Note that such stable synchronous motion can still be chaotic, i.e. any slight perturbation would lead to a transition from one synchronous collective orbit to another also synchronous orbit.

This result is quite remarkable. Indeed, it holds for any single-element dynamics (6.79), no matter how complicated it is. Hence, mutual synchronization is always possible in ensembles consisting of any identical dynamical elements. If the dynamics of an individual element is not chaotic and therefore its maximal Lyapunov exponent is not positive, the synchronization takes place at any coupling intensity, i.e. even for vanishingly small values of the coupling strength ε. In contrast to this, if the single-element dynamics is chaotic, the collective motion of elements becomes synchronous only when the coupling strength ε exceeds the threshold $\varepsilon_{\mathrm{c}} = \lambda$.

In Sect. 6.3 we analyzed the influence of noise on mutual synchronization in populations of identical phase oscillators (clocks) with simple periodic dynamics. We have seen that in the presence of noise the synchronization occurs only when the coupling exceeds a threshold determined by the noise intensity (cf. (6.56)). Comparing this with the above results for chaotic oscillators, it can be concluded that the intrinsic stochasticity of individual elements plays to a certain extent a role that is similar to that of the external noise.

Though the analysis presented above is fairly general, since it applies for elements with any internal dynamics, the coupling between the elements was introduced in (6.84) in a special form. Indeed, even if the coupling is linear, it may generally be different for different components $w_k(t)$ of the state variable $\mathbf{w}(t)$. In this case (6.84) is replaced by the equations

$$\dot{\mathbf{w}}_i = \mathbf{F}(\mathbf{w}_i) + \varepsilon \widehat{A} (\overline{\mathbf{w}} - \mathbf{w}_i), \tag{6.93}$$

where $\widehat{A}$ is some constant matrix other than unity. It turns out that such a small modification makes the problem much more difficult [305, 325]. For some choices of the matrix $\widehat{A}$ the synchronization of chaotic dynamics does not take place at any coupling strength, while for other choices it still occurs when the coupling is sufficiently strong. Moreover, the synchronization is sometimes found only inside a finite window of the coupling strength ε [329].

Another important deficiency is that even if the coupling is chosen in the form (6.84), the above analysis proves only the linear stability of synchronous motion. Based on it, nothing can still be said about the evolution of the system after a relatively strong perturbation. It may well be that such a strong perturbation destroys the synchronous collective motion and leads to a different dynamical attractor, where the synchronization is absent.

Zanette and Mikhailov [331] have proposed a nonlinear form of global coupling that ensures absolute synchronization in the system. It differs from (6.93) by the inclusion of an additional coupling term

$$\dot{\mathbf{w}}_i = \mathbf{F}(\mathbf{w}_i) + \varepsilon \widehat{A} (\overline{\mathbf{w}} - \mathbf{w}_i) + \varepsilon' (\mathbf{F}(\overline{\mathbf{w}}) - \mathbf{F}(\mathbf{w}_i)), \tag{6.94}$$

whose strength is characterized by a separate parameter ε'. When $\varepsilon' = 1$, equations (6.94) are reduced to

$$\dot{\mathbf{w}}_i = \mathbf{F}(\overline{\mathbf{w}}) + \varepsilon \widehat{A} (\overline{\mathbf{w}} - \mathbf{w}_i), \tag{6.95}$$

and we have

$$\dot{\overline{\mathbf{w}}} = \mathbf{F}(\overline{\mathbf{w}}). \tag{6.96}$$

Subtracting (6.96) from (6.95), we obtain the *exact* linear equation for transverse deviations $\delta \mathbf{w}_i^{\mathrm{T}} = \mathbf{w}_i - \overline{\mathbf{w}}$:

$$\dot{\delta \mathbf{w}}_i^{\mathrm{T}} = -\varepsilon \widehat{A} \delta \mathbf{w}_i^{\mathrm{T}}. \tag{6.97}$$

Hence, these deviations are damped if all eigenvalues of the matrix $\widehat{A}$ are either positive or have positive real parts. In this case, synchronous motion represents the only dynamical attractor and will be observed for any initial conditions. Though the absolute convergence to synchronous collective motion is proven for the system (6.94) only at $\varepsilon' = 1$, this property would probably be also found at least in a small neighborhood of this point, for smaller values of the coupling coefficient ε'.

A similar analysis can be performed for globally coupled maps representing dynamical systems with discrete time. If a single map is

$$\mathbf{w}(t+1) = \mathbf{f}\left(\mathbf{w}(t)\right), \tag{6.98}$$

an ensemble of such coupled maps can be defined as

$$\mathbf{w}_i(t+1) = (1-\varepsilon)\mathbf{f}\left(\mathbf{w}_i(t)\right) + \frac{\varepsilon}{N}\sum_{j=1}^{N}\mathbf{f}\left(\mathbf{w}_j(t)\right). \tag{6.99}$$

This form of coupling corresponds to (6.84). Repeating the above arguments, it can be shown that the ensemble of chaotic maps will always have a linearly stable synchronous collective motion for sufficiently strong coupling intensity ε. The synchronization threshold ε_c is determined by the condition

$$\lambda + \ln\left(1 - \varepsilon_c\right) = 0, \tag{6.100}$$

where λ is the maximal Lyapunov exponent of the single map.

Alternatively, global coupling between the maps can be introduced as

$$\mathbf{w}_i(t+1) = (1-\varepsilon)\mathbf{f}\left(\mathbf{w}_i(t)\right) + \varepsilon\mathbf{f}\left(\overline{\mathbf{w}}(t)\right), \tag{6.101}$$

where the global average is defined as

$$\overline{\mathbf{w}}(t) = \frac{1}{N}\sum_{j=1}^{N}\mathbf{w}_j(t). \tag{6.102}$$

Such a form of coupling is similar to (6.94). It is easily seen that absolute synhronization takes place here at $\varepsilon = 1$; this collective behavior can also be expected in an interval of ε below $\varepsilon = 1$.

6.6 Further Reading

Synchronization phenomena in populations of coupled oscillators have been the subject of intensive research since the pioneering works of Winfree in the field of biology [298] and of Kuramoto on lattices of locally (i.e. diffusively) coupled oscillators [299]. The motivation for this research can be found in the broad variety of phenomena that can be modeled in this framework, such as synchronous flashing in swarms of fireflies [308, 309], crickets that chirp in unison [310], epileptic seizures in the brain [311], and electrical synchrony of cardiac pacemaker cells [312]. The origins of oscillations in biological and biochemical systems are considered in the books by Murray [300] and Goldbeter [301] and in the review by Hess [313]. An interesting and readable account of biological sleep rhythms is provided in the book by Coleman [307]. Synchronization phenomena in neural systems and related typical models will be discussed in *Further Reading* in the next Chapter. At the physical level, oscillator synchronization is possible, e.g. in the arrays of Josephson junctions [315]. Herz and Hopfield [314] have shown that synchronization phenomena may play a significant role in the development of earthquakes.

The Kuramoto model of coupled phase oscillators (which we also called "interacting clocks") represents the canonical form for any population of weakly coupled limit-cycle oscillators. The reduction to phase equations is described in the book [299]. When the coupling is not weak, such reduction is not possible and the amplitudes of the oscillations are as important as their phases. The behavior of strongly coupled osillator populations is not universal and depends on the choise of a particular system.

Near the Hopf bifurcation, leading to the emergence of a stable limit cycle from a stable fixed point, the oscillations are nearly harmonic and can be described by simple dynamical equations. The collective behavior of such globally coupled oscillators has been extensively investigated [316–321]. In addition to synchronization and asynchronous oscillations, these systems show dynamical clustering (discussed in the next chapter) and collective synchronous chaos. Oscillator systems with both local and global coupling between elements have also been considered [302].

Investigations of synchronization in coupled chaotic dynamical systems were pioneered in 1983 by the paper of Fujisaka and Yamada [305], who analytically determined the conditions, that ensure the linear stability of the synchronous state and numerically analyzed that synchronization of coupled Lorenz systems. An extensive numerical analysis of mutual synchronization in populations of globally coupled logistic maps has been performed by Kaneko [322]. Mutual synchronization of chaotic dynamical systems with continuous time, such as the Rössler oscillators, has been analytically, numerically and experimentally studied [323–325]. Weaker forms of chaotic syhronization, such as phase and lag synchronization [332], have been considered.

The synchronization of chaos is closely related to the problem of controlling it. The control means stabilization of an unstable orbit by applying an appropriate small perturbation [345–348]. Work on the synchronization and control of chaos has led many researchers to focus on applications in communications [323, 334–339], chemistry [340], laser physics [341–343], and electronic circuits [344] (for a review, see Pecora et al. [329], Ditto and Showalter [330], and Pikovsky et al. [333]).

An important application of synchronized chaotic systems for secure communications was proposed by Pecora and Carroll. A message coupled to a chaotic drive system is passed through a communication channel to a receiver. If the receiver has a copy of the chaotic drive system, then the message can be subtracted form the chaotic part [323].

The broadband properties that make chaotic signals interesting also make the signals difficult to use in communication schemes. Normally, when signals are transmitted through a channel, the properties of the medium vary with frequency. Broadband signals such as chaos are greatly distorted by transmission through a channel. Some work has concentrated on how to adapt chaotic receivers to correct for distortion due to transmission through the channel [326, 327]. An alternative way is to ask how much information about the chaotic drive system is necessary in order to synchronize the receiver. One may be able to avoid the channel distortion problems by sending only this necessary information. Carroll and Johnson could show that it is possible to reduce the amount of information about the chaotic system by sending the drive signal through a bandpass filter and still synchronize the receiver system [328].

Synchronization of spatiotemporal chaos in coupled extended chaotic dynamical systems has been investigated [349, 350]. A related problem is the synchronization of stochastic dynamics in coupled cellular automata [351, 352].

The role of time delays in the synchronization phenomena in the model of globally coupled phase oscillators has beed discussed by Yeung and Strogatz [353]. Izhikevich [354] has derived such phase models for populatitions of weakly coupled limit-cycle oscillators with interaction delays.

7. Dynamical Clustering

Large ensembles of globally coupled chaotic elements can spontaneously organize themselves into a set of coherently operating groups – dynamical clusters. Inside a cluster, the dynamical states of all the elements are (nearly) identical. The collective dynamics of interacting clusters may still be very complex and chaotic. This transition can be viewed as an onset of functional structuralization in an initially uniform population.

The best-studied example of dynamical clustering is provided by the discrete-time system of logistic maps. Similar phenomena are found when chaotic Rössler oscillators with continuous time are considered. Under the conditions of dynamical clustering, these systems usually possess a large number of attractors corresponding to different cluster partitions. The application of weak noise leads to wandering over this attractor network, accompanied by successive structural reorganizations.

Individual elements, forming an ensemble, may have very rich internal dynamics. As an example, we shall consider an ensemble whose single elements represent entire neural networks with chaotic internal dynamics. When these networks are cross-coupled, they display both complete mutual synchronization and dynamical clustering, depending on the interaction strength.

Dynamical clustering is also found in the systems formed by stochastic elements. This phenomenon may play a significant role in molecular biology. A living cell is essentially an ensemble of interacting protein machines whose coherent operation is possible. To illustrate this, a simple example of an allosterically regulated enzymic reaction in a small spatial volume is considered.

7.1 Logistic Maps

Many insects live just one year, so that their number in the next summer is determined by how many insects are born annually. Moreover, some species reproduce only once per year. If t is the discrete time, measured in years, $n(t)$ is the number of insects in year t and α is the reproduction rate, the number of individuals in the next year may be written as

$$n(t + 1) = \alpha n(t). \tag{7.1}$$

A mathematician would say that this equation defines a *map* $n(t) \mapsto n(t+1)$, putting into correspondence a certain number $n(t+1)$ to any given number $n(t)$.

The linear map (7.1) does not actually constitute a good model of discrete population dynamics. Its solution is $n(t) = \alpha^t n_0$, where n_0 is the initial population number. Hence, the population either grows indefinitely with time if $\alpha > 1$ or it undergoes extinction if $\alpha < 1$. To make the model more realistic, we can take into account that the reproduction rate can itself depend on the population number. Indeed, if more insects are present, less food is available and therefore the reproduction chances are lower. For instance, we can assume $\alpha = \alpha_0 - \beta n(t)$ so that (7.1) is replaced by

$$n(t+1) = (\alpha_0 - \beta n(t))\, n(t), \qquad (7.2)$$

Introducing a new dynamical variable $x = (n-B)/A$ with $A = (\alpha_0/4\beta)\,(\alpha_0-2)$ and $B = \alpha_0/2\beta$ and defining the parameter $a = (\alpha_0/4)\,(\alpha_0 - 2)$, we can write (7.2) as

$$x(t+1) = f(x(t)), \qquad (7.3)$$

with the nonlinear function

$$f(x) = 1 - ax^2. \qquad (7.4)$$

Equations (7.3) and (7.4) define the *logistic map*. When the control parameter a lies in the interval from 0 to 2, the map projects the segment $[-1, 1]$ into itself. This map displays complex behavior. Its applications to population biology have been extensively discussed by May [355]. The dynamics of the logistic map becomes chaotic for $a > 1.40155\ldots$. Grossmann and Thomae [356] and Feigenbaum [357] have shown that the transition to chaos in this system is preceded by an infinite sequence of period-doubling bifurcations with interesting scaling properties. Today, the logistic map represents one of the most extensively investigated discrete-time systems with an intrinsically chaotic dynamics. Because of its simplicity, the system (7.3) and (7.4) is often chosen as an example and as a basic element in the construction of more complex abstract models. The logistic map is the *Drosophila fly* of nonlinear dynamics.

We consider a system of N globally coupled logistic maps

$$x_i(t+1) = (1 - \varepsilon)f(x_i(t)) + \frac{\varepsilon}{N}\sum_{j=1}^{N} f(x_j(t)), \qquad (7.5)$$

where $i = 1, 2, \ldots, N$ and the function $f(x)$ is given by equation (7.4). The properties of globally coupled logistic maps have been extensively investigated by Kaneko [20, 358].

Mutual synchronization is observed in this system. When the dynamics of an individual map is periodic, the synchronization takes place at any coupling strength ε. If the dynamics is chaotic, it occurs only when the coupling strength exceeds a certain threshold ε_c. This agrees with the general analysis by Fujisaka and Yamada [359] (see Sect. 6.5) and could well have been expected.

Surprisingly, Kaneko has found [358] that mutual synchronization is preceded in globally coupled chaotic logistic maps by a regime of *dynamical clustering*. This new regime is observed in an interval of coupling strengths below the synchronization threshold ε_c. It is characterized by the formation of synchronous groups (clusters) of elements.

As an example, Fig. 7.1 shows the evolution of an ensemble of $N = 5000$ chaotic maps ($a = 1.6, \varepsilon = 0.1$) with randomly chosen initial conditions. At the beginning, the states of individual elements are greatly dispersed, so that the points representing them in the plot merge to produce a uniformly black area. Later on, the elements start to organize into clusters. At the end, only four synchronous clusters (corresponding to four different collective trajectories in Fig. 7.1) are found.

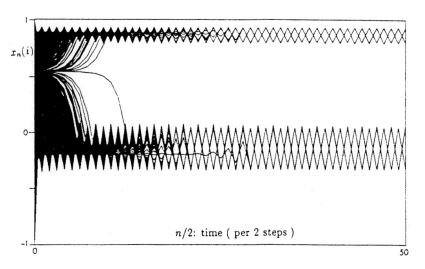

Fig. 7.1. Evolution of an ensemble of $N = 5000$ chaotic maps ($a = 1.6, \varepsilon = 0.1$) with randomly chosen initial conditions. (From [358])

By carrying out many computations, an approximate phase diagram of globally coupled logistic maps has been constructed (see [358]). When $a = 1.6$, the asynchronous ("turbulent") phase is found below $\varepsilon = 0.075$. Dynamical clustering is observed inside the interval $0.32 > \varepsilon > 0.075$. Mutual synchronization of the entire ensemble (the "coherent" phase) develops for $\varepsilon > 0.32$. If $a = 1.8$, dynamical clustering takes place inside the interval $0.37 > \varepsilon > 0.14$, and so on.

In the middle of the clustering interval, the system usually forms two clusters with various sizes N_1 and N_2. When clusters are almost equally large ($N_1 \approx N_2$), the collective motion is periodic or quasiperiodic. As the difference in cluster sizes increases, the motion becomes more complex and, through a sequence of period-doubling bifurcations, it transforms into chaos. A similar dependence on cluster sizes is observed in regimes with three dynamical clusters.

A characteristic property of dynamical clustering is that it is extremely sensitive to variations in the initial conditions. The same system can evolve to very different cluster partitions, depending on the initial states of its elements. Some of these partitions correspond to periodic motions (though the dynamics of an isolated single map is chaotic), while collective motions for other partitions are quasiperiodic or chaotic.

Since all elements are identical, any cluster partition is defined by indicating the total number K of clusters and the numbers N_k of elements in the kth cluster ($k = 1, 2, \ldots, K$). It is convenient to order the clusters according to their sizes, so that $N_1 \geq N_2 \geq \ldots \geq N_k \geq \ldots \geq N_K$. Generally, there would also be some elements which are not entrained and do not belong to any cluster. In this section we shall consider them, following [358], as single-element "clusters" with $N_k = 1$. Thus, any state of the ensemble can be described as a certain partition $(K; N_1, N_2, \ldots, N_k, \ldots, N_K)$. In these notations, the partition $(1; N)$ corresponds to complete mutual synchronization, whereas $(N; 1, \ldots, 1)$ represents an asynchronous state of the entire ensemble. The notation $(K; N - K + 1, 1, 1, \ldots, 1)$ means that the system has a single cluster with $N - K + 1$ elements and $K - 1$ non-entrained single elements. Since the latter ones are also counted as "clusters", this partition is viewed as having K clusters.

Any asymptotic dynamics, corresponding to a particular cluster partition, represents an *attractor* of the considered dynamical system. When dynamical clustering takes place, the system has a great number of different attractors. Suppose that a particular partition $(K; N_1, N_2, \ldots, N_k, \ldots, N_K)$ is realized. Since all N elements in the ensemble are identical, we can choose the particular N_k elements, forming any kth cluster, by many different ways. For instance, if only two clusters of sizes N_1 and N_2 are formed ($N_1 + N_2 = N$), this partition corresponds to

$$M(N_1, N_2) = \frac{N!}{N_1! N_2!} \tag{7.6}$$

different distributions of elements. Because of the intrinsic symmetry of the system, the attractors corresponding to all such distributions should have the same properties and must be characterized by equal basin volumes. Furthermore, since clusters may have various sizes, many different partitions with the same number of clusters exist. Even if only two clusters are found ($K = 2$), this can correspond to $N/2$ different partitions $(2; N_1, N_2)$.

The *basin of attraction* is the set of all phase space points such that, if a trajectory starts from a given point, it leads to a given attractor. The volume of the basin is an important property of attractors. When the volume is large and fills almost the entire phase space, the attractor is strong: it is approached for almost any initial condition. Attractors with small basin volumes are weak: to find such an attractor, special initial conditions must be chosen. Therefore, it might be difficult to detect these attractors in numerical simulations.

For instance, in the middle of the region of dynamical clustering strong attractors, corresponding to two or three clusters, are typically observed. However, even in this region the system may possess many other weak attractors. It is well known that an individual logistic map has a large number of periodicity windows in the chaotic region. Inside any such interval of the parameter a and in its small neighbourhood, the ensemble of globally coupled maps should have an attractor corresponding to the synchronous motion of all elements. But such an attractor may be weak and thus almost undetectable. Weak attractors corresponding to other cluster partitions are also possible.

In addition to the basin volume, another important property can be defined for any attractor. Suppose that we have allowed the system to evolve for a long enough time and it has reached one of its attractors. Let us now add small random perturbations $\delta x_i = \sigma \xi_i$ to all variables x_i (here the coefficient σ specifies the perturbation intensity and ξ_i are independent random numbers, selected from the interval $[-0.5, 0.5]$) and again allow the system to follow its own dynamics (without the noise term). Sometimes the system will return to the original attractor, but for sufficiently strong perturbations it is also possible that the system leaves its original attractor and switches to a different one. The minimal perturbation intensity σ_c which is necessary to induce a switching is an important property of any attractor. It characterizes the minimal distance between the attractor and the boundary of its attraction basin.

Since many attractors coexist in the dynamical clustering regime, their basins often have quite complicate shapes, i.e. they are *riddled*. Generally, weak attractors can be expected to have much smaller σ_c than strong attractors with large basins. However, if a basin is riddled, its boundaries may well come extremely close to the attractor, though the basin occupies a large volume of phase space. We would have then a situation where evolution from many different initial conditions leads to the considered attractor, but even small perturbations would suffice to push the trajectory out of it.

Milnor has presented examples [360] of dynamical systems with quasi-attractors whose basins occupy almost all phase space and which are still unstable with respect to infinitely weak perturbations ($\sigma_c \to 0$). Such objects are often found when a true attractor disappears. An attractor ruin, which remains, is still globally attracting but the local dynamics is already modified. As a result, the trajectory is attracted by a region in phase space, where a true attractor previously existed, but escapes later from this region after spending some time inside it.

In the context of dynamical clustering, Milnor attractors correspond to metastable cluster partitions. If we follow the system evolution in this case, we would see that elements tend to condense into a certain number of clusters. However, these clusters actually exist only for a finite time and later spontaneously break down, leading to other cluster partitions.

Because of the finite lifetime of such metastable clusters, the elements inside them never have identical states, though they may come very close to one another. This presents a serious problem when numerical simulations of such systems are undertaken. Indeed, any real number is encoded in a computer by a string of bits of a fixed length and thus can be specified only up to a certain precision. Two real number that differ less than the computer precision, would be encoded by the same strings and are treated as being identical. As metastable clusters are formed, the distances between their elements may become shorter than the computer precision. Once this has occurred, the elements would never separate in the running computation and a conclusion might be made that a stable cluster has been found. To eliminate such pseudoattractors, representing computation artifacts, very small noise should be continuously added during the entire simulation.

Various attractors and numerous attractor ruins, which are found in the dynamical clustering regime for globally coupled maps, may form a complex network in the phase space of this dynamical system. In the presence of noise the system may wander for a long time along such a network until a robust attractor, stable with respect to perturbations, is finally approached. It may also happen that such robust attractors are absent for particular parameter values and the system indefinitely wanders through the attractor network.

These effects are indeed observed near the boundary, separating the region with dynamical clustering from the region of asynchronous dynamics [361]. Close to this boundary, in addition to coherent groups many individual non-entrained elements are typically observed. Since they are counted as single-element "clusters" in the employed classification, partitions with large numbers of clusters $K = O(N)$ are observed in this region. Figure 7.2 shows the computed dependence of the average σ_c on the control parameter a of logistic maps for the coupling strength $\varepsilon = 0.1$ and three different sizes ($N = 10, 50$ and 100) of the ensemble. The averaging is performed over 10^4 random initial conditions, leading to various attractors. We see that in the interval from $a \approx 1.63$ to $a \approx 1.7$ the average values of σ_c become very small, indicating strong sensitivity to perturbations. This interval is located at the boundary between asynchronous ("turbulent") dynamics and dynamical clustering. Further examination reveals that many attractors in this region actually represent Milnor attractors ($\sigma_c \to 0$), i.e. they are destabilized even by extremely weak noise.

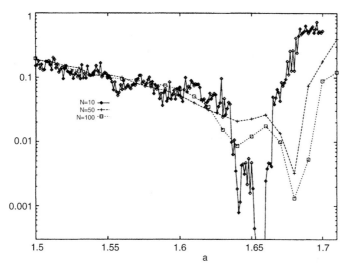

Fig. 7.2. Dependence of the average σ_c on the control parameter a of logistic maps for the coupling strength $\varepsilon = 0.1$ and three different sizes ($N = 10, 50$ and 100) of the ensemble. The averaging is performed over 10^4 random initial conditions leading to various attractors. (From [361])

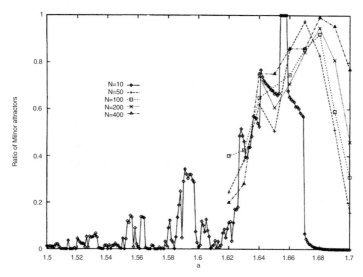

Fig. 7.3. The basin volume ratio of Milnor attractors as a function of the parameter a for the coupling strength $\varepsilon = 0.1$ and five different ensemble sizes $N = 10, 50, 100, 200$ and 400. (From [361])

Figure 7.3 displays the basin volume ratio of Milnor attractors as a function of the parameter a for a coupling strength $\varepsilon = 0.1$ and five different ensemble sizes $N = 10, 50, 100, 200$ and 400. To construct this figure, Kaneko took 1000 different initial conditions for each a and N. The evolution starting from each initial condition was followed over 10^4 time steps to get an attractor. To check whether an orbit returns to its original attractor after a perturbation, each attractor was perturbed by $\sigma = 10^{-7}$ over 100 trials. If the orbit has not returned at least in one of the trials, the attractor was counted as a Milnor attractor. The basin volume ratio has been calculated as the fraction of randomly chosen initial conditions leading to Milnor attractors. It is clearly seen that in the interval $1.67 > a > 1.63$ for $N = 10$ and in the interval $1.69 > a > 1.63$ for the larger ensembles the basins of Milnor attractors occupy a large fraction of the available phase space. For some parameters a this fraction becomes very close to unity or even reaches it, so that practically any initial condition would lead to a Milnor attractor.

After a long wandering through attractor ruins – a network of Milnor attractors – the system may finally end its evolution in a rare robust attractor, stable with respect to small perturbations. In this case the Milnor attractors would essentially represent only transients, preceding the establishment of a robust dynamical behavior. The existence of long transients in some parameter regions is an important property of coupled logistic maps.

Examining the properties of coupled maps in the interval separating the region of dynamical clustering from the synchronous ("coherent") phase, Kaneko has noted [358, 362] that in this interval the system also usually forms a large number K of clusters, comparable with the total size N of the ensemble. Recent simulations [363], however, show that after long transients the system always approaches inside this interval a state with just a few large clusters. During a long transient, the system slowly evolves by gradually decreasing the number of clusters (Fig. 7.4) until only a few of them (mostly two or three) are present. The analysis reveals [363] that the average number of clusters decreases exponentially with time and the exponent does not significantly depend on the system size N. Figure 7.5 shows the average length of a transient for $\varepsilon = 0.3$ in an ensemble of size $N = 100$ as a function of the control parameter a (the vertical bars indicate the maximal and the minimal values of the transient length for each value of a). We see that the transient can be extremely long, reaching 10^7 time steps at the boundary between dynamical clustering and complete synchronization. When weak noise is added, it does not prevent the system from relaxing to partitions with a few large clusters. The difference is only that, instead of settling in one of such partitions, the system may finally wander through a small network formed by them.

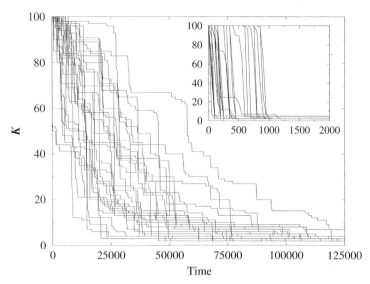

Fig. 7.4. Number of clusters as function of time for $a = 1.6$, $\epsilon = 0.3$, $N = 100$. For comparison, the inset shows the respective dependence for the same system at $a = 1.8$, where short transients are found. (From [363])

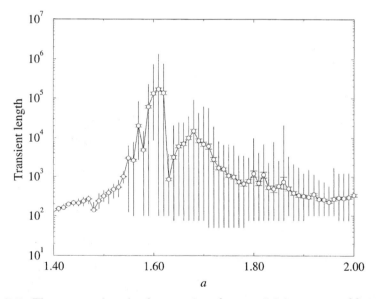

Fig. 7.5. The average length of a transient for $\varepsilon = 0.3$ in an ensemble of size $N = 100$ as a function of the control parameter a (the vertical bars indicate the maximum and the minimum values of the transient length for each value of a). (From [363])

7.2 Rössler Oscillators

Dynamical clustering, considered in the previous section for globally coupled logistic maps, represents a general property of coupled chaotic systems. Below we show that a similar behavior is found in large ensembles of globally coupled chaotic Rössler oscillators. The dynamics of a single such oscillator is described by three variables x, y and z satisfying the evolution equations [364]

$$
\begin{aligned}
\dot{x} &= -y - z, \\
\dot{y} &= x + ay, \\
\dot{z} &= b - cz + xz,
\end{aligned}
\tag{7.7}
$$

where a, b and c are positive parameters. In a similar way to the logistic map, this dynamical system shows a transition to chaos through a sequence of period-doubling bifurcations. We shall fix the parameters as $a = 0.15, b = 0.4$ and $c = 8.5$, to ensure that the motion of an individual oscillator is intrinsically chaotic.

Global coupling between such oscillators can be introduced in different ways. We shall assume that an ensemble of N identical Rössler oscillators $i = 1, 2, \ldots, N$ obeys the equations

$$
\begin{aligned}
\dot{x}_i &= -y_i - z_i, \\
\dot{y}_i &= x_i + ay_i + \varepsilon\left(\overline{y} - y_i\right), \\
\dot{z}_i &= b - cz_i + x_i z_i + \varepsilon\left(\overline{xz} - x_i z_i\right),
\end{aligned}
\tag{7.8}
$$

where ε is the coupling parameter and $\overline{x}, \overline{y}, \overline{z}$ are defined by equations of the form

$$
\overline{x}(t) = \frac{1}{N} \sum_{i=1}^{N} x_i(t).
\tag{7.9}
$$

Note that this is a special case of nonlinear coupling (6.94) with $\varepsilon' = \varepsilon$ and the matrix $\widehat{A}$ whose elements are all zero except for $A_{xy} = A_{xz} = -A_{yx} = 1$, $A_{zz} = c$ and $A_{yy} = 1 - a$.

The eigenvalues of the matrix $\widehat{A}$ are $\Lambda_1 = c$ and $\Lambda_{2,3} = \frac{1}{2}(1 - a) \pm \frac{1}{2}\sqrt{(1-a)^2 - 4}$. All three eigenvalues (or their real parts) are positive if $a < 1$. Therefore, when this condition is satisfied and $\varepsilon = 1$, all deviations from the global averages exponentially decrease with time (see Sect. 6.5) and the states of all elements in the ensemble asymptotically converge. Thus, at $\varepsilon = 1$ this system has a single global attractor that corresponds to the synchronous motion of all elements. This attractor coincides with the attractor of a single Rössler oscillator, which is chaotic for the considered choice of the parameters a, b and c.

When the coupling strength is gradually increased towards $\varepsilon = 1$, the system (7.8) must undergo a transition to complete synchronization. To investigate such a transition, numerical simulations of the system (7.8) with

$N = 1000$ Rössler oscillators have been performed [365]. These simulations have shown that synchronization is observed in the considered ensemble already starting from $\varepsilon \approx 0.1$ and it is preceded by dynamical clustering of the elements.

In our analysis of dynamical clustering in logistic maps in Sect. 7.1 we have counted, following Kaneko, individual non-entrained elements as one-element "clusters". This definition might be misleading. Indeed, when mutual synchronization of phase oscillators was considered in Chap. 6, we distinguished between the synchronous group of elements and the rest, which were non-entrained oscillators. The synchronization transitions, studied in Sects. 6.2 and 6.3, were expressed by the formation of a synchronous group of elements and could be interpreted as condensation taking place in this system. The size of the condensate, i.e. the fraction of elements belonging to the synchronous group, represented the order parameter of these transitions. When the condensate first appeared at the transition point, it was small and its size gradually increased with the growing coupling strength.

For logistic maps, we found in Sect. 7.1 that complete synchronization was preceded by the formation of synchronous groups. In addition to such groups, the system could have individual non-entrained elements. The number of non-entrained elements was found to be large near the boundary separating the dynamical clustering region from the region with asynchronous dynamics. The phenomenon of dynamical clustering resembled the condensation of water vapor, where many small water drops are initially formed from the gas phase. In the analysis of the dynamical clustering of chaotic Rössler oscillators [365] we shall pay special attention to this aspect of the synchronization behavior.

To prepare the initial condition, the system (7.9) is first allowed to evolve up to $t = 100$ without global coupling ($\varepsilon = 0$) so that its elements get uniformly distributed over the Rössler attractor. Global coupling is then introduced, time is reset to zero, and the evolution of the coupled population is started. As time goes on, the orbits of some elements converge and become identical (within the computer precision). These elements form synchronous groups and belong to the dynamical condensate.

To describe the condensation, we define distances d_{ij} between dynamical states of elements i and j,

$$d_{ij} = \sqrt{(x_i - x_j)^2 + (y_i - y_j)^2 + (z_i - z_j)^2}, \qquad (7.10)$$

and compute these distances for all pairs of elements (i, j) at a fixed moment of time $t = 2000$. Figure 7.6 shows typical histograms of distributions over pair distances for four different coupling strengths. Such histograms are constructed by counting pairs with distances d lying within the subsequent intervals of width $\Delta d = 0.25$. The number of pairs inside each interval, divided by the total number of pairs $N(N - 1)/2$, yields the height of the respective bar.

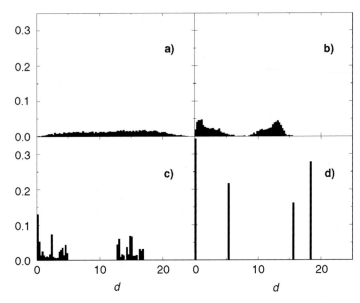

Fig. 7.6. Histograms of distributions over pair distances for four different coupling strengths **(a)** $\varepsilon = 0$, **(b)** $\varepsilon = 0.01$, **(c)** $\varepsilon = 0.02$, **(d)** $\varepsilon = 0.05$ at $a = 0.15$, $b = 0.4$ and $c = 8.5$ for the system of size $N = 1000$. (From [365])

In absence of global coupling the elements are relatively uniformly distributed over the single-oscillator attractor and a smooth distribution over distances d is thus observed (Fig. 7.6a, $\varepsilon = 0$). When weak global coupling is introduced, the distribution is modified (Fig. 7.6b, $\varepsilon = 0.01$). It remains smooth, but now has two maxima. Note that a significant fraction of pairs with small distances is already present. As the coupling strength ε gets larger, a strongly nonuniform distribution with narrow peaks develops (Fig. 7.6c, $\varepsilon = 0.02$). Further increase of the coupling strength produces distributions that consist of several distinct lines (e.g. Fig. 7.6d for $\varepsilon = 0.05$). When the coupling strength exceeds $\varepsilon = 0.1$, the histogram has only one line located at $d = 0$ indicating complete synchronization (not shown).

Distributions formed by several lines are characteristic for the situations where the entire ensemble breaks down into a few coherent groups. All elements in a group follow the same trajectory and, hence, distances between pairs of elements belonging to the same group are zero. When elements from different groups are chosen, their distances are all equal. If M such groups are present, the distribution over pair distances would consist of $M(M-1)/2+1$ individual lines. For instance, the four lines seen in Fig. 7.6d correspond to a distribution with three synchronous groups. Hence, the analysis of distributions over pair distances reveals that the ensemble of globally coupled Rössler oscillators undergoes dynamical clustering.

While providing valuable insight into the mutual organization of the ensemble, distributions over pair distances are not permanent. Indeed, the elements forming the clusters continue to move (typically, this motion is chaotic, in contrast to coupled logistic maps, where periodic motions inside clusters are often observed). Therefore, the histograms of such distributions depend on time.

To quantitatively characterize the dynamical clustering, we now introduce two *order parameters*. The first of them represents the ratio r of the number of pairs with zero distances to the total number of pairs. In the absence of synchronous groups, $r = 0$. On the other hand, $r = 1$ when mutual synchronization has taken place and the states of all elements are identical. The second order parameter s represents the fraction of elements belonging to synchronous groups. It is defined as the relative number of elements that have at least one other element at zero distance. Note that $1 - s$ yields the fraction of non-entrained elements in the ensemble. If the system has only one synchronous group of elements, the order parameter s has the same meaning as the order parameter σ of the synchronization transition in Chap. 6.

Figure 7.7 shows the computed dependence of both order parameters on the coupling strength ε, obtained by averaging over 20 realizations at $t = 2000$. The bars display the statistical dispersion (mean-square deviation) of the data at some selected points. The synchronous groups of elements first

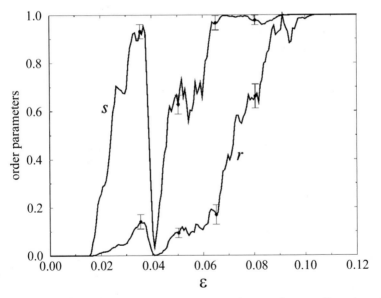

Fig. 7.7. Dependence of order parameters r and s on the coupling strength ε, obtained by averaging over 20 realizations at $t = 2000$. The bars display the statistical dispersion (mean-square deviation) of the data at some selected points; $a = 0.15$, $b = 0.4$, $c = 8.5$, and $N = 1000$. (From [365])

appear at $\varepsilon_c \approx 0.017$. The number of elements, belonging to synchronous groups, grows with ε and at $\varepsilon \approx 0.03$ almost any element of the ensemble belongs to one of such groups, so that the order parameter s becomes close to unity. At higher values of the coupling strength both order parameters sharply decrease and reach a minimum at $\varepsilon \approx 0.04$. After that they start to grow and eventually approach unity. Note that the order parameter s becomes close to unity already at $\varepsilon \approx 0.06$, when the order parameter r is still relatively small.

The two order parameters can be used to estimate the number of synchronous groups in the ensemble. By the definition of s, the total number of elements belonging to synchronous groups is sN. Assuming that all these groups have equal sizes and there are K such groups, this would imply that each group contains $n = sN/K$ elements. The distances between the elements within each group are zero. Therefore, the number of identical pairs of elements in a single group is approximately $n^2/2$ for $n \gg 1$. Hence, the number of identical pairs is $Kn^2/2$. Dividing this by the total number of pairs $N^2/2$, we find that

$$r = \frac{Kn^2}{N^2} = \frac{s^2}{K}. \tag{7.11}$$

The relationship (7.11) can be used to estimate the number of synchronous groups,

$$K = \frac{s^2}{r}, \tag{7.12}$$

and the mean size of such a group,

$$n = \frac{r}{s}N. \tag{7.13}$$

It should be remembered that equations (7.12) and (7.13) provide only rough estimates, since they were derived under the assumption that the sizes of all synchronous groups are equal. In reality, groups of different sizes are typically found and the size distribution varies from one realization to another. Figure 7.8 displays the exact mean number (bold curve) of synchronous group, obtained by direct counting and averaging over 20 independent realizations, and its estimate (dashed curve) given by equation (7.12). The bars show the statistical dispersion of the data. Though there are some significant deviations, both curves show the same basic behavior. We see that dynamical clustering is observed in a wide interval of the coupling strength and the number of synchronous groups is usually relatively large. Figure 7.9 shows the relative mean size of a synchronous group, estimated as $n/N = r/s$, as function of the coupling strength ε. The general trend is that the relative size grows from very small values near the onset of clustering and approaches unity when full synchronization is reached. The growth is not, however, strictly monotonous.

Figure 7.10 shows the dependence of the order parameter r on the system size N at a fixed intensity $\varepsilon = 0.08$ of global coupling in the regime of

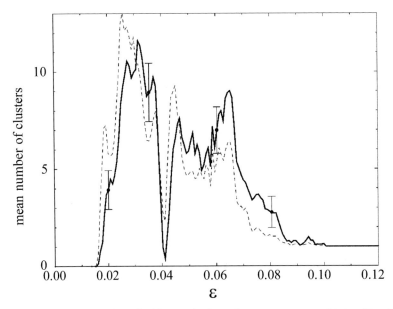

Fig. 7.8. The mean number (bold curve) of synchronous groups, obtained by direct counting and averaging over 20 independent realizations, and its estimate (dashed curve) given by equation (7.12). The bars show the statistical dispersion of the data. The parameters are the same as in Fig. 7.7. (From [365])

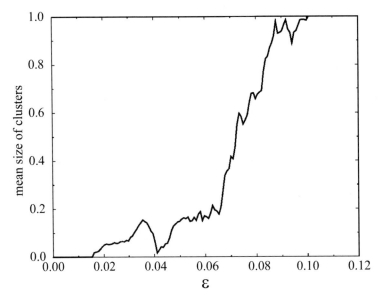

Fig. 7.9. The relative mean size $n/N = r/s$ of a synchronous group as function of the coupling strength ε. The parameters are the same as in Fig. 7.7. (From [365])

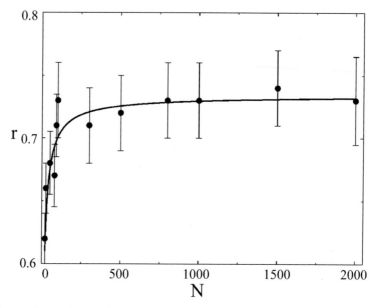

Fig. 7.10. Dependence of the order parameter r on the system size N at a fixed intensity $\varepsilon = 0.08$ of global coupling in the regime of dynamical clustering. Each point has been obtained by averaging over 20 independent realizations; the bars indicate the statistical dispersion of the data. The parameters are the same as in Fig. 7.7. (From [365])

dynamical clustering. Each point has been obtained by averaging over 20 independent realizations; the bars indicate the statistical dispersion of the data. A strong size dependence is characteristic of relatively small ensembles ($N < 500$). Starting from an ensemble size of about $N = 1000$, the dependence displays saturation, with remaining small variations lying within the statistical dispersion range. Hence, the results for an ensemble of 1000 chaotic Rössler oscillators are already representative for the limit $N \to \infty$ of large globally coupled systems.

The above analysis has been performed in the absence of noise and, therefore, some of the observed asymptotic regimes of dynamical clustering may actually correspond to pseudoattractors. Recently, numerical similations of a population of globally coupled Rössler oscillators with a different form of coupling in the presence of noise have been performed [366]. Weak added noise does not significantly influence the dynamical clustering, while strong noise destroys the clusters.

7.3 Neural Networks

Dynamical clustering and mutual synchronization are characteristic of a broad class of dynamical systems. We have already seen that these phenomena are observed in such different systems as ensembles of globally coupled logistic maps and chaotic Rössler oscillators. Below in this section we shall consider the synchronization effects in ensembles whose individual elements represent entire neural networks.

The human brain is a giant network that consists of billions of neural cells, called neurons. On the other hand, a small animal such as a worm might have a nervous system made of only a hundred neurons. The properties of individual neural cells are similar in all animals. McCulloch and Pitts [367] proposed in 1943 a simple model of a neural network. They suggested the description of neurons as automata that can be found in two states, active or passive. Thus, each neuron i is characterized by a time-dependent binary variable s_i, taking values $s_i = 0$ (passive state) and $s_i = 1$ (active state).

A neural network is formed by N connected neurons. Time is discretized into small steps. The next state of the neuron i is determined by the signal h_i which it receives from all other neurons forming the network:

$$s_i(t+1) = H\left(h_i(t) - b_i\right), \qquad (7.14)$$

where $H(z)$ is the step function, $H(z) = 1$ for $z \geq 0$ and $H(z) = 0$ if $z < 0$, and b_i is the threshold (or bias) of neuron i. Thus, the neuron is active at the next moment of time $t+1$ only if the total signal $h_i(t)$ which it has received at time t exceeds the threshold b_i.

A connection extending from an element i to an element j is characterized by the connection weight G_{ij}. The signal h_i received by neuron i at time t is given by the sum

$$h_i(t) = \sum_{j=1}^{N} G_{ij} s_j(t). \qquad (7.15)$$

This connection is activatory if $G_{ij} > 0$ and inhibitory if $G_{ij} < 0$. If $G_{ij} = 0$, the connection is absent. If a neuron j is active ($s_j = 1$), it sends a positive signal to neuron i, when the connection is activatory, and a negative signal in the case of an inhibitory connection.

Of course, the McCulloch–Pitts model represents only a caricature of biological neural networks. Real neurons are much more complex than the idealized two-state automata. However, the dynamical behavior of networks described by (7.14) and (7.15) is surprisingly rich and the model apparently grasps some essential properties characteristic of neural systems. A discussion of the information-processing properties of McCulloch–Pitts networks can be found, for instance, in the book [368]. Note that artificial neural networks, implementing (7.14) and (7.15), can easily be constructed to perform various tasks of information processing.

When the connection matrix is symmetric, i.e. $G_{ij} = G_{ji}$ for any elements i and j, the dynamics of such networks is relaxational: starting from any initial condition, they approach some stationary state described by a certain activity pattern of neurons. This final state, representing an attractor of such a dynamical system, is generally different for different initial conditions.

In contrast to such simple relaxational dynamics, the behavior is much more complex when the connections are asymmetric, $G_{ij} \neq G_{ji}$. In this case the system can display persistent oscillations. These oscillations are characterized by periodic or irregular variation of the activity pattern formed by the network. Thus, the behavior of large neural network may also be chaotic.

Suppose that we have a population of simple animals (or *robots*) whose individual behavior is controlled by nervous systems representing McCulloch–Pitts networks. Let us assume furthermore that the animals (or robots) can communicate by effectively exchanging information about their neural activity patterns and this information can influence their own neural dynamics. Can such an interaction lead to the mutual synchronization of complex activity patterns in the "brains" of individual animals? Can dynamical clustering develop in the ensemble, so that synchronously operating groups of animals (or robots) spontaneously emerge? The answers to these questions are positive, as follows from the analysis of a simple model of cross-coupled neural networks introduced by Zanette and Mikhailov [369].

We take M networks with identical connection matrices G_{ij} and zero biases ($b_i = 0$) and introduce coupling between them. The activity state of a neuron i in the network α (where $\alpha = 1, 2, \ldots, M$) at discrete time t is characterized by the variable $s_i^\alpha(t)$. The next state of this neuron is

$$s_i^\alpha(t+1) = (1-\varepsilon) H\left(h_i^\alpha(t)\right) + \varepsilon H\left(\overline{h}_i(t)\right), \qquad (7.16)$$

where

$$h_i^\alpha(t) = \sum_{j=1}^{N} G_{ij} s_j^\alpha(t) \qquad (7.17)$$

is the signal received by the neuron i from all other neurons in its own network α and $\overline{h}_i$ is the average signal received by neurons occupying the same position i in all M networks,

$$\overline{h}_i(t) = \frac{1}{M} \sum_{\beta=1}^{M} h_i^\beta(t). \qquad (7.18)$$

The coefficient ε specifies the strength of the cross-network coupling. When $\varepsilon = 0$, the networks forming the ensemble are independent. On the other hand, if $\varepsilon = 1$, the first term in (7.16) vanishes and the states of all neurons i, occupying the same positions in different networks, are identical since they are determined by the same average signal $\overline{h}_i$. Hence, the ensemble dynamics must be synchronous at $\varepsilon = 1$. In the interval $0 < \varepsilon < 1$ the dynamics is governed by an interplay between local coupling inside the networks and global coupling across them.

The further analysis [369] is based on numerical simulations. For technical reasons, it is convenient to treat s_i^α as continuous variables and replace the step function $H(z)$ in (7.16) by the sigmoidal function

$$H(z) = \frac{1}{2}\left[1 + \tanh\left(\chi z\right)\right]. \tag{7.19}$$

When $\chi = 10$, this function is a good approximation of the discontinuous step function. Most of the simulations have been performed for ensembles of $M = 100$ identical networks, each consisting of $N = 50$ neurons.

First we set the connection weights between neurons in an individual network. Each matrix element G_{ij} is chosen at random with equal probability from the interval between –1 to 1. The weights G_{ij} and G_{ji} of forward and reverse connections are independently selected and are therefore generally different. We run a simulation of a single network with the chosen connection matrix and accept this matrix only if the network dynamics is irregular and is not characterized by the relaxation to a stationary pattern or by periodic alterations of the activity patterns. After that the connection weights are fixed and remain unchanged during the whole simulation series involving the network ensembles. The initial conditions for all neurons in each simulation are randomly chosen.

Since subsequent states of all neurons in all networks are recorded, each simulation yields a large volume of data that should be further analyzed. This analysis is facilitated by monitoring the integral activities $u^\alpha(t)$ of all networks, which are defined as

$$u^\alpha(t) = \sum_{i=1}^{N} s_i^\alpha(t). \tag{7.20}$$

If the synchronization of the network activity has taken place, the activity patterns of some networks coincide and their integral activities $u^\alpha(t)$ must then also be identical.

When the cross-network coupling is absent or very weak, all signals $u^\alpha(t)$ are different. However, as the coupling strength ε is increased, the behavior is changed. Since the initial conditions are random and independent for all networks, the integral signals of different networks are at first not correlated. However, starting from some moment, some of the networks in the ensemble generate identical (within the computer precision) activity signals, indicating the onset of synchronization in the system.

Figure 7.11 shows typical integral activity signals when the global coupling is relatively strong. As an illustration, we display here only the signals $u^\alpha(t)$ for 10 selected networks in the ensemble of the total size $M = 100$ (the indices α of these selected networks are given on the left-hand side of the respective plots). The signals are complex and apparently chaotic. Nonetheless, a closer examination shows that, starting from certain moments, some of them are actually identical! We have marked such identical plots in Fig. 7.11

by short vertical bars with horizontal arrows. The synchronization begins at the moment corresponding to the bar location and is maintained at all later moments. When several synchronous groups (network clusters) are present, they are identified by letters A, B, C at the corresponding arrows. Thus, when $\varepsilon = 0.35$ (Fig. 7.11a), the entire ensemble breaks down into several synchronous clusters. At a higher intensity of cross-network coupling ($\varepsilon = 0.5$, Fig. 7.11b) the activity of all networks in the ensemble is synchronous.

As a quantitative measure, characterizing the degree of mutual synchronization in the ensemble, we can use the time-dependent statistical dispersion $D(t)$ of activity patterns, defined as

$$D(t) = \frac{1}{M} \sum_{\alpha=1}^{M} \sum_{i=1}^{N} [s_i^\alpha(t) - \bar{s}_i(t)]^2 , \qquad (7.21)$$

where

$$\bar{s}_i(t) = \frac{1}{M} \sum_{\alpha=1}^{M} s_i^\alpha(t) \qquad (7.22)$$

is the average activity of neurons occupying the same positions i in all networks of the ensemble. The dispersion (7.21) must be zero when the states of all such neurons are identical, i.e. complete mutual synchronization has taken place.

Figure 7.12 shows, on a logarithmic scale, how the statistical dispersion D evolves with time in a typical simulation at a fixed intensity $\varepsilon = 0.5$ of the global cross-network coupling (cf. Fig. 7.11b). To understand the observed time dependence of $D(t)$, we should take into account that the synchronization begins with the formation of a nucleus of a few synchronous networks. This nucleus grows by an aggregation process, where further networks are subsequently added to the existing synchronous cluster. While non-entrained networks remain in the ensemble, the dispersion D is relatively large, though it gradually decreases with time. When the last network has approached the synchronous cluster, it gets strongly attracted and its distance from the cluster begins to exponentially decrease. This rapid decrease is reflected in the final steep linear fall seen in Fig. 7.12.

Though the dispersion serves as a good indicator of full synchronization, it is not sensitive to dynamical clustering where several different synchronous groups are formed. To investigate clustering, we have calculated pair distances between activity patterns of all networks in the ensemble. The pair distance $d^{\alpha\beta}(t)$ between activity patterns of networks α and β at time t is defined as

$$d^{\alpha\beta}(t) = \sqrt{\sum_{i=1}^{N} \left(s_i^\alpha(t) - s_i^\beta(t) \right)^2}. \qquad (7.23)$$

By counting at a fixed moment the number of network pairs (α, β) that have pair distances lying within subsequent equal intervals, a histogram of the distribution over pair distances can be constructed.

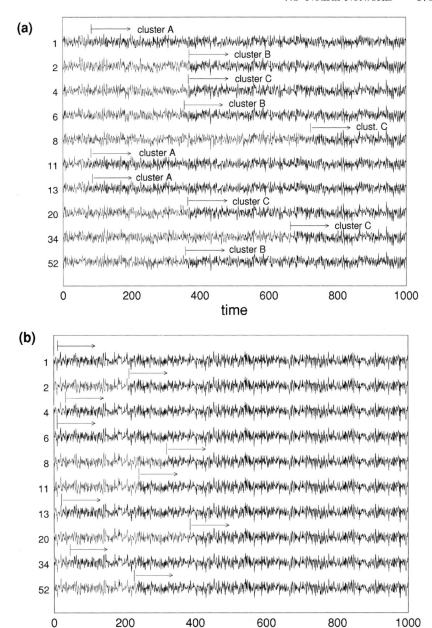

Fig. 7.11. Time-dependent integral activity of ten selected neural networks with $N = 50$ in an ensemble of size $M = 100$ for (**a**) $\varepsilon = 0.35$ and (**b**) $\varepsilon = 0.5$. Synchronization of each signal begins at the corresponding bar. In (**a**), clusters are identified by different letters. (From [369])

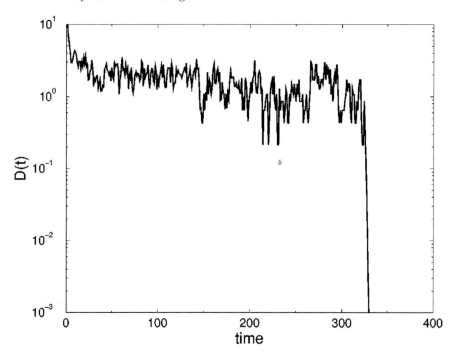

Fig. 7.12. Dispersion D of the activity patterns of all neural networks in the ensemble as a function of time ($\varepsilon = 0.5$, $M = 100$, $N = 50$). (From [369])

Figure 7.13 shows such distributions over pair distances for four different intensities ε of the cross-network coupling. When the coupling is weak (Fig. 7.13a, $\varepsilon = 0.15$) the histogram has a single maximum at a typical distance between the activity patterns of noncorrelated networks. Increasing the coupling intensity, we find that above a certain critical point ($\varepsilon \approx 0.17$) some pairs of networks in the ensemble begin to have the same activity patterns, so that the distance between them is zero. This corresponds to the development of a peak at $d = 0$ in the respective histograms. This new peak is already visible in Fig. 7.13b for $\varepsilon = 0.28$. When the coupling is further increased, this peak grows, indicating a large number of synchronous pairs (Fig. 7.13c, $\varepsilon = 0.34$). Moreover, we can already see in Fig. 7.13c that the histogram contains additional peaks at nonzero distances, so that the distribution is jagged. At a slightly higher coupling intensity (Fig. 7.13d, $\varepsilon = 0.35$), the distribution consists of several well-defined lines. Such histograms are characteristic for dynamical states with a few synchronous clusters. When the coupling intensity is further increased, complete mutual synchronization is established in the ensemble for $\varepsilon \approx 0.4$ (this final regime is not shown in Fig. 7.13).

It is interesting to compare Fig. 7.13 with Fig. 7.6 for an ensemble of globally coupled chaotic Rössler oscillators. We see that dynamical clustering

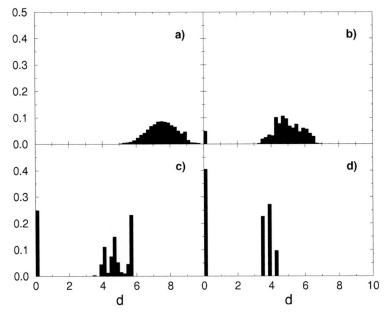

Fig. 7.13. Histograms of distributions over pair distances d between activity patterns of all neural networks for **(a)** $\varepsilon = 0.15$, **(b)** $\varepsilon = 0.28$, **(c)** $\varepsilon = 0.34$, **(d)** $\varepsilon = 0.35$. The other parameters are the same as in Fig. 7.12. (From [369])

has similar properties in both systems. It begins with the formation of small synchronous groups on the background of a large number of non-entrained elements. The number of elements inside coherent groups increases with the coupling intensity and, in the middle of the dynamical clustering interval, almost all elements belong to one of the synchronous groups. At still higher coupling intensities, complete synchronization takes place. Thus, though neural networks are much more complex dynamical systems than individual Rössler oscillators, their collective behavior is not significantly different.

The model (7.16) involves a relatively high amount of communication between individual neural networks. Indeed, each neural cell in any network must receive information about the average activity state of all other cells occupying the same positions in other networks in the ensemble. It would be more realistic to assume that only a certain fraction of neurons in each network takes part in the interactions. This situation is described by the model where the evolution of individual neurons is governed by the equations

$$s_i^\alpha(t+1) = (1 - \varepsilon\zeta_i)\, H\left(h_i^\alpha(t)\right) + \varepsilon\zeta_i H\left(\overline{h}_i(t)\right), \tag{7.24}$$

where ζ_i are independent random variables, taking the values 1 or 0 with the probability p or $1 - p$. In this modification of the model, only a subset of neurons (with $\zeta_i = 1$) in each network is affected by cross-network coupling. Note that this subset is the same for all networks in the ensemble.

The numerical simulations of the model (7.24) show that the system undergoes mutual synchronization and displays dynamical clustering even when a relatively high fraction of possible cross-network interactions is absent. Based on a large number of simulations, an approximate synchronization diagram has been constructed (Fig. 7.14). Complete synchronization is found inside the dark-gray region. Dynamical clustering is observed inside the light-gray region. The latter region is defined by the condition that at least two networks in the ensemble develop identical activity patterns. Some of the networks in this region may be non-entrained. Examining Fig. 7.14, we notice that dynamical clustering persists even when only about 10% of cross-network connections remain in the system.

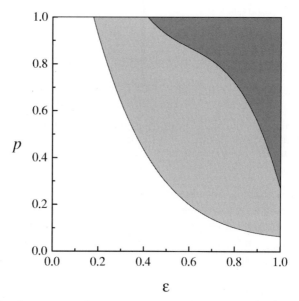

Fig. 7.14. Synchronization diagram in the parameter plane (ε, p) for an ensemble of $M = 100$ neural networks, each consisting of $N = 50$ neurons with randomly chosen asymmetric synaptic connections. Full synchronization is found inside the dark-gray area, partial synchronization is observed inside the light-gray area. (From [369])

Synchronization of network activity and dynamical clustering have been found [369] under a random choice of connection weights for networks and ensembles of various sizes. This suggests that such effects may represent a generic property of neural systems.

Though only networks with random connection matrices have been investigated so far, similar synchronization behavior should apparently also be expected when connection patterns are not arbitrary, but are designed or selected in such a way that a neural network is able to perform certain operations of information processing or to control the activity of motor units. Dynamical clustering in a population of animals or robots with such neural networks would mean that the whole populations breaks down into a number of coherently operating groups, executing different functional tasks. Hence, it can be viewed as describing a spontaneous onset of a functional structure in an originally uniform population.

7.4 Protein Machines

In the previous section we have considered dynamical clustering in large systems, whose individual elements represented entire neural networks. Now we shall discuss the behavior on the much shorter length scales characteristic of an individual biological cell. A living cell is a tiny chemical reactor where tens of thousands of chemical reactions can simultaneously go on. The very fact that these reactions proceed in a regular and predictable manner, despite thermal fluctuations and variations in environmental conditions, already indicates a high degree of organization in this system.

The biochemical activity of a cell can be compared with the operation of a large industrial factory where certain parts are produced by a system of machines. Products of one machine are then used by other machines to manufacture their products or to regulate their functions.

Two possible modes of operation of such a factory can be imagined. In the asynchronous mode, parts produced by all machines are first deposited and accumulated in a common store. They are taken back from this store by other machines, when the parts are needed for further production. This kind of organization is not however optimal, since it requires large storage facilities and many transactions. It becomes deficient when the intermediate products are potentially unstable and can easily be lost or damaged during the storage process.

When the synchronous operation mode is employed, intermediate products, required for a certain operation step in a given machine, are released by other machines and become available exactly at the moment when they are needed. Hence, large storage facilities are eliminated and the entire process may run much faster.

The synchronous manufacturing process implies much more complicated management than the asynchronous operation mode. In a real factory, this is achieved by careful planning and detailed external control of the production. Such a rigid external control cannot, however, be achieved at a molecular level inside an individual cell. The coherence, underlying the synchronous production, should emerge in this case as a natural consequence of the interactions between the individual elements. Remarkably, the predictable coherent operation of the complex molecular machinery must be maintained in a living cell under very stringent conditions, when strong thermal fluctuations are present and the parts needed in its assembly lines are transported by the random Brownian motion of the molecules. Below in this section we show that a spontaneous transition to coherent collective operation is indeed possible in biochemical systems. It represents the effect of dynamical clustering in a population of interacting protein machines.

Proteins are macromolecules formed by long heteropolymer chains. Their important property is that, starting from any initial condition, these chains fold into a unique compact spatial configuration corresponding to a deep minimum of the free energy. The final stable configuration is sensitive to the sequence of amino acids forming the chain. Different sequences would generally fold into different spatial configurations. Furthermore, even minor chemical modifications, such as the binding of a small additional molecule, can trigger a conformation transition.

The ability of proteins to change their spatial configurations in response to weak control perturbations allows their use as *molecular machines*. These machines can perform mechanical work and therefore proteins are often used by biological cells as motors. There is, however, also a large group of proteins whose function inside a cell is to serve as a "chemical machine". These proteins are *enzymes*.

Enzymes are single-molecule catalysts. A certain chemical reaction, such as, for example, the splitting of a small substrate molecule S into two product molecules P_1 and P_2, might be practically impossible under natural conditions because the energy barriers are too high. It can, however, be realized with the help of an enzyme. The substrate molecule binds at a special binding site on the surface of the enzyme and thus a *turnover cycle* is initiated:

Since the conditions have changed, the protein machine begins a sequence of conformational transformations, relaxing to a new equilibrium. In the process of such transformations the captured substrate molecule moves together with the surrounding machine parts. In this way, it is brought to an active center inside the enzyme where a special chemical environment is created that facilitates the reaction (i.e. the splitting of S into two products P_1 and P_2 in the case considered). Once this has occurred, the conditions are again changed and a new sequence of transformations is initiated. As it proceeds, the product molecules are expelled from the machine, which then gradually returns to its original free state, ready to bind another substrate molecule and repeat the whole turnover cycle.

Many examples of biochemical reactions catalyzed by various enzymes are known. The enzymes help not only to split the molecules, as just described. In a similar way, they can facilitate chemical reactions where two small substrate molecules S_1 and S_2 merge into a single product P.

An important further property of some enzymes is that their operation is *allosterically* regulated. Different mechanisms of such regulation are known. In the simplest case, an enzyme has an additional binding site on its surface where a special regulatory molecule can sit. If the regulatory molecule is present, this modifies the protein configuration. The modification involves also the surrounding of the substrate binding site and thus influences the capture probability of a substrate molecule. When the presence of a regulatory molecule enhances the substrate binding probability, allosteric activation takes place. If the presence of a regulatory molecule makes substrate binding more difficult, allosteric inhibition is achieved.

Biochemical reactions in a living cell are complex. They comprise several different types of enzymes and many intermediate products. Typically, some of these by-products would play the role of regulatory molecules of the same reaction. These regulatory molecules diffuse through the solution and, by binding to the regulatory sites of other enzymes, can control their operation. This makes the entire reaction a system of communicating protein machines.

Conformational transformations are usually slow. The folding of a protein to its native configuration may take several seconds. The turnover times of enzymes can vary greatly, but typically the duration τ of a turnover cycle lies in the range of tens or hundreds of milliseconds. In contrast to this, the characteristic transport times related to the diffusion of small regulatory molecules inside a biological cell can be much shorter.

Suppose that the reaction proceeds inside a volume of linear size L containing N enzymes of a particular kind. A regulatory molecule is released into the solution somewhere inside this volume and begins its diffusive motion through it, characterized by the diffusion constant D. The enzymes are heavy and therefore their own mobility can here be neglected. When during its diffusive motion a regulatory molecule comes into contact with a certain atomic target group of radius R on the surface of an enzyme, it will be captured with a high probability by this group and thus bound to the enzyme molecule.

The mean transit time t_{transit} needed for a regulatory molecule to find one of N targets which are randomly distributed in the reaction volume can roughly be estimated [370] as

$$t_{\text{transit}} = \frac{L^3}{NDR}. \tag{7.25}$$

The mean mixing time t_{mixing} during which the regulatory molecule would diffuse over the distance of order L and thus cross the entire volume, forgetting the initial position inside it, is given by

$$t_{\text{mixing}} = \frac{L^2}{D}. \tag{7.26}$$

When the typical numerical values $L = 1$ μm, $R = 1$ nm, $N = 1000$ and $D = 10^{-5}$ cm/s^2 are taken, this yields $t_{\text{transit}} \sim t_{\text{mixing}} \sim 1$ ms.

Both characteristic transport times t_{transit} and t_{mixing} in micrometer volumes can be significantly shorter than the turnover time ($\tau \sim 10$ ms) of a single enzyme molecule. This is a remarkable result. Usually, all characteristic kinetic times of a chemical reaction are much longer than the duration of a single molecular reaction event and therefore these events are treated as instantaneous. This forms the basis of the kinetic theory formulated in terms of random Markov processes. For biochemical reactions in small volumes, an opposite limit can easily be reached, where the processes inside individual reacting molecules are much slower than the diffusion kinetics.

When the condition $\tau \gg t_{\text{transit}} \geq t_{\text{mixing}}$ is satisfied, the reaction develops a *molecular network* [371]. It can be viewed as a population of slow cyclic protein machines that interact by releasing small regulatory molecules which rapidly move through the solution, find other machines and influence their cycles. Through this communication, the internal dynamics of individual machines becomes coupled. Since $t_{\text{transit}} \geq t_{\text{mixing}}$, a regulatory molecule would typically cross the whole reaction volume until it finds a target. Thus, it can bind with equal probability to any of N protein machines, no matter how far are they located from the point where the regulatory molecule was released. This means that coupling in the considered network is global.

Like globally coupled oscillator ensembles, molecular networks can show mutual synchronization and dynamical clustering. To demonstrate this, we consider below an example of a simple hypothetical reaction

$$S + E \rightarrow E + P + P^*, \quad P \rightarrow 0, \tag{7.27}$$

where binding of the substrate S by enzymes E is allosterically activated by the product P. The other product P^* does not further participate in the reaction. We shall assume that the substrate concentration is maintained constant and that the product P decays at a certain rate.

When the conditions of a molecular network hold, the internal dynamics of individual enzyme molecules is important and must constitute an essential part of the theory. This dynamics is complex and its details are not well enough known. Therefore, one can try to model the operation of a single enzyme molecule in terms of an *automaton*.

As in Sect. 6.1, any enzyme i will be modelled as a clock whose dynamics represents deterministic motion along a certain internal phase coordinate. The phase $\Phi_i = 0$ corresponds to the free state of the enzyme, where it is waiting to bind a substrate. The cycle begins when the substrate is bound. Inside the cycle, the phase Φ_i gradually increases. At a certain point $\Phi_i = \Phi_c$ the products are released. After this event the enzyme continues its motion until the end point $\Phi_i = \Phi_{\text{max}}$ of the cycle is reached, and the enzyme returns to its initial state $\Phi_i = 0$. Note that, though the motion inside the cycle is deterministic, the waiting time in the state $\Phi_i = 0$ is a random variable because the binding of a substrate molecule is a stochastic event. The clock stops after each rotation of the cycle and waits until a random cycle initiation has occurred.

For convenience, we introduce a short time interval Δt and consider the system at discrete moments $t_n = n\Delta t$, where $n = 0, 1, 2, 3, \ldots$. The integer phase Φ_i is defined as the number of time steps needed to reach a particular state. Hence, $\Phi_{\text{max}} = \tau/\Delta t$, where τ is the cycle duration. The dynamics of an individual enzyme i is therefore described by the evolution algorithm

$$\Phi_i(n+1) = \begin{cases} \Phi_i(n) + 1, & \text{if } 0 < \Phi_i(n) < \Phi_{\max} \\ 0, & \text{if } \Phi_i(n) = \Phi_{\max} \\ 1, & \text{with probability } w(n), \text{ if } \Phi_i(n) = 0 \\ 0, & \text{with probability } 1 - w(n), \text{ if } \Phi_i(n) = 0 \end{cases} \qquad (7.28)$$

A product molecule P is released each time that one of the enzymes passes through the state $\Phi_i = \Phi_c$. Moreover, each of these product molecules can decay; we assume that the probability of decay per one time step is g. Hence, the number $m(n)$ of product molecules evolves with time according to the algorithm

$$m(n+1) = m(n) + \sum_{i=1}^{N} \Delta(\Phi_i(n) - \Phi_c) - \sum_{j=1}^{m(n)} \xi_j. \qquad (7.29)$$

The second term on the right-hand side of this equation describes the release of new product molecules. Here $\Delta(\Phi)$ is the discrete delta-function, i.e. $\Delta(\Phi) = 1$ for $\Phi = 0$ and $\Delta(\Phi) = 0$ otherwise, and N is the total number of enzymes. The last term describes the spontaneous decay of the product molecules. The binary random variables ξ_j take values 1 and 0 with the probabilities g and $1 - g$, respectively.

The probability $w(n)$ of the cycle initiation (i.e. the binding of a substrate molecule) at time n depends on the number $m(n)$ of product molecules in the system. Any of m product molecules can trigger the cycle of a given enzyme i with a certain probability w_1 per one time step. Moreover, there is also a certain small probability w_0 of spontaneous cycle initiation. To estimate the total probability of the cycle initiation, we note that the initiation *does not* take place at a given time step, either spontaneously or induced by any of the present product molecules, with the probability $q = (1 - w_0)(1 - w_1)^m$. Therefore, the probability of the cycle initiation is $w = 1 - q$ or, explicitly,

$$w(n) = 1 - (1 - w_0)(1 - w_1)^{m(n)}. \qquad (7.30)$$

If the probabilities w_0 and w_1 are small (as can be expected for short time intervals Δt), this equation approximately yields

$$w = w_0 + m w_1. \qquad (7.31)$$

Equations (7.28), (7.29) and (7.30) define the stochastic evolution algorithm of the automata ensemble. They can be used to perform numerical simulations of the considered chemical reaction [372]. A population of $N = 200$ enzymes is taken and time is discretized in such a way that each cycle consisted of $\Phi_{\max} = 100$ steps. As the initial condition, the state with a random distribution over cycle phases and a small number of products is chosen. The probability w_0 of spontaneous cycle initiation is small ($w_0 = 0.01$). The decay probability g of product molecules is so high ($g = 0.2$) that the characteristic lifetime $1/g$ of a product molecule in the system is much shorter than the cycle duration ($1/g \ll \Phi_{max}$). Two parameters that are varied in the simulations are the probability w_1 and the moment Φ_c inside the cycle when the product is released.

The parameter w_1 can be viewed as specifying the strength of the coupling between the machines. Indeed, if $w_1 = 0$, the cycles of all enzymes are independent. Figure 7.15 displays the behavior of the system when w_1 is small. The distribution over cycle phases at a fixed moment of time (Fig. 7.15a) reveals that the cycles of individual enzymes are not in this case correlated. Because allosteric regulation is not effective, the binding of substrate molecules occurs mainly by spontaneous, nonactivated events. The waiting times of the substrate binding are relatively long and therefore a significant fraction ($N_0 > 45$) of the enzymes is found in the free state ($\Phi = 0$). The rest of the enzymes are almost uniformly distributed over various phase cycles. The dependence of the total number m of product molecules on time shows (Fig. 7.15b) only irregular fluctuations, described by the Poisson distribution.

The kinetic regime of the reaction is, however, completely changed when the allosteric regulation is stronger (i.e. the parameter w_1 is increased). This is seen in Fig. 7.16. Examining the distribution over the cycle phases (Fig. 7.16a), we see that it is strongly nonuniform. The entire enzymic population is effectively divided into two synchronously operating groups whose turnover cycles are shifted by half a cycle period. Moreover, the number of enzymes in the state with $\Phi = 0$, waiting to bind a substrate, is now much smaller ($N_0 < 10$). The dependence of the number of product molecules on time shows a sequence of spikes (Fig. 7.16b). Its period is shorter by a factor of two than the molecular turnover time τ. Each of the spikes is generated by the synchronous release of products from one of the two enzymic groups. Since the lifetime of the product molecules is short, all the products generated by one group die before the appearance of the next spike, generated by a different group. Because the binding of the substrate and the decay of the product are stochastic processes, the spikes are not identical and show some stochastic variation. Figure 7.17 shows the transient process, by which synchronization is established starting from an initial state with a random distribution of enzymes over the cycle phases.

The synchronization properties depend on the details of the turnover cycle. The simulations, presented in Fig. 7.18 and showing two synchronous groups, are performed at $\Phi_c/\Phi_{max} = 0.5$, i.e. when the product is released in the middle of the turnover cycle. By varying the phase Φ_c when the product is released, regimes with different numbers of synchronous groups can be observed. For instance, when $\Phi_c/\Phi_{max} = 0.8$, four synchronous groups are found (Fig. 7.18). Below the synchronization threshold (Fig. 7.18a), the distribution shows irregular variations. As the intensity w_1 of allosteric activation is increased, strong correlations between the phases of individual enzymes develop and four distinctive enzymic groups are already seen (Fig. 7.18b). Further increase of the activation intensity makes the maxima, corresponding

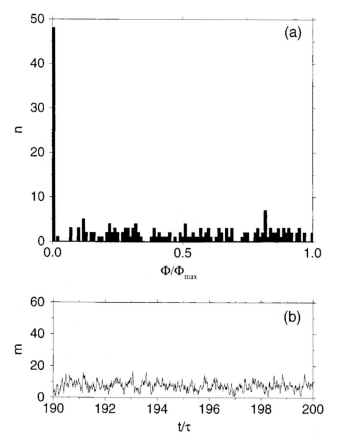

Fig. 7.15. Distribution over cycle phases (*a*) and time dependence of the number of product molecules (*b*) for the asynchronous reaction regime. The reaction parameters are $w_0 = 0.01, w_1 = 0.002, \Phi_c = 50, \Phi_{max} = 100, g = 0.2$ and $N = 200$. (From [372])

to each enzymic group, more narrow (Figs. 7.18c,d). Hence, the statistical phase dispersion of enzymes inside the groups gets smaller and the synchronization effect is more pronounced.

By varying the phase Φ_c, kinetic regimes with one and three synchronous groups have also been found [372].

To analytically investigate the observed clustering behavior, the mean-field approximation can be used. This approximation in chemical kinetics corresponds to neglecting fluctuations in concentrations of reactants that are due to the atomistic stochastic nature of the reaction processes. If $n(t)$ is the number of enzymes in the ground state at time t and $\nu(t)$ is the probability rate of substrate binding, then, on average, $\nu n(t)\Delta t$ of these enzymes will bind the substrate within a short time interval Δt, and hence the number of

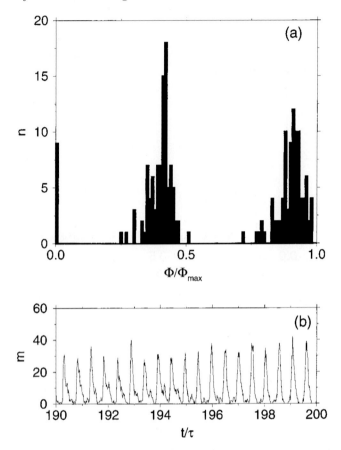

Fig. 7.16. Distribution over cycle phases (**a**) and time dependence of the number of product molecules (**b**) for the synchronous reaction regime with two coherent enzymic groups. The reaction parameters are $w_0 = 0.01, w_1 = 0.01$, $\Phi_c = 50, \Phi_{max} = 100, g = 0.2$ and $N = 200$. (From [372])

enzymes in the ground state will be decreased by this amount. On the other hand, during the same time interval some enzymes will complete their cycles and return to the ground state. To estimate the number of such returning enzymes, we note that, because fluctuations in turnover times are neglected, all these enzymes must have started their cycles at a delayed time moment $t-\tau$. Therefore, the number of such returning enzymes is equal to the number of enzymes that have bound the substrate at this time $t - \tau$, i.e. it is given by $\nu(t - \tau)n(t - \tau)$.

   If the total number of enzymes in the considered population is large and the change in the number of enzymes in the ground state within a time interval Δt is small, a continuous description can be used. In this description, the internal state of an enzyme is specified by the (continuous) phase $\phi = \Phi\Delta t$.

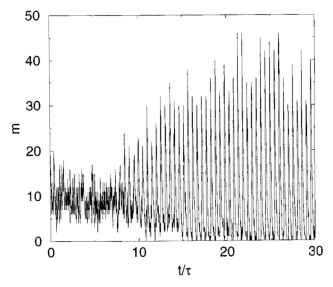

Fig. 7.17. The transient leading to synchronous two-group spiking. The parameters are the same as in Fig. 7.16. (From [372])

The rate of change in the mean number of enzymes, found in the ground state, is then given by the equation

$$\frac{dn(t)}{dt} = -\nu(t)n(t) + \nu(t - \tau)n(t - \tau). \tag{7.32}$$

Furthermore, according to (7.31), the binding probability per short time Δt is linearly dependent on the number $m(t)$ of product molecules present at time t in the system. Therefore, for the probability rate $\nu = w/\Delta t$ at time t we get

$$\nu(t) = \nu_0 + \nu_1 m(t), \tag{7.33}$$

where $\nu_0 = w_0/\Delta t$ and $\nu_1 = w_1/\Delta t$.

To construct the respective evolution equation for the mean number $m(t)$ of product molecules, we note that on average $\gamma m(t)\Delta t$ of these molecules will die within a short time Δt. On the other hand, during the same time interval new product molecules will be released by the enzymes. The number of such new product molecules is given by the number of enzymes that have passed within interval Δt through the state with $\phi = \tau_c$. But the latter is equal to the number of enzymes that have started their cycles at time $t - \tau_c$, i.e. to $\nu(t - \tau_c)n(t - \tau_c)$. Therefore, we obtain the following equation

$$\frac{dm(t)}{dt} = -\gamma m(t) + \nu(t - \tau_c)n(t - \tau_c). \tag{7.34}$$

Equations (7.32), (7.33) and (7.34) form a closed system. We see that in contrast to the usual kinetic equations of chemical kinetics, which are

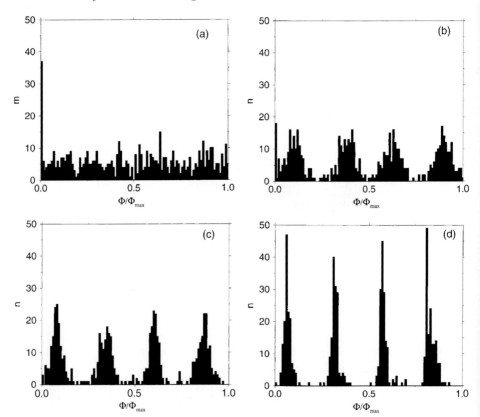

Fig. 7.18. Distribution over cycle phases in the asynchronous (**a**) and synchronous (**b, c, d**) regimes with four coherent groups in a population of $N = 600$ enzymes at different intensities of allosteric activation: (**a**) $w_1 = 0.002$, (**b**) $w_1 = 0.004$, (**c**) $w_1 = 0.06$, and (**d**) $w_1 = 0.2$. Other parameters are $w_0 = 0.01, \Phi_c = 80, \Phi_{max} = 100$ and $g = 0.2$. (From [372])

also formulated in the mean-field approximation, this system of equations includes *delays*. The correspondence between these equations and the original automata model is established by the relationships $\Phi_{max} = \tau/\Delta t, \Phi_c = \tau_c/\Delta t,$ $w_0 = \nu_0/\Delta t, w_1 = \nu_1/\Delta t,$ and $\gamma = g/\Delta t$.

To test the agreement between the mean-field and automata descriptions, we can numerically integrate (7.32), (7.33) and (7.34) with the parameter values corresponding to previous automata simulations. The time dependences of the number of product molecules in the system, obtained by two such simulations, are shown in Fig. 7.19.

The parameters in Fig. 7.19a correspond to the parameters chosen in Fig. 7.15, where an asynchronous regime with random Poisson fluctuations in the number of product molecules was observed. We see that the mean-field approximation yields in this case relaxation to a steady state of kinetic

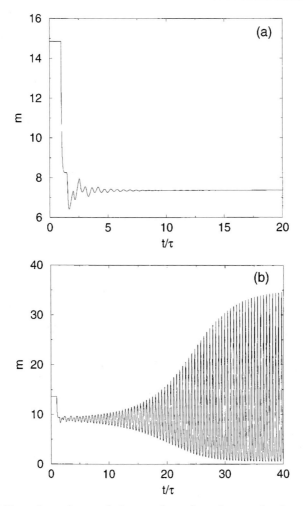

Fig. 7.19. Time dependence of the number of product molecules, obtained by integration of the mean-field equations, for the parameters (**a**) $\nu_0 = 1$, $\nu_1 = 0.2$, $\gamma = 20$, $\tau_c = 0.5$, $\tau = 1$ and $N = 200$ (as in Fig. 7.15) and (**b**) $\nu_0 = 1$, $\nu_1 = 1$, $\gamma = 20$, $\tau_c = 0.5$, $\tau = 1$ and $N = 200$ (as in Fig. 7.16). Transients leading to the stable steady state (**a**) and periodic two-group spiking (**b**) are shown. (From [372])

equilibrium with a stationary product concentration that is close to the mean number of product molecules in the automata model. On the other hand, when the parameters corresponding to synchronous spiking with two enzymic groups (as in Fig. 7.16 and Fig. 7.17) are chosen, the mean-field equations show the development of stable periodic oscillations (Fig. 7.19b). By comparing Fig. 7.19b and Fig. 7.17, we see that both descriptions yield similar periods and amplitudes of oscillations, though, as can be expected, fluctuations in the spike amplitudes are absent in Fig. 7.19b.

Thus, the mean-field approximation provides satisfactory agreement with the stochastic automata model and it can be used to analyze the synchronization phenomena.

The equilibrium steady states $n = n_0$ and $m = m_0$ of the system are given by the fixed points of equations (7.32), (7.33) and (7.34). It can be noted that (7.32) is satisfied identically for any steady state, whereas the two other equations give in the steady state

$$\nu = \nu_0 + \nu_1 m_0, \tag{7.35}$$

$$-\gamma m_0 + \nu n_0 = 0. \tag{7.36}$$

The total number N of enzymes is equal to the sum of their number n_0 in the ground state, waiting to bind a substrate, and the number N_1 of enzymes that are currently found inside the catalytic cycles. At equilibrium, N_1 can be determined from the following simple considerations. At a given time t we would find inside their cycles all enzyme molecules whose cycles have been initiated at time $t-\tau < t' < t$. Under steady state conditions, $\nu n_0 \Delta t$ enzymes start their cycles within each short time interval Δt. Therefore, the number of enzymes inside the cycles is $N_1 = \nu n_0 \tau$. Hence, we have

$$N = n_0 + \nu n_0 \tau. \tag{7.37}$$

Solving (7.35), (7.36) and (7.37) we find the steady state of the system as

$$n_0 = \frac{N}{1 + \bar{\nu}\tau}, \quad m_0 = \frac{\bar{\nu}}{\gamma} n_0, \tag{7.38}$$

where

$$\bar{\nu} = \frac{1}{2}\left(\nu_0 + \frac{\nu_1 N}{\gamma\tau} - \frac{1}{\tau}\right) + \sqrt{\frac{1}{4}\left(\nu_0 + \frac{\nu_1 N}{\gamma\tau} - \frac{1}{\tau}\right)^2 + \frac{\nu_0}{\tau}}. \tag{7.39}$$

The steady state may become unstable when the intensity ν_1 of allosteric activation is increased. To perform the stability analysis, we introduce small perturbations $m(t) = m_0 + \delta m(t)$, $n(t) = n_0 + \delta n(t)$. Substituting this into (7.35) and (7.37) and linearizing these equations, we obtain

$$\frac{d\delta m}{dt} = -\gamma\delta m + \bar{\nu}\delta n(t - \tau_c) + \nu_1 n_0 \delta m(t - \tau_c), \tag{7.40}$$

$$\frac{d\delta n}{dt} = -\bar{\nu}\delta n - \nu_1 n_0 \delta m + \bar{\nu}\delta n(t - \tau) + \nu_1 n_0 \delta m(t - \tau). \tag{7.41}$$

The solution of these linear differential delay equations can be sought in the form $\delta m \propto e^{\lambda t}$, $\delta n \propto e^{\lambda t}$. Substituting these expressions, we obtain the characteristic equation

$$\lambda^2 + \left[\gamma + \bar{\nu}\left(1 - e^{-\lambda\tau}\right) - \nu_1 n_0 e^{-\lambda\tau_c}\right]\lambda + \gamma\bar{\nu}\left(1 - e^{-\lambda\tau}\right) = 0. \tag{7.42}$$

The nonpolynomial nature of the characteristic equation is typical for dynamical systems with delays.

The roots of equation (7.42) are generally complex, i.e. $\lambda = \kappa + i\omega$. The steady state is stable if the real part is negative. The instability boundary is thus determined by the condition $\kappa = 0$. At this boundary, the imaginary part ω is nonzero and therefore we have a Hopf bifurcation.

Putting $\lambda = i\omega$ into (7.42), we find

$$-\omega^2 + i\left[\gamma + \bar{\nu}\left(1 - e^{-i\omega\tau}\right) - \nu_1 n_0 e^{-\lambda\tau_c}\right]\omega + \gamma\bar{\nu}\left(1 - e^{-i\omega\tau}\right) = 0, \quad (7.43)$$

which is equivalent to the two equations (for the real and imaginary parts)

$$-\omega^2 - \bar{\nu}\omega\sin\omega\tau + \gamma\bar{\nu}\left(1 - \cos\omega\tau\right) = 0, \quad (7.44)$$

$$\gamma + \bar{\nu}\omega\left(1 - \cos\omega\tau\right) + \gamma\left(1 - \cos\omega\tau\right)\sin\omega\tau = 0. \quad (7.45)$$

Their solution determines the stability boundaries of the steady state in the parameter space and yields the frequency ω of the oscillations that start to grow as these boundaries are crossed.

The analysis of (7.44) and (7.45) reveals that they have an infinite number of solutions that correspond to the development of oscillations with frequencies lying near $\omega_k = 2\pi k/\tau$ with $k = 1, 2, 3, \ldots$. These oscillations correspond to spiking regimes with different numbers k of coherent molecular groups. Equations (7.44) and (7.45) have been solved numerically for different parameter values. Figure 7.20 displays the computed bifurcation diagram in the parameter plane (τ_c, ν_1).

At low values of the allosteric activation intensity ν_1, the steady state corresponding to absence of oscillations is stable. If we increase the parameter ν_1, going along the vertical direction in the bifurcation diagram, one of the instability boundaries will be crossed. Above this boundary, the steady state is unstable with respect to the development of periodic oscillations. Note that, once the first boundary is crossed, the system is no longer found in the steady state. Therefore, other instability boundaries of the steady state which lie above the first boundary are not relevant.

The different curves in Fig. 7.20 correspond to the onset of oscillations with different numbers k of coherent groups. The number of coherent groups in the first unstable mode depends on the parameter τ_c. Inside the intervals $0 \leq \tau_c/\tau \leq 0.34$ and $0.955 \leq \tau_c/\tau \leq 1$ the first crossed boundary corresponds to $k = 1$ and hence synchronous oscillations of the entire enzymic population are expected. Inside the intervals $0.34 \leq \tau_c/\tau \leq 0.4$ and $0.65 \leq \tau_c/\tau \leq 0.78$ spiking with three coherent groups ($k = 3$) is expected, etc.

The frequencies ω of the first unstable modes with different k at the respective bifurcation boundaries are shown in Fig. 7.21. We see that they are not exactly equal to $\omega_k = 2\pi k/\tau$. The actual frequencies ω are always somewhat larger than ω_k because the enzymes spend part of time in the ground state, waiting to bind a substrate molecule.

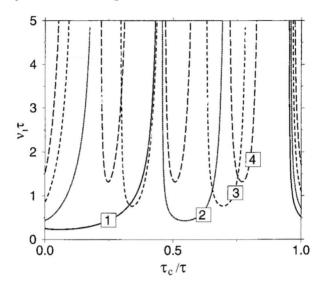

Fig. 7.20. The bifurcation diagram in the parameter plane (τ_c, ν_1). The steady state becomes unstable with respect to the onset of oscillations with k coherent groups when the curve marked as "k" $(k = 1, 2, 3, \ldots)$ is crossed while moving in the vertical direction. Other reaction parameters are $\nu_0 = 1, \gamma = 20$, and $N = 200$. Only the instability boundaries for $k = 1, 2, 3$ and 4 are plotted. (From [372])

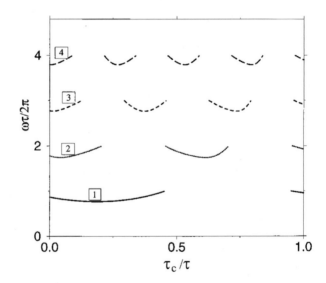

Fig. 7.21. Frequencies ω of oscillatory modes with different numbers k of enzymic groups. The parameters are the same as in Fig. 7.20; the parameter ν_1 is taken on the respective bifurcation boundaries shown in Fig. 7.20. (From [372])

The bifurcation diagram in Fig. 7.20 has two gaps near $\tau_c/\tau = 0.45$ and $\tau_c/\tau = 0.95$. A more detailed examination reveals that the oscillations develop inside these regions too, but they set on at much higher intensities of allosteric activation ν_1 and correspond to regimes with large numbers of coherent enzymic groups.

The behavior of the system above the instability boundary has been studied by numerical integration of the full nonlinear delay equations of the mean-field approximation. This investigation has shown that the instability indeed leads to the development of persistent oscillations with the frequencies and numbers of coherent groups determined by the linear stability analysis. However, the bifurcation is often subcritical and hysteresis is observed [372].

7.5 Further Reading

Dynamical clustering is possible in populations of globally coupled phase oscillators ("clocks") [373, 374]. Golomb et al. [375] have considered a system with the equations of motions

$$\dot{\phi}_i = \omega + f(\phi_i) - \frac{1}{N}\sum_{j=1}^{N} g(\phi_j), \qquad (7.46)$$

where f and g are continuous periodic functions, $f(\phi + 2\pi) = f(\phi)$ and $g(\phi + 2\pi) = g(\phi)$. In contrast to the Kuramoto model of phase oscillators (see Chapt. 6), the interaction terms in these equations do not depend on the phase ϕ_i of the driven oscillator (this situation is, for example, characteristic of Josephson-junction arrays). Numerical investigations have shown that such a system has coexisting dynamical attractors corresponding to various cluster partitions. Interestingly, it was noted that the apparently simplest case with $f(\phi) = A\sin\phi$ and $g(\phi) = \sigma\sin\phi$ is not generic and has special properties resulting from the presence of additional symmetries. The influence of noise has been further analyzed. Okuda [376] has considered the Kuramoto phase oscillators model containing the higher harmonics in the coupling function and found clustering in this system.

Populations of globally coupled limit-cycle oscillators, described by the amplitude equations, can also show dynamical clustering [377, 378]. The total number of clusters and their relative sizes depend on the initial conditions and the parameters of the system. However, in a remarkable contrast to the results of Kaneko for globally coupled logistic maps (Sect. 7.1), the maximum number of clusters is three here. The dynamical properties of clustered states in this system have been studied [379].

Globally coupled logistic maps represent the most extensively investigated system where dynamical clustering and complex collective dynamics are possible (see also [380–385]). Other examples of globally coupled chaotic systems that have been investigated include circle maps [386], sawtooth maps [387] and coupled Hamiltonian systems [388].

Recent experiments on arrays of globally coupled chaotic electrochemical oscillators [389] show that the dynamical clustering in such systems is similar to the clustering behavior in large ensembles of globally coupled Rössler oscillators and can be quantitatively characterized using the order parameters introduced in Sect. 7.2.

Mutual synchronization and dynamical clustering are observed in the brain. The synchronized firing activity of neurons in the central nervous system has been found by Singer and co-workers [390]. Buzsaki et al. [391] have experimentally observed that a network of neurons in the hippocampus of the rat oscillates at a multiple of the single-oscillator frequency, indicating dynamical clustering.

The brief discussion of a neural network model in Sect. 7.3 aimed only to illustrate the possibility of synchronization and clustering in ensembles with very complex individual elements, each representing a network. We used there an idealized model where the activity of an individual neuron is described by a continuous variable. In real biological neurons, sequences of short electric pulses (spikes) are generated. The activity $x(t)$ is understood as the spike rate, i.e. as the number of spikes generated per unit time by a neuron.

Though the models of neural networks, formulated in terms of activity rates $x_i(t)$ of individual neurons, are popular and provide some insight into the organization of information processing, they have serious deficiencies. The problem is that the communication code based on a temporal average is intrinsically slow. As recently pointed out by Gerstner [392], to perform an efficient time-averaging a neuron needs to count between 5 to 10 arriving spikes. Given the typical rates of cortical neurons, this would take a time between 100 and 500 ms. Experimental evidence [393] is however available that humans can classify complex visual scenes and make decisions already within 200 ms, i.e. within a time interval that may be shorter than the time needed by a single neuron to read the code!

This means that, in order to adequately describe the brain operation, models should be employed where the spiking activity of individual neurons is temporally resolved. Currently, much attention is attracted to neural models with integrate-and-fire (IF) neurons [394–396]. A network of IF neurons is described by the equations (e.g., see [397])

$$\dot{u}_i = a_i - b_i u_i + \varepsilon \sum_{j=1}^{N} g_{ij} \delta(t - t_j^*), \qquad (7.47)$$

where a_i is a positive constant and the coefficients b_i are positive for leaky neurons and zero for non-leaky neurons. If the matrix element $g_{ij} > 0$, the connection from neuron j to neuron i is excitatory; if $g_{ij} < 0$, the connection is inhibitory. When the variable $u_i(t)$ reaches 1, the ith neuron is said to fire a spike, and u_i is reset to the new value $u_i = 0$. This moment is marked as t_i^*.

The dynamics of an IF neuron ($b_i = 0$) is that its state variable steadily rises from $u_i = 0$ to $u_i = 1$, when the neuron fires. The firing can be accelerated or retarded, depending on the total signal arriving from other neurons in the network and described by the last term in (7.47). The coefficient ε specifies the strength of the interactions between the neurons.

The collective dynamics of IF neurons can be formulated in terms of equations for the population activity, obtained by instantaneous averaging of the states of all neurons. The evolution equations for such average activities have been suggested in different forms by several authors [398–401]. An important property of large populations of IF neurons is that they can quickly (within one firing cycle) respond to abrupt changes in the input [402–405].

Populations of interacting IF neurons with excitatory connections can undergo spontaneous transitions into locked synchronous states with simultaneous firing of all neurons [395, 396, 406]. Populations of IF neurons with inhibitory connections and time delays were considered by Ernst et al. [407], who have shown that several synchronous clusters are typically formed in such systems.

While the collective dynamics of IF neurons apparently captures important properties, characteristic of populations of real biological neurons, this model still represents a rough mathematical idealization and its predictions cannot be directly experimentally verified. Detailed models of "realistic" neurons have therefore been proposed (e.g., see [408]). These models incorporate important physical processes, responsible for the operation of a neural cell, and use several ordinary differential equations to describe a single neuron.

Golomb and Rinzel [409] have analyzed the collective dynamics of a large population of excitable "realistic" neurons with inhibitory coupling. The spontaneous formation of coherent phase-shifted clusters of neurons that sequentially excite the cycles of each another has been found in this study. For a recent review of synchronization phenomena in populations of "realistic" neurons, see Abarbanel et al. [410].

While neurons communicate by sending and receiving electrical pulses, other types of biological cells may communicate chemically, through the release of various chemical substances into the intercellular solution. In turn, the modification of the chemical composition of this solution can influence the chemical processes inside the cells. If chemical oscillations are taking place inside the cells and the solution is stirred, the global coupling of the chemical oscillators in individual cells is achieved.

Kaneko and Yomo [411, 412] have considered general models where cells represent chemical microreactors, each containing a complex network of chemical reactions. Such microreactors are coupled through the common medium. The reaction kinetics in a single cell is typically chaotic. The investigations have shown that dynamical clustering may easily occur in this uniform system, so that coherently acting groups of cells emerge. This phenomenon was used to explain the biological effect of isologous differentiation. From a general viewpoint, these studies provide a proof that dynamical clustering may develop in ensembles of elements with complex individual dynamics and thus confirm the results of our discussion of coupled chaotic neural networks in Sect. 7.3.

Despite its impressive advances, the molecular biology of the cell has for a long time remained based on the concepts of classical chemistry. The situation is, however, changing. In his recent article "The cell as a collection of protein machines: preparing the next generation of molecular biologists" [413] Alberts writes that "the entire cell can be viewed as a factory that contains an elaborate network of interlocking assembly lines, each of which is composed of a set of large protein machines". Like macroscopic mechanical machines invented by humans, they are made of highly coordinated moving parts. The effects of fluctuations are reduced and the operation of such machines can be described by equations of motion. Furthermore, "underlying this highly organized activity are ordered conformational changes in one or more proteins driven by nucleoside triphosphate hydrolysis (or by other sources of energy, such as an ion gradient)".

Thus, the cell is populated by an ensemble of active dynamical units whose operation is powered by the chemical energy. Their collective coordinated dynamics is responsible for the main cell functions. Hess and Mikhailov [371] have pointed out that self-organization inside living biological cells cannot represent a reduced copy of nonequilibrium pattern formation in macroscopic reaction–diffusion systems. The principal role is played here by collective coherent dynamics in molecular networks formed by communicating protein machines. Though the communication is based on the release and diffusion of small molecules, it can be very efficient. The estimates show that, as a result of pure diffusion, any two molecules inside a micrometer volume would collide each second!

The protein machines can execute various functions inside a cell. Some of them represent molecular motors moving along a filament or a nuclear acid strand. Other machines are involved in the proofreading of genetic information and in the production of new protein molecules. Enzymes are protein machines whose function is to perform chemical operations with single small molecules. Typically, they would split a molecule into two parts or sew two different substrates into a new product molecule. The concept of enzymes as molecular machines was introduced and discussed by McClare [414] and Blumenfeld and Tikhonov [415].

The internal dynamics of protein machines can be synchronized. In the experiment of Gruler and Müller-Enoch [416] (see also [417]) the turnover cycles of individual enzymes of the cytochrome P-450 dependent monooxygenase system were externally synchronized by applying repeated illumination pulses. Hess and Mikhailov have suggested [418] that mutual synchronization of the turnover cycles of enzymes can take place. In addition to the case of allosterical product-activated reactions, considered in Sect. 7.4, other principal reaction schemes were also analyzed. They included allosteric reactions with product inhibition [419] and non-allosteric reaction loops where a fraction of the products is converted back to the substrate [420]. Synchronization of turnover cycles is possible in all these reactions. It remains sufficiently robust against the influence of thermal intramolecular noise.

The reader might have already noticed analogies between the simple models, used to describe a single enzyme and an integrate-and-fire neuron. Indeed, in both cases the active units are modeled as automata. The difference is that in the case of an allosteric enzyme the arrival of a regulatory molecule does not speed up or decelerate the motion inside a cycle, but changes the probability of the stochastic cycle initiation influencing the waiting time to bind a substrate. The communication is therefore effective only when the target machine is in a certain internal state. This leads to significant differences in the collective coherent behavior in the two classes of models. The activatory enzymic interactions usually result in dynamical clustering, while product inhibition yields synchronous oscillations of the entire enzymic population.

8. Hierarchical Organization

Hierarchical organization is typical of complex living systems. Hierarchies of two kinds should be distinguished. Evolutionary hierarchies, such as genealogical trees, are memories of a branching process. The degree of kinship represents a natural measure of distance between elements in such trees, leading to the concept of an ultrametric space. Dynamical hierarchies are formed by coexisting structures of different levels that interact to determine the system's dynamics.

The simplest example of a system that builds up evolutionary hierarchies is provided by the Sherrington–Kirkpatrick model. This model describes an ensemble of interacting Ising spins, or a spin glass. When a control parameter is varied, a hierarchical structure develops through a sequence of bifurcations of equilibrium states in the model. Its statistical properties can be analyzed by considering the distribution of overlaps between realizations (replicas).

The stochastic dynamics of the Sherrington–Kirkpatrick model represents relaxation to one of its equilibrium states. In contrast to this, dynamical glasses have hierarchically organized multiple attractors which correspond to regular or chaotic oscillations. We show that an ensemble of globally coupled logistic maps forms a dynamical glass in its clustering regime. Hence, it provides another example of an evolutionary hierarchy.

The classical illustration of a dynamical hierarchy is fluid turbulence. The turbulent state is constituted by a set of interacting vortices of various sizes. Larger vortices can be viewed as consisting of many smaller enslaved ones. The intrinsic property of all such systems is the existence of a hierarchy of timescales corresponding to different levels. The structures of a higher hierarchical level have slower dynamics and thus do not directly interfere with the rapid internal dynamics of subordinate units. However, they can effectively guide their dynamics. As an example, we consider a dynamically structured swarm formed by a population of actively moving particles.

8.1 Hierarchies

Hierarchies are ubiquitous in nature. Quarks form elementary particles, which combine to produce atoms. Atoms give rise to molecules, which in turn constitute solid bodies or fluids. The interactions between them produce all the systems of the macroscopic world. This physical hierarchy extends further to planets, stars and galaxies.

The lowest level of the biological hierarchy is formed by elementary biochemical reactions that combine to produce complex chains. A sophisticated system of such reaction chains builds up a living cell. Interacting cells constitute a biological organism. Animals and plants form populations whose interactions determine all the processes in the biosphere.

The elementary unit of social structural hierarchies is a human individual. The hierarchical structure of a traditional society is based on kinship relations. Parents and their progeny form families. Groups of families with common ancestors build clans or tribes. Related tribes produce nations. In contrast to this, in a modern society kinship plays a less important role. Social groups and political parties reflect professional, cultural and economic differentiation in a society.

When economics is considered, we can again notice a complex hierarchical organization. A large industrial company is a conglomerate of various production divisions. Each of them has a number of manufacturing plants, which in turn consist of smaller manufacturing units, assembly lines and work teams.

The above examples refer to gross hierarchies whose existence is obvious. A detailed analysis would, however, also often reveal the hierarchical structure of various particular processes in biological organisms, ecological systems and human societies. Apparently, the hierarchical organization is not accidental. It must be essential for the functioning of complex living systems.

To understand the origins and the role of natural dynamic hierarchies, let us return to physics. Subsequent structural levels correspond here to the operation of different forces responsible for the interactions between elements. It can be noted that the forces acting at lower levels are significantly stronger. The hierarchy of structures has its parallel in a hierarchy of interactions – from extremely strong nuclear forces to relatively weak electromagnetic interactions and further to very weak gravitational effects.

The hierarchy of interactions makes physical systems *nearly decomposable.* If we could switch off the electromagnetic forces, we would see that the atomic nuclei would still exist, but the atoms and molecules would already cease to form. If gravitational forces are eliminated, the planets, stars and galaxies would be absent, but all lower levels of the structural physical hierarchy would be left intact.

The decomposability implies that, to describe structure formation at a certain hierarchical level, one needs to take into account only the forces operating at this particular level. For example, atomic nuclei are produced by strong nuclear interactions between protons and neutrons. The influence of electromagnetic forces can here be neglected. If we move one step higher and consider atoms and molecules, the dominant role is played by electromagnetic interactions. Such interactions operate in a system made of nuclei and electrons. Though the nuclei actually have complex internal structure, they can be viewed as simple particles at such a higher structural level. This becomes possible because electromagnetic forces are too weak to interfere with the internal organization of nuclei.

Note that the same forces can give rise to interactions of varying strength. The interactions between atoms and molecules are also essentially electrostatic, though they are much weaker than the interactions between the particles inside an atom. This is explained by the fact that the total electric charge of such composite particles is zero, and hence the principal electrostatic forces are almost compensated. The residual forces give rise to weak dipole–dipole or van der Waals interactions, which are responsible for the formation of liquids or solids.

The decomposability of physical systems plays a fundamental role. If all the interactions had the same strength, the separation into different structural levels would have not been possible. Then the whole Universe would have had just a single huge nondifferentiated structure.

Apparently, a similar decomposability underlies various biological and social hierarchies. The problem is how to compare and define the "strength" of biological or social interactions. Indeed, all physical forces are quantitatively well-defined by the respective laws. Their intensities are measured in the same units and can easily be compared. In contrast to this, no universal dynamical laws are available for biological and social systems. At most, we have here various phenomenological models that describe particular aspects of their behavior.

The strength of the interactions in a physical system determines the timescale of the processes resulting from such interactions. The stronger the interactions, the shorter the characteristic timescale of the respective process. Indeed, if an elastic string is more stiff and a stronger force is needed to expand it, the oscillation period of this string is smaller. The structural hierarchy of physical systems corresponds to a hierarchy of their timescales.

This suggests that we can estimate and compare the strength of chemical, biological or social interactions by looking at their characteristic timescales. Viewed from this perspective, the interactions between the biochemical reactions in a living cell are strong, since they lead to characteristic times of a millisecond. The physiological processes in a human body correspond to weaker interactions, since their characteristic times would typically lie in the range of seconds or minutes. The social dynamics of small human groups proceeds on the scale of hours or days, whereas large social groups evolve only on the scale of months or years.

The separation of timescales is important. Because of it, the slow processes of a higher structural level cannot directly interfere with the dynamics at lower levels. However, they can effectively guide this dynamics by setting conditions under which it takes place. The variables of a higher structural level can play a role as control parameters for the subordinated systems of a lower level. Their slow evolution induces instabilities and bifurcations in such systems, switching their dynamical behavior. On the other hand, low-level processes also cannot interfere with the high-level dynamics. These processes are so rapid that their influence on the slow dynamics of a high-level structure is simply averaged out.

Thus, biological and social systems are also nearly decomposable. The self-organization of structural units at any hierarchical level is determined by interactions with the timescale corresponding to this level. When higher levels are considered, the elements of a previous level can be viewed as simple objects described by a small number of relevant properties. Consequently, the theoretical understanding and mathematical modeling of emerging hierarchically organized patterns becomes possible.

From a general perspective, hierarchical organization provides a solution of an apparent contradiction between the complexity of dynamics and its stability and predictability. At a first glance, the collective dynamics of more complex systems, consisting of a larger number of various interacting components, should generally be expected to be more complicated and less predictable. If, however, a complex system is appropriately hierarchically designed, its behavior can still be quite regular. This aspect will be further discussed in Sect. 8.4.

So far we have been talking about *dynamical hierarchies*, all of whose structural levels correspond to actually existing dynamical objects. Looking around, we can, however, notice that apparently there are also other kinds of hierarchies, which are not directly related to any dynamics. Rather they reflect the evolutionary history of a particular system.

A good example of such an *evolutionary hierarchy* is provided by genealogical trees. If you wish to reconstruct the genealogy of your family, you must use your memory, ask relatives, work in archives and so on. In other words, you cannot deduce the genealogy by just examining your current social environment. Moreover, though several family generations often overlap, your more distant ancestors will not be physically present.

A genealogical tree establishes a relation of kinship, i.e. shows how genetically close to you are the family members. The kinship does not itself constitute any social interaction. In a modern society, you would probably not work together with your brothers and cousins in the same team. Neither are you expected to belong to the same political party or a cultural group. Hence, this hierarchy establishes a certain classification, which is not necessarily playing a role in the system's dynamics.

Of course, the genetical similarity can determine to a certain extent the physical properties of an individual. Some diseases are inherited and the occurrence of many others is genetically biased. Hence, by examining your family history you may deduce how probable would be a certain illness in your case. Nonetheless, an even better result would yield a straight genetic analysis of your chromosomes.

The structure of genealogical trees is complex. Its branches can meet and intersect. Indeed, you may marry one of your cousins. Other evolutionary hierarchies are usually more simple. An important example is the evolutionary hierarchy of biological species whose branches can only split or undergo extinction.

The knowledge of an evolutionary history leads to a natural classification in a system. However, sometimes the records are missing and our classification can only be based on a physical examination of the individual elements. Indeed, a genetic analysis can establish a degree of kinship between two individuals, even when they do not know their history or try to hide it. On the other hand, when Linneus produced in the 18th century a detailed hierarchical classification of plants, all his conclusions were based on a systematic comparative analysis of their biological structures. Only later was the theory of biological evolution constructed that explained the origin of this empirically established hierarchy.

Generally, even if a pattern of similarity between elements forms a hierarchy, this cannot be considered as a proof that the system has actually chronologically evolved through a sequence of structural bifurcations. However, a control parameter can often be identified whose gradual variation would lead the system through such a sequence and build up the observed hierarchical classification.

Sets endowed with a hierarchical classification structure are interesting mathematical objects. We can define *ultrametric distance* $d(A, B)$ between any two elements A and B of such a set as the minimal number of steps (nodes) one should go up the hierarchy to find a common ancestor of these two elements. For example, in the hierarchy shown in Fig. 8.1 the distance between elements A and B is equal to one, whereas the distance between A and C is equal to two. Thus defined, the ultrametric distance reflects the kinship relation between elements.

The properties of the ultrametric distance are different from the usual (Euclidean) distance between points. If we take three points A, B and C on a plane, these points form a triangle. The distances $l(A, B), l(B, C)$ and $l(A, C)$ between these points are just the lengths of the respective sides of this triangle. As we know from geometry, they always satisfy the inequality

$$l(A, C) \leq l(A, B) + l(B, C). \tag{8.1}$$

In contrast to this, the ultrametric distances between any three elements A, B and C satisfy a stronger inequality

$$d(A, C) \leq \max \{d(A, B), d(B, C)\}. \tag{8.2}$$

If we take elements A, B and C in the tree shown in Fig. 8.1, we find that the ultrametric distances between them are $d(A, C) = 2, d(A, B) = 1$ and $d(B, C) = 2$, so that the inequality (8.2) is indeed valid. It can easily be

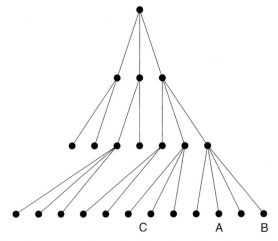

Fig. 8.1. Graphical representation of a hierarchy. The ultrametrical distance between elements A and B is equal to one, the distances between elements A and C or B and C are equal to two

checked that such an inequality will hold for any three elements from the lowest level of any branching tree.

Sets of elements where a distance is introduced are called metric spaces. Generally, we can choose as a distance any real nonnegative property defined for any pair of elements and obeying the triangular inequality (8.1). A metric space where the distance obeys the stronger inequality (8.2) is called an *ultrametric space*.

Historically, the concept of ultrametric spaces was first proposed in algebraic number theory and only later has its connection to hierarchical structures become clear. A good review of the related mathematical aspects has been given by Rammal et al. [421]. In an ultrametric space, all triangles are either equilateral or isosceles with a small base (third side shorter that the two equal ones, see (8.2)). Equivalently, we can say that the diameter of any ball is equal to its radius.

While this formal mathematical reasoning is nice, it does not have many practical implications. There is a one-to-one correspondence between ultrametric spaces and hierarchies. Any ultrametric set can be displayed as a tree in which the vertical axis reveals the ordering of objects into nested clusters of increasing radius. Hence, whenever one talks about studying physical processes in ultrametric spaces, this simply means that some processes taking place in a set of hierarchically ordered elements shall be investigated.

A different mathematical question related to hierarchies is how to specify the complexity of a given tree. It is clear that, for instance, the regular tree shown in Fig. 8.2 is less complex than the tree in Fig. 8.1. Intuitively, the complexity of a hierarchy seems to be related to its diversification, i.e. to the number of non-isomorphic subtrees found at each level. Ceccatto and Huberman [422] have formalized this property and suggested several quantitative complexity measures of hierarchically organized systems.

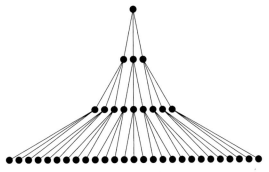

Fig. 8.2. A regular tree

8.2 The Sherrington–Kirkpatrick Model

The hierarchy of species is a result of a long and complex biological evolution. It was generally assumed that the development of hierarchies could only be observed in extremely complex systems, comprising a great number of various components. Today we know that this is not true. A simple automata model, proposed in 1975 by Sherrington and Kirkpatrick [423], already reveals an emergent hierarchy.

The model is formulated in terms of Ising spins σ_i, which can be viewed as automata with two discrete states $\sigma_i = \pm 1$. The energy of a system of N interacting spins is

$$E = -\frac{1}{2} \sum_{i,j=1}^{N} J_{ij} \sigma_i \sigma_j. \tag{8.3}$$

The interaction matrix is symmetric ($J_{ij} = J_{ji}$) and its elements can be positive or negative. They are chosen independently at random for any pair (i, j) with a Gaussian probability distribution,

$$p(J_{ij}) = \sqrt{\frac{N}{2\pi}} \exp\left(-\frac{N}{2} J_{ij}^2\right). \tag{8.4}$$

Once the elements of the interaction matrix are randomly chosen, they remain fixed. Hence, the system is characterized by the presence of quenched disorder. Self-interaction is absent (i.e. $J_{ii} = 0$ for any i).

We now introduce the stochastic dynamics of the Sherrington–Kirkpatrick (SK) model, assuming that the states of its N automata can randomly change at discrete time moments t. The updating algorithm can be described as follows. At any moment t we choose at random one of the spins i and flip it, so that the modified state of this automaton is $\sigma'_i = -\sigma_i(t)$. Next, we determine the energy difference $\Delta E = E' - E$ between the modified and the old patterns,

$$\Delta E = -\sum_{j=1}^{N} J_{ij}(\sigma'_i - \sigma_i(t))\sigma_j(t). \tag{8.5}$$

Depending on the sign and the magnitude of this difference, we either accept or reject such a flip. Namely, the modified state σ'_i is taken as the actual state $\sigma_i(t+1) = \sigma'_i$ of the spin i at the next moment $t+1$ with the probability

$$p(\Delta E) = \begin{cases} 1, \text{if } \Delta E < 0 \\ \exp\left(-\Delta E/\theta\right), \text{ if } \Delta E \geq 0 \end{cases}. \tag{8.6}$$

Hence, an attempted flip is always accepted if it leads to a decrease in the energy E of the total system. However, it is also accepted with a certain small probability if it leads to a higher energy. The parameter θ is interpreted as the temperature.

When $\theta \to 0$, the transitions leading to higher energies are forbidden and, therefore, the evolution of this system ends when a stationary state corresponding to a minimum of energy (8.3) is reached. Hence, its dynamics in this case represents relaxation to one of its stable stationary states. At a finite temperature θ, fluctuations are present and the system may also climb up the energy landscape. As a result, even after a long evolution there will be a certain probability $P(\{\sigma_i\})$ of finding this system in any state $\{\sigma_i\}$. It can be shown (see [424]) that the asymptotic equilibrium probability distribution reached under this stochastic dynamics is the Boltzmann thermal distribution

$$P(\{\sigma_i\}) = Z^{-1} \exp\left(-\frac{1}{2}\sum_{i,j=1}^{N} J_{ij}\sigma_i\sigma_j\right), \tag{8.7}$$

where Z is a normalization constant. (See, however, the ergodicity discussion below.)

Actually, the Metropolis evolution algorithm (8.6) is just one of the possible ways to introduce a stochastic dynamics that would ensure relaxation to thermal equilibrium with the probability distribution (8.7). Below in this section we focus our attention on the equilibrium statistical properties of the SK model, where the kinetics is irrelevant.

Originally, the SK model was introduced as a mean-field approximation for spin glasses with a finite interaction radius. However, its significance goes beyond particular applications in material science. For instance, this model may also be interpreted as describing a certain neural network (e.g. see [425]).

To begin the analysis of this model, let us first briefly discuss its behavior if all interactions are identical and attractive, i.e. $J_{ij} = J > 0$ for any i and j

(such spin systems are called *ferromagnetic*). In this case there are only two minima of energy, corresponding to the states where all spins are looking in the same direction, i.e. either $\sigma_i = 1$ for any i or $\sigma_i = -1$ for any i. At zero temperature $(\theta \to 0)$ they represent the actual equilibrium configurations. The system will evolve to one of such *pure configurations*, depending on its initial conditions.

At a low temperature, thermal fluctuations are present. They lead to occasional spin flips against the force direction (i.e. with $\Delta E > 0$). A sufficiently long sequence of such random flips may reverse all spin directions and move the system from one pure configuration to another. The length of this sequence will be of the same order of magnitude as the system size N. If $p \ll 1$ is the typical probability of a single flip, the probability of a transition between pure configurations can roughly be estimated as p^N. Hence, this probability vanishes in the limit $N \to \infty$ of an infinite system. This means that at low temperatures a large system will still be found in one of the states that are close to its pure configurations. We can picture this system as staying in one of its two energy valleys, corresponding to pure configurations and separated by infinite energy barriers. In the first valley the average magnetization is $< \sigma_i > \approx 1$, while in the second we have $< \sigma_i > \approx -1$.

When the interactions J_{ij} have random signs and magnitudes, the system has a large number of pure configurations corresponding to minima of energy (8.3). In these configurations the spins are not parallel and the total magnetization $\sum_i \sigma_i$ is close to zero. Such configurations are separated by energy barriers of various heights. When the limit $N \to \infty$ of an infinite system is considered, many of these barriers remain low. However, there is also a certain fraction of barriers whose height goes to infinity in this limit. The energy landscape is formed by deep valleys, such that to go from one valley to another an infinite barrier must be overcome. Inside each valley a great number of pure configurations, corresponding to different energy minima and separated by finite barriers of various heights, is found.

If thermal fluctuations are present, the system wanders between different spin configurations. However, it cannot leave a valley where it was initially located. This means that thermal equilibration takes place only within this particular valley. Thus, if we start from a certain initial condition α,

$$\sigma_i(0) = \sigma_i^{(\alpha)} \text{ for } i = 1, 2, \dots, N, \tag{8.8}$$

the system relaxes to an equilibrium probability distribution $P^{(\alpha)}(\{\sigma_i\})$ that generally depends on this initial condition. Indeed, the dynamical phase space of our system turns out to be divided into a set of disconnected domains (valleys). Since during its evolution the system remains confined to the domain that is fixed by the initial condition, it does not know what would happen outside it and is not influenced by this.

Note that many different initial conditions α correspond to the same valley. Ideally, it would have been better to enumerate the energy valleys and assign these indices to the respective probability distributions. The problem is, however, that we do not explicitly know the shapes and the sizes of these phase domains. Moreover, they would be different for different random choices of the interaction matrix J_{ij}.

Let us consider the time average

$$< \sigma_i >_t^\alpha = \lim_{T\to\infty} \frac{1}{T} \sum_{t=0}^{T} \sigma_i^\alpha(t), \qquad (8.9)$$

where $\sigma_i^\alpha(t)$ is the series of states of the spin i at subsequent discrete times t if the evolution starts from the initial condition α. After a transient, the stochastic motion of the system would represent random thermal wandering through the energy valley determined by the initial condition. Therefore, this time average will coincide with the statistical average $< \sigma_i >^\alpha$ that should be calculated using the equilibrium probability distribution $P^{(\alpha)}(\{\sigma_i\})$ which corresponds to the chosen initial condition. The memory of an initial condition implies that the stochastic dynamics of the SK model is *non-ergodic*.

It must be emphasized that the absence of ergodicity is found only in the limit $N \to \infty$. If the system size N is finite, the barriers between different valleys have finite heights. Therefore, occasional transitions between the valleys are possible. If we wait long enough, the system will visit at random all energy valleys and the initial condition will be forgotten. In this case time averages coincide with statistical averages calculated with the common probability distribution (8.7). However, the time needed for a system to forget its initial condition exponentially increases with the system size and becomes astronomically long for relatively small sizes. On realistic timescales, the above-described nonergodic behavior will actually be found.

A characteristic statistical property of a ferromagnetic system is its magnetization $m =< \sigma_i >$. In spin glasses, this property is not useful, because the quantity $< \sigma_i >$ is different for different spins i and, furthermore, it is sensitively dependent on the choice of a particular random realization of the interaction matrix J_{ij}. Thus, $< \sigma_i >$ vanishes after averaging over all realizations of J_{ij}.

The statistical description of spin glasses is constructed in terms of *overlaps* between different replicas. Let us take two identical copies of the system (8.3) of large size $N \to \infty$ and choose different initial conditions α and β for both of them. If we start a parallel stochastic evolution of these two systems, this would generate two different "trajectories" $\{\sigma_i^\alpha(t)\}$ and $\{\sigma_i^\beta(t)\}$. We shall call them two different *replicas* α and β of the same system. The overlap $q^{\alpha\beta}$ between two replicas is defined as

$$q^{\alpha\beta} = \lim_{T\to\infty} \frac{1}{NT} \sum_{t=0}^{T} \sum_{i=1}^{N} \sigma_i^\alpha(t)\sigma_i^\beta(t). \qquad (8.10)$$

If the two replicas coincide, i.e. $\sigma_i^\alpha(t) = \sigma_i^\beta(t)$ for any i and t, the overlap between them is $q^{\alpha\beta} = 1$. On the other hand, if $\sigma_i^\alpha(t) = -\sigma_i^\beta(t)$ for any i and t, we have $q^{\alpha\beta} = -1$. Generally, overlaps lie in the interval $-1 \le q^{\alpha\beta} \le 1$. Note that (8.10) is often written in a different, but equivalent form, i.e. in terms of statistical averages (e.g. see [426]).

Suppose that we have generated at random many different replicas α and β and computed the overlaps $q^{\alpha\beta}$ between all of them. Then we can determine the distribution $P_J(q)$, such that $P_J(q)dq$ yields the probability of finding an overlap in the interval from q to $q + dq$. The index J refers here to the fact that all replicas belong to the same system with a particular interaction matrix J_{ij}. This matrix is randomly chosen and the respective probability distribution is given by (8.7). By additional averaging over all possible realizations of the interaction matrix, we obtain the mean overlap distribution $P(q) = \overline{P_J(q)}$.

What is the form of the overlap distribution $P(q)$? When Edwards and Anderson [427] introduced the replica description, they believed that all overlaps must be equal, so that $q^{\alpha\beta} = q_1$ for any α and β. Using this assumption of replica symmetry, Sherrington and Kirkpatrick have constructed a solution of their model [423]. The subsequent analysis has, however, revealed that, though the solution with replica symmetry exists, it does not always correspond to thermal equilibrium of the SK model. For instance, the calculations show that the entropy for this solution becomes negative at sufficiently low temperatures, which is unacceptable.

Thus, the true solution of the SK model must be characterized by a range of various possible overlaps. Note that this implies the breakdown of replica symmetry. Though all replicas seem to be statistically equivalent, some of them are "more similar" than the others! In 1980 Parisi constructed an approximate analytical solution of the SK model that was based on the idea of spontaneous replica symmetry breaking [428]. The replicas originally appeared in the solution only as formal mathematical constructions. Later, their connection to various statistical realizations of the system was established [429]. The detailed analysis of replica-symmetry breaking in the SK model can be found in the books [425, 426, 430].

Since the distribution of overlaps in absence of external fields is always symmetric, i.e. $P(-q) = P(q)$, it is sufficient to consider it only for positive overlaps q. The general result of Parisi is that the overlap distribution in the SK model below the critical temperature $\theta_c = 1$ has the form

$$P(q) = p(q) + p_1 \delta(q - q_1), \tag{8.11}$$

where $p(q)$ is a smooth function that vanishes for $q > q_1$. Thus, there is a finite probability p_1 of finding the maximum possible overlap q_1 between any two randomly chosen replicas. However, with a certain probability density $p(q)$ the overlap q between them may also take any smaller value.

Near to the critical temperature θ_c, the distribution parameters can be analytically calculated. One finds that $p(q) \approx \frac{1}{2}$, $q_1 \approx \tau$ and $p_1 = 1 - \frac{1}{2}\tau$, where $\tau = \theta_c - \theta$ and $\tau \ll 1$. On the other hand, it is known that $q_1 \approx 1 - \alpha_0 (\theta/\theta_c)^2$ at low temperatures (α_0 is a numerical constant). Some estimates are also available in the intermediate temperature range (see [428]).

The structure of the distribution (8.11) can be understood if we recall the existence of disconnected energy valleys in the SK model at $N \to \infty$. Suppose that we have a large set of "boxes", each containing an identical copy of the SK system. We choose independent random initial conditions in each of the boxes and simultaneously start the evolution in all of them. The trajectory of the system in one of the boxes yields one of the replicas of the SK model.

Some of the trajectories would belong to the same energy valley. All such trajectories correspond to relaxation to the same equilibrium statistical distribution. After a certain time, the difference in their initial conditions will be forgotten. Hence, overlaps between all replicas belonging to one valley must be equal. Moreover, they will be the same as the self-overlap of a replica in this valley, defined as

$$q_{\text{self}} = \lim_{T \to \infty} \frac{1}{NT} \sum_{t=0}^{T} \sum_{i=1}^{N} \sigma_i^\alpha(t) \sigma_i^\beta(t + T_1), \qquad (8.12)$$

where the time T_1 is chosen to be long enough, i.e. much longer than the correlation time during which the memory of the initial conditions is still present. Though self-overlaps might generally be different for different energy valleys, in the SK model they are all equal. Actually, q_{self} is also independent of a particular random choice of the interaction matrix J_{ij}. Hence, the self-overlap q_{self} is an important statistical property of the SK system which can depend only on its temperature θ. Self-overlaps for spin glasses were introduced by Edwards and Anderson [427]. Note that q_{self} yields the maximum value of overlaps q between any two replicas.

If we generate at random two replicas α and β, there is a certain finite probability that they both would belong to the same energy valley and, therefore, we would find $q^{\alpha\beta} = q_{\text{self}}$. Hence, the distribution of overlaps $P(q)$ will always contain a delta-function at $q = q_{\text{self}}$. This explains the presence of a delta-function term in (8.11). As implied by such an interpretation, p_1 represents simply the probability that two arbitrarily generated replicas would belong to the same valley. Furthermore, we see that $q_1 = q_{\text{self}}$.

The smooth part $p(q)$ of the overlap distribution (8.11) corresponds to the overlaps between replicas belonging to different energy valleys. It is the most striking property of the Parisi solution that the overlaps between such replicas can take *any* value in the interval $(0, q_{\text{self}})$. Though the replicas seem to be statistically equivalent (and indeed have equal self-overlaps), they are different with respect to the degree of similarity between them. If we fix a particular replica and start to randomly generate other replicas, we would be able to find replicas with any possible degree of similarity to the fixed one. Thus, the set of all possible replicas can be ordered, based on their overlaps.

These theoretical conclusions are supported by direct Monte Carlo simulations of the SK model. Figure 8.3 shows overlap distributions $P(q)$ computed for systems of different sizes $N = 32, 64, 128$ and 192 at temperature $\theta = 0.4\theta_c$ [431]. For comparison, the dotted line displays the approximate analytical solution at this temperature in the limit $N \to \infty$.

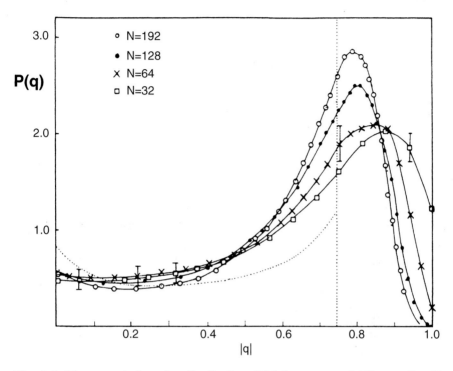

Fig. 8.3. The computed overlap distributions $P(q)$ for systems of different sizes N and the analytical prediction (dotted line) for the SK model at temperature $\theta = 0.4\theta_c$ (from [431])

We see that the computed distributions have maxima near the largest possible overlaps. As the system size increases, these maxima become more narrow and their center moves to the left, closer to the location $q = 0.744$ of the delta function in the analytical solution. Moreover, it can be noticed that the distribution includes a plateau at intermediate overlaps, whose height is in good agreement with the analytical prediction. An interesting special feature of the approximate analytical solution at $\theta = 0.4\theta_c$ (absent near the critical temperature) is the presence of an upturn at $q \to 0$. The numerical data shows a similar behavior at small overlaps, though the upturn is less pronounced here.

The presence of a continuous component in the overlap distribution $P(q)$ indicates that the set of replicas belonging to different energy valleys can be gradually ordered with respect to their overlaps. However, we do not yet know whether this set is hierarchically organized.

If some set is hierarchically organized, one can always define an ultrametric distance $d(A, B)$ between any two elements A and B (see Sect. 8.1). This distance must satisfy the inequality (8.2), i.e. for any three elements A, B and C the condition $d(A, C) \leq \max \{d(A, B), d(B, C)\}$ should be realized. Thus, to prove that replicas are hierarchically ordered it would be enough to show that some distance between replicas satisfying the ultrametric condition (8.2) can be defined.

The overlaps $q^{\alpha\beta}$ between replicas measure the degree of similarity between them. In contrast to this, the distance $d(\alpha, \beta)$ between two replicas α and β must measure how dissimilar they are. We can define the distance as some monotonously decreasing function of the overlaps, i.e.

$$d(\alpha, \beta) = f(q^{\alpha\beta}). \tag{8.13}$$

The actual form of function f is irrelevant. We should only require that $f(q_1) = 0$, so that the distance between any two replicas belonging to the same valley is zero.

If the replicas are hierarchically organized, the distance between any three replicas α, β and γ will satisfy the condition $d(\alpha, \gamma) \leq \max \{d(\alpha, \beta), d(\beta, \gamma)\}$. Therefore, if the distances are uniquely determined by respective overlaps, these overlaps must obey the condition

$$q^{\alpha\gamma} \geq \min \{q^{\alpha\beta}, q^{\beta\gamma}\}. \tag{8.14}$$

This inequality means that *the two smallest overlaps between any three replicas should always be equal.*

Thus, to prove the hierarchical organization one should consider the joint probability distribution[1]

$$P(q_1, q_2, q_3) = \sum_{\alpha,\beta,\gamma} \delta(q^{\alpha\gamma} - q_1)\delta(q^{\alpha\beta} - q_2)\delta(q^{\beta\gamma} - q_3), \tag{8.15}$$

which specifies the probability density of finding overlaps q_1, q_2 and q_3 between any three arbitrarily chosen replicas α, β and γ. If replicas are hierarchically organized, the probability density $P(q_1, q_2, q_3)$ must always be zero if two minimal overlaps are not equal.

It can be shown that the replica-symmetry breaking solution of the SK model yields the distribution $P(q_1, q_2, q_3)$ which indeed satisfies this ultrametric condition (see [421]). Numerical tests of ultrametricity in the SK model have also been performed. Bhatt and Young [432] made Monte Carlo simulations of the SK model at temperature $\theta = 0.6\theta_c$. The statistics of

---

[1] We use here q_1 to denote one of the arguments of the joint probability distribution. It should not be confused with a similar notation used in (8.11) to denote a parameter of the overlap distribution.

overlaps in randomly generated replica triads was accumulated. This allowed the determination of the probability distribution $H(\Delta q)$ of the difference Δq between the two smallest overlaps ($\Delta q = q_{\text{mid}} - q_{\text{min}}$) for a fixed larger overlap value $q_{\text{max}} = 0.5$.

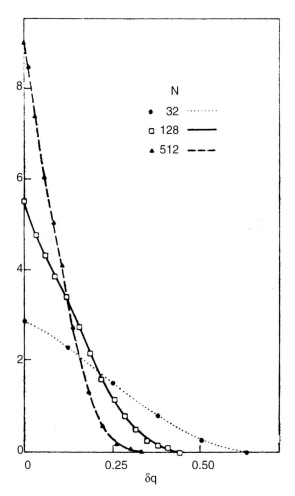

Fig. 8.4. Probability distributions $H(\Delta q)$ of the difference between the two smaller overlaps for a fixed larger overlap $q = 0.5$ in systems of different sizes N at temperature $\theta = 0.6\theta_c$ (from [432])

Figure 8.4 shows such computed distributions for three different system sizes $N = 32, 128$ and 512. We see that the distributions have maxima at $\Delta q = 0$ and the probability $H(\Delta q)$ vanishes for relatively large differences Δq. As the size N increases, the maxima become more narrow. They seem to approach in the limit $N \to \infty$ the singular distribution $H(\Delta q) = \delta(\Delta q)$, which is expected if the ultrametricity condition is satisfied.

Thus, replicas of the SK model are arranged into a hierarchical tree. The ultrametrical distance between these replicas can be defined as a function (8.13) of the overlaps between them. The distance $d^{\alpha\beta}$ can further be interpreted as the height to which one should go vertically up in this hierarchy to find common "ancestors" of two replicas α and β. Thus, the hierarchical organization of the SK model may be viewed as resulting from a certain branching process (see [421] for a detailed discussion).

The hierarchy which we find in the SK model is not principally different from the hierarchy of species, which is a consequence of biological evolution. The difference is that in biology the ancestors of all currently observed species have actually existed some time ago and the branching process has indeed taken place. The branching process of the SK system is virtual and, generally, has no relation to its dynamics (which represents relaxation to equilibrium states). Under certain conditions, this evolution can nonetheless be realized as some temporal development. If we take temperature θ as a control parameter, start from a temperature above θ_c and gradually decrease it, we would see that the hierarchical organization emerges in the course of time.

The hierarchical organization of replicas reflects a hierarchical structure of the energy landscape of the SK model. Any replica belongs to one of the energy valleys which are separated by infinite barriers in the limit $N \to \infty$. As already noted, all replicas from the same valley have the maximum possible overlap q_1 between them and are therefore separated by a zero ultrametric distance $d = f(q_1) = 0$. Hence, all these replicas correspond to the same node of the hierarchy. We see that the physical elements forming the hierarchy are, in effect, different energy valleys of the system.

The existence of a hierarchically organized set of valleys, separated by infinite barriers, tells us that the energy landscape of the SK model must be extremely complex. A detailed analysis shows, furthermore, that the structure of every single energy valley is also very complicated. It turns out that it is split into a set of smaller subvalleys that are, however, separated only by *finite* barriers of various heights. Many of these barriers will still be much larger than the typical energy of thermal fluctuations. Therefore, if a system falls into a subvalley separated by such a barrier, it would spend a long time sitting inside it. Only when a sufficiently strong fluctuation occurs may the system leave such a subvalley and move to another one, which corresponds to a deeper energy minimum. Consequently, the relaxation process taking place in a single copy of the SK system at $\theta < \theta_c$ should be very slow. Before reaching thermal equilibrium inside a valley, it would wander for a long time over a set of subvalleys. Such slow nonexponential relaxation is another characteristic property of the considered model.

8.3 Replica-Symmetry Breaking in Dynamical Glasses

A glass is a solid that lacks crystalline structure. In contrast to crystals, glasses have many equilibrium states, all of which are disordered. Thus, the actual spatial arrangement of atoms in a glass depends on its history, i.e. on the initial conditions. Spin glasses are solids with frozen magnetic disorder. Atoms form here a periodic crystalline structure, but their magnetic moments (spins) are irregularly oriented. The Sherrington–Kirkpatrick model, which we have considered in the preceding section, describes an idealized spin glass where the radius of interactions between spins is infinite. All glasses are equilibrium physical systems. Therefore, their dynamics consists of relaxation to some equilibrium stationary states that represent the only possible attractors. A special property of glasses is that they have a great number of various stable stationary states whose attraction basins are interwoven into a complex web.

Generally, attractors of dynamical systems may also correspond to periodic, quasiperiodic or chaotic oscillations. Moreover, the same system can have a large number of attractors. Thus, depending on the initial conditions, the system will approach different asymptotic dynamical states corresponding to various attractors. These dynamical states represent possible persistent motions of the system.

Systems with such properties can be described as dynamical glasses. Below in this section we show that replica symmetry may be broken in dynamical glasses. Our analysis is based on the numerical studies [433, 434] that were performed for a system of globally coupled logistic maps.

The system of globally coupled logistic maps has been already introduced in Sect. 7.1. It is described by the set of equations

$$x_i(t+1) = (1 - \epsilon)f(x_i(t)) + \frac{\epsilon}{N} \sum_{j=1}^{N} f(x_j(t)) , \qquad (8.16)$$

where $f(x) = 1 - ax^2$, the parameter ϵ specifies the coupling strength, $i = 1, 2, \ldots , N$ and N is the system size. The control parameter a is chosen below in such a way that the dynamics of an individual map is chaotic.

When coupling is absent ($\epsilon = 0$), the motions of single maps are not correlated. On the other hand, when ϵ is sufficiently high, the dynamics of all maps becomes strictly synchronous while remaining chaotic. We have shown in Sect. 7.1 that the transition to full chaotic synchronization is preceded by a wide interval of ϵ where dynamical clustering is observed. In this interval, the system builds a number of coherent clusters. The states of elements inside a cluster are identical. Depending on the initial condition, the system can evolve to different cluster partitions at the same parameter values. These partitions differ not only in the number of clusters, but also in their relative sizes. Every cluster partition has its own dynamics, which may be chaotic, quasiperiodic or periodic with different periods. These cluster partitions correspond to various attractors of the system.

The replicas in this system represent its dynamic realizations corresponding to different initial conditions. Thus, any replica is simply a certain trajectory $\{x_i(t)\}$ of the entire ensemble. To compare the replicas, it is convenient to transform them into binary sequences. At any discrete time t for any element i we assign a binary number $\sigma_i(t)$, such that $\sigma_i(t) = 1$, if $x_i(t) \leq x^*$ and $\sigma_i(t) = -1$, if $x_i(t) > x^*$, where

$$x^* = \frac{1}{2a}(-1 + \sqrt{1 + 4a}) \tag{8.17}$$

is the fixed point of a single logistic map. Hence, any replica α is encoded as a certain spin chain configuration $\{\sigma_i^{(\alpha)}(t)\}$. The overlap $q^{\alpha\beta}$ between two replicas α and β, yielded by two different randomly chosen sets of initial conditions $\mathbf{x}^{(\alpha)}(0)$ and $\mathbf{x}^{(\beta)}(0)$ for the same parameters a and ϵ, can then be computed as

$$q^{\alpha\beta} = \frac{1}{NT} \sum_{t=0}^{T} \sum_{i=1}^{N} \sigma_i^{(\alpha)}(t_0 + t)\sigma_i^{(\beta)}(t_0 + t) . \tag{8.18}$$

The time t_0 is chosen large enough to assure that the transient is discarded.

Two periodic orbits can be identical up to a certain time lag (i.e. a phase shift). If we directly apply the definition (8.18), different overlaps depending on the phase shift will be in this case detected. To eliminate this effect, we can additionally maximize $q^{\alpha\beta}$ with respect to possible phase shifts. It was empirically found that it was sufficient to consider only possible phase shifts in the range from 2 to 32, because periodic trajectories with higher periods were practically absent. For chaotic trajectories such optimization was not actually needed if T was long enough. Using the above definition, it was found that in the case of one-band chaos (which appears for the individual map near $a^* = 1.56$), the binary sequence for a chaotic replica changes sign in an apparently uncorrelated fashion with respect to a second replica (be this periodic or chaotic) and always returns an almost zero value of the overlap in the limit of long averaging times T. For two-band chaos and for periodic orbits, the overlap (8.18) is not vanishing and is well defined, provided that the maximization with respect to phase shifts is performed for the periodic orbits. Quasiperiodic trajectories are rare and were not observed for the parameter values considered.

Special attention should be paid to the preparation of initial condition. They must be chosen at random, in such a way that their ensemble is uniformly spanning the phase space of the dynamical system. At first glance, such "maximally random" choice of initial conditions is done if the initial state $x_i(0)$ of any map $i = 1, \ldots, N$ is chosen independently at random from the interval between -1 and 1. However, the resulting ensemble of initial conditions would be strongly biased.

To demonstrate this, we note that the set (8.16) of evolution equations for the maps can also written as

$$x_i(t+1) = 1 - a[(1-\epsilon)x_i^2(t) + \epsilon m(t)] , \qquad (8.19)$$

where

$$m(t) = \frac{1}{N} \sum_{i=1}^{N} x_i^2(t) \qquad (8.20)$$

is the synchronizing field that acts on any given map and is collectively produced by the whole system. Let us consider statistical properties of the synchronizing field at the initial time moment $t = 0$. If the initial states $x_i(0)$ of all maps are random and independent, this field represents a sum of a large number of independent random variables and should obey for $N \to \infty$ a Gaussian probability distribution

$$p(m) = \frac{1}{2\pi\sigma} \exp\left[-\frac{(m-\overline{m})^2}{2\sigma}\right] \qquad (8.21)$$

where $\overline{m}$ is the mean value of the field $m(0)$ and σ is its mean-square statistical variation,

$$\overline{m} = \frac{1}{N} \sum_{i=1}^{N} < x_i^2(t) > = \frac{1}{2N} \sum_{i=1}^{N} \int_{-1}^{1} x^2 dx = \frac{1}{3} , \qquad (8.22)$$

$$\sigma = < (m-\overline{m})^2 > = \frac{1}{N^2} \sum_{i=1}^{N} (< x_i^4(0) > - < x_i^2(0) >^2) = \frac{4}{45N} . \qquad (8.23)$$

Thus, in the limit $N \to \infty$ the initial synchronizing field $m(0)$ approaches a constant value, independent of the realization. For large N it shows fluctuations of order $1/\sqrt{N}$. We see that, though the initial conditions for individual maps were randomly generated, the initial collective properties of the ensemble become very similar for large system sizes for any statistical realization. Therefore, this set of initial conditions can be described as the *fixed-field ensemble*.

To eliminate the bias, the procedure used for random generation of initial conditions must be modified. For each realization, we first choose at random a parameter ξ from the interval between 0 and 1. Then the initial states $x_i(0)$ of all individual maps in the system are independently drawn from the interval $(-\xi, \xi)$. The resulting set of initial conditions is called the *random-field ensemble*.

To analyze the distribution of overlaps, their histograms were constructed. To obtain such a histogram, the interval $[0,1]$ of possible overlaps q was divided into equal boxes of width 0.004 and the numbers of overlaps $q^{\alpha\beta}$ between randomly generated pairs (α, β) of replicas, which were found lying in different boxes, have been counted. Averages over 2×10^4 replicas have been performed. Systems with sizes of up to $N = 8192$ have been analyzed [433, 434].

Figure 8.5 shows normalized histograms $P(q)$ computed for $a = 1.3$ and $\epsilon = 0.15$ and two system sizes, using the random-field ensemble of initial

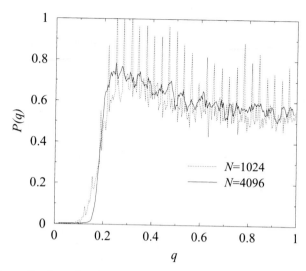

Fig. 8.5. Normalized overlap distributions $P(q)$ for globally coupled logistic maps with $a = 1.3$, $\epsilon = 0.15$ and different system sizes. (From [434])

conditions. For these parameter values, all attractors represented different two-cluster partitions. We see that orbits with greatly varying degrees of similarity, corresponding to the overlaps in the interval from approximately 0.1 to 1, can be found here. Moreover, the increase in the size N of the system from 1024 to 4096 leads only to some smoothing of the overlap distribution, retaining its general properties. Note that when $a = 1.3$ an individual logistic map is periodic.

In Fig. 8.6 we give another example of the overlap distributions, corresponding to the parameters $a = 1.55$ and $\epsilon = 0.1$. Now the behavior of an individual map is chaotic and we are close to a boundary of transition from dynamical clustering to the "turbulent" regime where the states of individual maps are no longer synchronized. In contrast to Fig. 8.5, the overlap distribution has a more complicated form in this case and replicas with overlaps inside the entire variation interval from 0 to 1 can be found here. Again, we see that the distribution does not strongly depend on the size of the system.

Thus, this numerical study provides evidence that the system of globally coupled logistic maps in the dynamical clustering regime indeed has a broken replica symmetry – a property that is characteristic of equilibrium spin glasses, such as the SK model. Note that this result essentially depends on the choice of the ensemble of initial conditions. If, instead of the random-field ensemble, a different set of initial conditions corresponding to the fixed-field ensemble is chosen, the replica symmetry apparently recovers [433, 434] in the limit of large system sizes N. Of course, this is simply a consequence of the fact that all initial conditions become increasingly similar in the statistical sense in the limit $N \to \infty$ when the fixed-field ensemble is chosen.

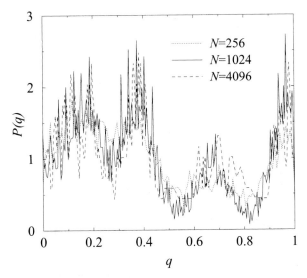

Fig. 8.6. Normalized overlap distributions $P(q)$ for $a = 1.55$ and $\epsilon = 0.1$ for different system sizes. (From [434])

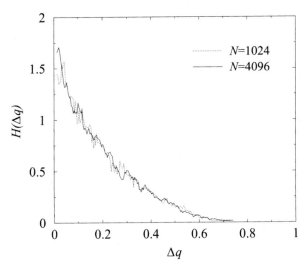

Fig. 8.7. Distributions $H(q)$ for systems of different sizes. The same parameters as in Fig. 8.6. (From [433]).

To test ultrametricity in the dynamical clustering regime of globally coupled logistic maps, the parallel evolution of many triads of replicas α, β and γ was then realized and overlaps between each pair of them were computed. The obtained numerical data was used to construct the probability distribution $H(\Delta q)$ over differences between two minimal overlaps in a triad. Figure 8.7

displays the distributions $H(\Delta q)$ for the same parameter values and system sizes as in Fig. 8.6.

We see that the distributions are broad and almost do not depend on the system size. Hence ultrametricity is absent in the considered system.

Though replica-symmetry breaking is a necessary condition for nontrivial overlap distributions, it does not imply exact ultrametricity, which is a more demanding condition. Deviations from exact ultrametricity are known and have been discussed for spin glasses [457].

8.4 Fluid Turbulence

The dynamical glasses, which we have considered in the previous section, have a large number of attractors that can be hierarchically ordered. The dynamics, corresponding to a particular attractor of a glass, can, however, be simple (e.g. periodic). Examination of a single trajectory of this system would not reveal any sign of its hierarchical organization.

There is, however, a different class of systems where the dynamics corresponding to an individual attractor is hierarchically organized. This dynamics is accompanied by the formation of a hierarchy of interacting structures with largely different timescales. The structures of a higher level are slow and control (or *enslave*) the dynamics of lower-level structures in the system.

Such dynamical hierarchies play an important role in social and biological processes. However, the internal complexity of a system is not a prerequisite for the development of a hierarchical organization of its dynamics. A classical example of hierarchical dynamics is provided by the phenomenon of hydrodynamical turbulence. The viscous flows of incompressible fluids are described by the Navier–Stokes equation for the local velocity field $\mathbf{v}(\mathbf{r}, t)$, i.e.

$$\frac{\partial \mathbf{v}}{\partial t} + (\mathbf{v}{\cdot}\nabla)\,\mathbf{v} = -\frac{1}{\rho}\nabla p + \nu\nabla^2\mathbf{v} + \mathbf{f}(\mathbf{r}, t), \qquad (8.24)$$

where ρ is the fluid density, ν is the viscosity constant, $\mathbf{f}(\mathbf{r}, t)$ is an external force field and p is the local pressure. Without the viscous term, (8.24) had already been derived in 1755 by L. Euler. The viscosity effects were incorporated into (8.24) in 1827 by C. L. Navier and a systematic derivation of this equation was given in 1845 by G. G. Stokes.

The second term on the left-hand side of (8.24) can be transformed using the identity

$$(\mathbf{v}{\cdot}\nabla)\,\mathbf{v} = \frac{1}{2}\nabla\left(\mathbf{v}^2\right) - [\mathbf{v}\,\mathbf{rot}\,\mathbf{v}], \qquad (8.25)$$

where the brackets denote a vector product. This transformation yields

$$\frac{\partial \mathbf{v}}{\partial t} + \frac{1}{2}\nabla\left(\mathbf{v}^2\right) = [\mathbf{v}\,\mathbf{rot}\,\mathbf{v}] - \frac{1}{\rho}\nabla p + \nu\nabla^2\mathbf{v} + \mathbf{f}(\mathbf{r}, t). \qquad (8.26)$$

By applying the differential operator **rot** to both sides of (8.26) and taking into account the incompressibility condition $\nabla \cdot \mathbf{v} = 0$, the pressure p is excluded and thus a closed dynamical equation for the velocity field is obtained,

$$\frac{\partial}{\partial t}\mathbf{rot\ v} = \mathbf{rot}\ [\mathbf{v\ rot\ v}] + \nu\nabla^2\left(\mathbf{rot\ v}\right) + \mathbf{rot\ f}(\mathbf{r}, t). \qquad (8.27)$$

This simple dynamical equation, which contains only one nonlinear term (of convective origin), describes the vast range of complex hydrodynamical flows and is of immense importance. It has been investigated over the last two centuries.

Taking a tube and starting to pump fluid through it, one finds that, beginning from a certain critical pumping rate, the fluid flow becomes strongly irregular, i.e. *turbulent*. Alternatively, the turbulent state can be reached by stirring the fluid. Such stirring can be achieved, for instance, by the application of a randomly varying volume force field $\mathbf{f}(\mathbf{r}, t)$ with some large length scale L and a large time scale T.

A developed turbulent flow is characterized by the presence of velocity vortices with many different length and timescales. The vortices of smaller sizes are involved in the motion produced by larger vortices. The system of interacting vortices forms a hierarchy. The dynamics of this system is chaotic and only its statistical description is therefore meaningful.

A complete theory of developed hydrodynamical turbulence is not available. It was, however, shown in 1941 by A. N. Kolmogorov that the principal statistical properties of such hierarchically organized spatiotemporal chaos can be determined by simple scaling considerations [435].

Every level in this dynamical hierarchy is associated with a certain characteristic length l that represents the typical size of a vortex belonging to the respective level. The velocity field $\mathbf{v}(\mathbf{r}, t)$ is a sum of contributions corresponding to vortices of various sizes. Hence, it includes random pulsations $\mathbf{v}_l$ with different lengths l, i.e.

$$\mathbf{v}(\mathbf{r}, t) = \sum_l \mathbf{v}_l(\mathbf{r}, t). \qquad (8.28)$$

Note that $\mathbf{v}_l(\mathbf{r}, t)$ can be viewed as the component of the velocity field generated by Fourier modes with the characteristic wavenumber $k = 2\pi/l$. Our task is to estimate typical magnitudes of the velocity pulsations $\mathbf{v}_l$.

The analysis shows that viscosity effects become important only at very short length scales $l \ll l_0$ (we shall determine l_0 below). At larger length scales, the fluid viscosity can be neglected. Since viscosity is responsible for the dissipation of energy, this implies that energy will be dissipated only at small scales $l \ll l_0$.

To maintain turbulence, energy must be continuously supplied. This is done through stirring, resulting from the action of the forces $\mathbf{f}(\mathbf{r}, t)$. We assume that these forces are characterized by a characteristic length scale $L \gg l_0$. They induce large-scale fluid motions with a characteristic velocity V and length L.

Through stirring, energy is pumped into the system. The pumping rate can be specified by introducing the property ϵ that represents the amount of energy supplied per unit time to a unit mass element. Note that, in a steady-state regime, this property must coincide with the amount of energy dissipated per unit time per unit mass element. Hence, ϵ is at the same time the energy dissipation rate.

To roughly estimate the energy pumping rate which must be applied to induce large-scale motions with the characteristic length L and velocity V, we note that the dimensionality of ϵ is $\mathrm{erg/g \cdot s} = \mathrm{cm^2/s^3}$. Because the viscous effects are not important at such scales, ϵ should be expressed only in terms of the parameters L and V. There is only one combination of these two parameters with the correct dimension, and therefore the energy pumping rate can be estimated as

$$\epsilon \sim \frac{V^3}{L}. \tag{8.29}$$

The energy supplied at the highest hierarchical level, corresponding to the largest length scale L, is expended to induce motions at the lower levels characterized by smaller lengths l. Within a wide inertial interval $L \gg l \gg l_0$ both dissipation and external supply of energy are absent, and thus energy is only transferred from one mode to another. Because direct interactions between modes corresponding to strongly different characteristic lengths are difficult, it seems natural to assume that energy can be transferred only between the modes that occupy adjacent hierarchical levels. This leads to the picture of an *energy cascade*, with the energy supplied at the top level and gradually transferred down the dynamical hierarchy.

In a steady regime, the amount of energy passing per unit time within the inertial interval through any level of the hierarchy must be the same and equal to the energy supply (and dissipation) rate ϵ. Hence, the characteristic velocity v_l of patterns at a level with the characteristic length l may depend only on the properties ϵ and l. Since the dimension of v_l is cm/s, while those of ϵ and l are $\mathrm{cm^2/s^3}$ and cm, we can construct only one combination with the correct dimension, i.e.

$$v_l \sim (\epsilon l)^{1/3}. \tag{8.30}$$

This important relationship tells us that in any developed turbulent flow the change of velocity over a short distance is proportional to the cubic root of this distance.

After passing through all hierarchical levels, corresponding to the inertial energy, the energy flux ϵ arrives at a level with the characteristic length l_0, where it is eventually dissipated. The length l_0 can depend only on the energy flux ϵ and the viscosity constant ν, whose dimension is $\mathrm{cm^2/s}$. Again, only one combination of these two properties with the dimension of length is possible, and thus we obtain the estimate

$$l_0 \sim \left(\nu^3/\epsilon\right)^{1/4}. \tag{8.31}$$

We see that the dissipative length l_0 depends only weakly on the energy supply rate and is mainly determined by the viscous constant.

The relationship (8.30) can be expressed in several different forms. For instance, we can introduce the distribution $E(k)$ of kinetic energy (per unit mass element), such that $E(k)dk$ yields the kinetic energy accumulated in the modes with wavenumbers from k to $k + dk$. Inside the inertial interval, this distribution can depend only on the energy flux ϵ and the wavenumber $k \sim 1/l$. Since the dimension of $E(k)$ is cm^3/s^2, the only possible combination is

$$E(k) \sim \frac{\epsilon^{2/3}}{k^{5/3}}. \tag{8.32}$$

This is the famous scaling law derived in 1941 by Kolmogorov [435] (see the monograph [436] for further discussions).

The hierarchy of length scales of hydrodynamical turbulence is intrinsically related to a hierarchy of timescales in these processes. Two characteristic times are associated with any vortex. Since the vortex is involved in the flow at a certain mean velocity u, it will be transported by this flow over a distance comparable to its size l within the time $\Delta t \sim l/u$. A more interesting property is, however, a different timescale τ that specifies the internal dynamics inside a vortex, which can be seen if we eliminate the mean motion by going to a coordinate frame moving with the local flow velocity u.

Inside the inertial range, the characteristic velocity v_τ of the internal motions in a vortex with timescale τ can depend only on ϵ and τ. The dimensionality considerations lead to an estimate $v_\tau \sim (\epsilon\tau)^{1/2}$. The characteristic length scale associated with such a pattern is therefore $l \sim v_\tau\tau \sim \epsilon^{1/3}\tau^{2/3}$. Inverting this relationship, we determine the characteristic timescale of a pattern in terms of its size l as

$$\tau \sim \epsilon^{-1/3}l^{2/3}. \tag{8.33}$$

It can be checked that, if τ is thus expressed through l, the characteristic velocity v_τ coincides with v_l given by (8.30), as should have been expected.

Let us introduce the distribution $\mathcal{E}(\omega)$, such that $\mathcal{E}(\omega)d\omega$ would represent the kinetic energy (per unit mass element) which is associated with internal motions in patterns whose frequencies $\omega \sim 1/\tau$ lie in the interval from ω to $\omega + d\omega$. Inside the inertial range the spectral energy density $\mathcal{E}(\omega)$ can depend only on the frequency ω and the energy flux ϵ. Since the dimension of $\mathcal{E}(\omega)$ is cm^2/s, using scaling arguments we obtain

$$\mathcal{E}(\omega) \sim \frac{\epsilon}{\omega^2}. \tag{8.34}$$

This law is just a different expression of the basic Kolmogorov distribution (8.32).

The predictions yielded by the above dimensionality considerations have been subsequently tested both experimentally and numerically. Though some corrections have been detected, these scaling laws have generally been confirmed [437].

Developed hydrodynamical turbulence provides an example of a system with a dynamical hierarchy. The whole complex behavior is described here by the single nonlinear equation (8.27). All fluid elements are identical and interact in the same way with one another. Nonetheless, the system develops a hierarchy of dynamical structures with different timescales that are characterized by effective interactions of varying intensity.

This physical example is important. It shows that structural hierarchies can *emerge* through nonlinear interactions in an initially uniform system. The spontaneous emergence of hierarchies is typical of human societies. Social processes are, of course, different from fluid flows. However, the analysis of fluid turbulence already allows us to suggest that properly constructed mathematical models of economical and social phenomena may also have a tendency to build complex hierarchically organized structures.

8.5 Hierarchically Structured Swarms

Once a social or economic hierarchy is created, it can significantly influence the subsequent evolution of a system. Some links between individuals or companies would be intensified, while others would get weaker. Even if the initial system was relatively uniform, after a while the spontaneously developed hierarchy would be already imprinted in the modified parameters of the interactions between its units. In mature hierarchically organized systems, the constitutive elements and their interactions are strongly differentiated.

As the next example, we consider below a simple model of a structured swarm [438]. This system is formed by a population of N actively moving particles, like the swarm investigated in Sect. 6.4. In contrast to the previously discussed model, the particles are not identical now. Instead, they are divided into two groups of size $K = N/2$. Interactions between the particles belonging to the same group are much stronger than interactions between the particles from different groups. The dynamical equations of this system are

$$\ddot{x}_i + \left(\dot{x}_i^2 - 1\right)\dot{x}_i + \frac{a}{N}\sum_{j=1}^{N} w_{ij}(x_i - x_j) = f_i(t), \quad (i = 1, \dots, N), \qquad (8.35)$$

where

$$w_{ij} = \begin{cases} w, \text{ if } 1 \leq i \leq K \text{ and } 1 \leq j \leq K \\ w, \text{ if } K+1 \leq i \leq 2K \text{ and } K+1 \leq j \leq 2K \\ 1, \text{ otherwise} \end{cases} \qquad (8.36)$$

The parameter w is large ($w \gg 1$). The independent random forces $f_i(t)$ are modelled as white noise of intensity S, so that

$$\langle f_i(t) f_j(t') \rangle = 2S \delta_{ij} \delta(t - t'). \tag{8.37}$$

Thus, the system has two hierarchical levels. The lower level is formed by individual particles, while the higher level is formed by two strongly connected groups of them. We want to show that the dynamics of this system is nearly decomposable. It can be viewed as a slow motion of two compact objects. Each object consists of K particles which perform rapid internal motions inside it.

We define the coordinates X_1 and X_2 of the two groups as

$$X_1 = \frac{1}{K} \sum_{i=1}^{K} x_i \text{ and } X_2 = \frac{1}{K} \sum_{i=K+1}^{2K} x_i. \tag{8.38}$$

and introduce $\delta x_i = x_i - X_{1,2}$ for $1 \le i \le K$ and $K+1 \le i \le 2K$, respectively. Note that the coordinates δx_i describe the internal motions inside two groups.

The groups move according to the equations

$$\ddot{X}_1 + \frac{a}{2}(X_1 - X_2) + \dot{X}_1 \left(\dot{X}_1^2 - 1 \right) + 3 \dot{X}_1 \ll \delta \dot{x}_i^2 \gg_1 \tag{8.39}$$
$$+ \ll \delta \dot{x}_i^3 \gg_1 = \ll f_i(t) \gg_1 ,$$

$$\ddot{X}_2 + \frac{a}{2}(X_2 - X_1) + \dot{X}_2 \left(\dot{X}_2^2 - 1 \right) + 3 \dot{X}_2 \ll \delta \dot{x}_i^2 \gg_2 \tag{8.40}$$
$$+ \ll \delta \dot{x}_i^3 \gg_2 = \ll f_i(t) \gg_2 ,$$

which are derived by summing of (8.35). Here we use the notations $\ll \delta \dot{x}_i^2 \gg_1 \equiv \frac{1}{K} \sum_{i=1}^{K} \delta \dot{x}_i^2$, $\ll \delta \dot{x}_i^2 \gg_2 \equiv \frac{1}{K} \sum_{i=K}^{2K} \delta \dot{x}_i^2$, etc.

Rapid motions inside the first group are described by

$$\delta \ddot{x}_i + \frac{1}{2}(w+1) a \delta x_i + \delta \dot{x}_i (\delta \dot{x}_i^2 - 1) + 3 \dot{X}_1 \delta \dot{x}_i^2 + 3 \dot{X}_1^2 \delta \dot{x}_i \tag{8.41}$$
$$-3 \dot{X}_1 \ll \delta \dot{x}_i^2 \gg_1 - \ll \delta \dot{x}_i^3 \gg_1 = f_i(t) - \ll f_i(t) \gg_1 .$$

If the frequency $\omega_0 = \sqrt{(w+1) a/2}$ is large, i.e. $\omega_0 \gg 1$, the rapid oscillations are almost harmonic. Moreover, if their phases are randomly distributed, we can replace group averages by statistical means, i.e. put $\ll \delta \dot{x}_i^2 \gg_1 = \langle \delta \dot{x}_i^2 \rangle_1$, where $\langle \delta \dot{x}_i^2 \rangle_1$ denotes the statistical dispersion of the particles inside the first group. Therefore, we have in this case $\ll \delta \dot{x}_i^3 \gg = 0$. Furthermore, the average $\ll f_i(t) \gg$ is small and can be neglected. Finally, the term $(\delta \dot{x}_i^2 - \ll \delta \dot{x}_i^2 \gg_1)$ oscillates with the double frequency $2\omega_0$ and is therefore nonresonant, i.e. its influence will be small (of order $1/\omega_0$). Hence, it can also be neglected.

As a result, equations (8.41) are simplified to

$$\delta \ddot{x}_i + \omega_0^2 \delta x_i + \delta \dot{x}_i (\delta \dot{x}_i^2 - 1 + 3\, \dot{X}_1^2) = f_i(t). \tag{8.42}$$

In the same approximation, (8.39) takes the form

$$\ddot{X}_1 + \frac{a}{2}(X_1 - X_2) + \dot{X}_1 \left(\dot{X}_1^2 - 1 + 3 < \delta \dot{x}_i^2 >_1 \right) = 0. \tag{8.43}$$

Analogous equations hold for the second group of elements.

This system has three different dynamical regimes. In the first, both groups do not move, i.e. $X_1 = X_2 = $ const. In this case equation (8.42) is reduced to

$$\delta \ddot{x}_i + \omega_0^2 \delta x_i + \delta \dot{x}_i (\delta \dot{x}_i^2 - 1) = f_i(t). \tag{8.44}$$

and all elements perform small rapid self-oscillations approximately with frequency ω_0 and amplitude $1/\omega_0$.

In the other possible dynamical regimes, both groups move at a constant velocity in the same direction, i.e. $\dot{X}_1 = \dot{X}_2 = V = $ const. This regime is similar to what we have earlier considered in Sect. 6.4, with the only difference that the intrinsic frequency of random oscillations inside the moving cloud is ω_0 and therefore is relatively large. This regime becomes unstable when a certain critical noise intensity is exceeded.

In the above regimes the particles belonging to different groups are completely mixed. In contrast to this, in the third dynamical regime the two groups remain relatively compact and their centers X_1 and X_2 perform slow self-oscillations at a frequency close to $\Omega = \sqrt{a/2}$. To roughly analyze this dynamical regime, let us assume that the mass center oscillations are harmonic, i.e. $X_1 = A \sin \Omega t$ with $A \sim 1/\Omega$. This means that $\dot{X}_1^2$ oscillates at a double frequency in the range from zero to $\Omega^2 A^2 \sim 1$. If we look at (8.42), we note that internal oscillations are damped during intervals of time when $\dot{X}_1^2 > 1/3$, and are excited when $\dot{X}_1^2$ is smaller. Hence, $\ll \delta \dot{x}_i^2 \gg_1$ increases during time intervals when $\dot{X}_1^2 < 1/3$, and decreases down to the level determined by the noise between these excitation intervals. Thus, the slow motion of groups leads to their internal breathing. On the other hand, when $\ll \delta \dot{x}_i^2 \gg$ is higher, this slows down the collective group motion [see (8.43)]. This implies that the two groups will tend to spend more time in the state with low collective velocity $\dot{X}_1$, i.e. oscillations of their centers will be to a certain extent nonharmonic.

This regime is interesting because we observe here the coexistence of fast and slow motions in the same nonlinear system, with fast motions in its subsystems enslaved by slow collective motions. Figure 8.8 displays the results of numerical simulation[2] of the model (8.35) for a population of $N = 100$

---

[2] We have used here the same numerical procedure for the generation of noise, as in Sect. 6.4. See the footnote on p. 142.

particles and the parameters $a = 0.1, w = 1000$ under the action of relatively strong noise with intensity $S = 16.7$. The coordinates X_1 and X_2 of both group centers oscillate out of phase at a low frequency (Fig. 8.8a). The thin curve in Fig. 8.8b shows the time dependence of the coordinate δx_i for a single particle inside the first group, whereas the bold curve displays the respective time dependence of the dispersion $\ll \delta x_i^2 \gg^{1/2}$ of particles in the same group. We see that the motions inside the group are characterized by a much higher frequency and are irregular. Figures 8.9a and b show the frequency spectra of $X_1(t)$ and $\delta x_i(t)$ in the first group (note the logarithmic horizontal scale).

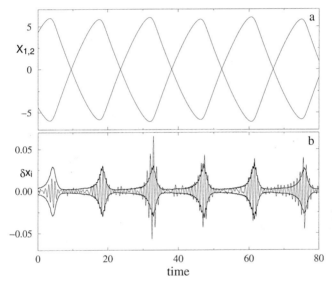

Fig. 8.8. Dynamics of a structured swarm described by equations (8.35). Time dependences of (**a**) coordinates X_1 and X_2 and of (**b**) the coordinate δx_i for a single particle inside the first group. The bold curve in (**b**) displays the respective time dependence of the dispersion $\ll \delta x_i^2 \gg^{1/2}$ of particles in the first group. The parameters are $a = 0.1, w = 1000, S = 16.7, N = 100$

At the considered parameter values, the oscillations of X_1 and X_2 are already nonharmonic. The groups move at an approximately constant velocity (corresponding to a constant slope in Fig. 8.8a) until they suddenly slow down and reverse their direction of motion. After that the groups move with the same velocity in the opposite direction, until the reversal again takes place. Examining Fig. 8.8b, we note that the reversals of the group motion are accompanied by bursts in the amplitude of internal motions in the group.

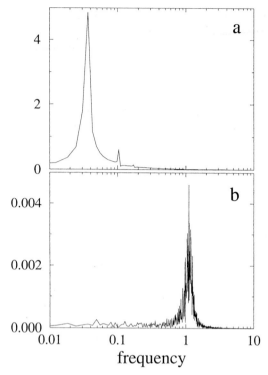

Fig. 8.9. The frequency spectra of (**a**) $X_1(t)$ and (**b**) $\delta x_i(t)$ in the first group for the same parameters as in Fig. 8.8

8.6 Further Reading

In this section we shall provide some suggestions for further reading related to biological, ecological or social aspects of hierarchical organization. This organization is already present at the level of a single biological cell. For example, ontogenesis involves a controlled and differentiated activity of genes and their expression. The genetic activity has to be complemented by special morphogenetic mechanisms, as the genes themselves code only for protein molecules (see Murray [439] for a discussion of chemical and mechanochemical morphogenetic mechanisms). The concept of genetic regulatory architectures, based on rigid hierarchical subordination, has been proposed [440, 441]. According to this concept a few master genes directly control overlapping downstream cascades of genes. The steering of developmental programs consists in unleashing the different possible cascades of activities by the master control gene at the top of each hierarchy. The overlapping activations spreading through the downstream cascades create different cell types, each with a unique battery of gene activities.

An alternative to the approach with direct one-way hierarchical control of gene activity is provided by models where gene regulation involves intrinsic feedback. Since the classical paper by Monod and Jacob [442] (for a model, see Goodwin [443]) it is known that DNA strands do not only contain genes that code regulatory and structural proteins. There are also regions which regulate the expression of a gene through negative feedback loops between the gene, the corresponding protein and a related substrate. These operator regions have either an inhibitory or an activatory effect and act like a switch that either turns a gene on or off. The relation between the regulatory and the coding part of a gene is a hierarchical one. For an overview of recent developments in genetic regulation, see Alberts et al. ([444]). Kauffman ([445]) compared and discussed experimental work and models in favor of either of these two concepts.

At the ecological level, interactions between individuals of different species build up a hierarchical trophical structure. Consumers (such as predators) are located at the next higher hierarchical level as compared to their prey. For a detailed discussion of involved models, see the books [439, 446–450].

A number of general conclusions can be drawn from the hierarchical organization of ecosystems and the limited amount of energy avilable. The ratio of energy influx of two adjacent trophic levels is called the ecological efficiency. Roughly speaking, the efficiency is independent of the position on the trophic ladder and is around 10%. As there is a certain minimal energy consumption per individual and a certain minimal population size, there is a tendency to find a decreasing number of individuals and a decreasing amount of biomass when moving up the trophic ladder. Moreover, there is a finite number of trophic levels. In fact, one rarely finds more than 4 or 5 trophic levels. These constraints are important for the sustainable use of biological resources.

Ludwig discussed the problem of sustainable harvesting ([451]). His analysis departed from the observation that populations usually fluctuate without our normally knowing the environmental factors that cause the fluctuations. Sometimes, although infrequently, catastrophes occur that have severe consequences leading to a breakdown. For populations subject to anthropogenic effects such as harvesting, Ludwig translated the constraint of sustainability into the problem of fixing a harvest rate that is low enough to ensure that the probability of early population breakdown is below a certain level. Ludwig's conclusion was that the knowledge that is normally available about an ecosystem hardly suffices to calculate the upper limits of safe exploitation without running the danger of having a system collapse. Giampietro analyzed the agricultural systems in various countries in terms of ecological constraints with respect to energetic flows ([452]). Ayres extended the concept of hierarchical organization and energetic throughput from the ecological level to include anthropogenic effects as well as links between ecology and economy [453, 454].

While some predators prey exclusively on one species, others feed on a number of prey species. Hence, the number of links to other predator or prey species can vary. It is clear that a predator preying only on one species is much more vulnerable to the availability of its prey than a predator that can choose among different prey species, some of which may become scarce. One could argue that ecosystems with a higher density of food web relations tend to be more stable. Much experimental work has been done to measure the connectivity in ecosystems and relate it to the system's stability, but a definite conclusion is not yet available. These experimental and theoretical topics are reviewed by May [446] and Holling [455].

Ecosystems are often used as metaphors for other complex systems where competition for resources and mutual support are important. In informatics, computational ecologies were introduced by Huberman and Hogg [456] to discuss the dynamical properties of distributed computer systems ensuing from sending and receiving messages, confidentiality requirements, technological change and other factors.

Spin glasses provide one of the central paradigms in modern statistical physics. The glass-like behavior is found in many different applications. It is typical of the problems of complex combinatorial optimization (see [457]). An example is the *traveling salesman problem*. Here one is asked to find the shortest closed tour that connects N given points ("cities"). Another example is the *graph bisection problem*, which is encountered in the design of computer chips. The task is to divide all nodes of a given graph of size $2N$ into two equal groups in such a way that the number of links between the nodes belonging to the opposite groups is minimal. The defining feature of a complex optimization problem is that the number of operations needed to compute its exact solution increases faster than any power of N. Hence, already for relatively small systems the exact solution cannot practically be constructed.

Putting into correspondence to any potential configuration (e.g. a tour in the salesman problem), the respective property (such as the tour length) and considering the space formed by all possible configurations, the similarity with spin glasses can clearly be seen. The optimized property is analogous to the free energy of a spin glass, whereas different configurations represent analogs of various spin orientation patterns. The optimization problem is to find a configuration with the lowest "energy". The difficulty lies in the fact that the system would have a great number of local "energy" minima. These minima may form a hierarchical structure.

An interesting application of the spin-glass theory is found in the financial problem of portfolio optimization. Given a set of financial assets, characterized by their average returns and their risk, the task is to determine the optimal weight for each asset, such that the overall portfolio provides the best return for a fixed level of risk. Galluccio et al. [458] have noticed that this optimization problem can be reduced to a certain statistical spin-glass problem. When the correlation matrix of assets is random, this problem has a great number of possible optimal solutions, corresponding to different portfolios. Actually, this number grows exponentially with the number of assets! All these solutions can be very far from each other, so that it can be rational to follow two totally different strategies. Furthermore, the solutions are usually "chaotic" in the sense that, say, the addition of an extra asset may completely change the risk associated with a particular portfolio. The authors conclude that in these situations "the very concept of rational decision making becomes questionable". Further analysis of this optimization problem has been given by Gábor and Kondor [459]. Spin-glass analogies in the behavior of social systems have also been considered by Axelrod and Bennett [460].

Applications of the methods of statistical physics to the analysis of hierarchically organized political systems have been discussed by Galam [461–463].

An important problem is to detect and reconstruct the hierarchical organization of a system through the analysis of its collective dynamics. Mantegna [464] has shown how this problem can be solved in the analysis of financial markets. The traded stocks are characterized by their daily prices and the only information available consists of the time series of prices of different stocks. Though the price variations are apparently random, the prices of some stocks may actually be correlated, indicating some kind of functional dependence on the performance of the respective companies. Using the daily price data, the correlation coefficient for any pair of stocks can be computed. In this manner, the correlation matrix of stocks is obtained. Mantegna [464] has used the correlation matrix to detect the hierarchical organization present in the portfolios of the stocks that are taken to determine the Dow Jones Index and the S&P 500 Index describing the performance of the New York Stock Exchange in the period from 1989 to 1995. Similar methods can apparently be used to analyze the hierarchical organization of systems of other origins, when only the time series are available.

9. Dynamics and Evolution of Networks

In a large class of complex living systems, a fundamental role is played by networks that are formed by coherently operating groups of elements. The examples range from large-scale social and economic structures of human societies to the brain, the immune system and the molecular biology of a cell. The dynamical networks spontaneously develop and interact with one another, leading to the patterns of cooperation and competition typical of all societies.

The connections between elements in a network constitute its graph. The theory of graphs is an advanced field of mathematics with important applications in transportation problems or the design of computer chips. When functional networks are considered, graphs set the scene for their dynamics.

When individual elements of a network represent chaotic dynamical systems, such as logistic maps, mutual synchronization and clustering are found at sufficiently high coupling strengths. The regime of dynamical clustering, preceding full synchronization, is especially interesting. Coherently operating groups of elements build up new effective dynamical networks, whose behavior depends on the number and sizes of the groups, as well as on the distribution of the elements over them. Therefore, the same population of elements may give rise to a huge variety of network structures with greatly different dynamics. The emerging networks are fluid: they have some lifetimes and evolve by transforming themselves into new network structures.

The underlying graph of connections between elements steers the dynamics by favoring certain cluster partitions and respective effective networks and forbidding the others. Thus, it provides a kind of "grammar" for this complex dynamical system.

On longer timescales, the graphs can evolve to improve the performance of their networks. In plastic graphs, some connections between elements are broken and new connections become established in the course of time. Through such random mutations and subsequent selection, the grammar of the networks evolves to optimize their dynamical expression. In this way a network may learn to generate particular temporal patterns or execute certain functional tasks.

9.1 Societies

Some time ago one of the authors of this book met a Japanese scientist, famous for his experimental and theoretical work in biophysics and nonlinear dynamics. We talked of various subjects including social psychology. With a touch of humor, he said that for the Japanese it might sometimes be difficult to understand what democracy means. As in other oriental cultures, a human has traditionally been viewed here as a void, acquiring its meaning through the pattern of social connections. Putting a stress on social interactions can indeed be justified if we note that, when large-scale social dynamics is considered, human personalities play only a limited role. At this level, the dominant structures of a society are various networks established by individuals and their groups.

A sentence in a language is a network made of words. These words may be spoken with different articulations or written in various scripts. They may have an interesting evolutionary history which can be successfully investigated. But the conveyed meaning of a sentence is independent of the factual properties of its words. We can even translate sentences into another language, replacing their words by different units, and the meaning will still be maintained. From the viewpoint of semantics, a word is defined through its pattern of connections to all other words in a language. When large-scale social dynamics is considered, a person is a "word" in a system of networks that collectively make up a society. Thus, to study a society one should identify its basic networks and consider their dynamics, evolution and interactions.

Human societies are not the only complex systems where the evolution and dynamics of networks are essential. Similar forms of organization are typical of social insects, such as ants (see Chap. 2). Moreover, networks also play a fundamental role in the functioning of individual living beings.

The brain is a giant physiological network formed by interconnected neural cells. Its dynamics is responsible for processing information about the surrounding world and generating the motor responses of an animal. In humans, the brain is also the seat of rationality and of the mind.

In contrast to modern computers, brain operation is not digital. When a cat jumps to catch a mouse, its brain does not numerically integrate a system of differential equations, as the computer would do in order to intercept a rapidly moving missile. Even the most careful examination will not reveal in the brain anything similar to a program operated by a computer in analogous situations. Yet, though the operational principles of the brain and of a computer are clearly different, they can both make efficient predictions about the outcomes of various processes in the external world, i.e. they can *emulate* it.

From this perspective, the brain can be viewed as a huge, incredibly flexible dynamical system that may tune itself to mimic a broad spectrum of processes taking place outside the animal. Although our knowledge of the brain is still very limited, there are good reasons to expect that the emulation is performed here through the emergence and development of special dynamical networks which can capture and repeat almost any targeted dynamics.

The neural system is plastic and the pattern of synaptic connections between neurons evolves with time. However, such neural plasticity is much too slow to explain the emergence and evolution of the dynamical networks needed to emulate a rapidly changing environment. It is more plausible that a new network is generated through the formation of coherently acting groups of neurons that belong to a given neural net. These coherent groups would represent elementary units of the emergent dynamical network. The number of such groups and the distribution of particular neurons over them would determine the pattern of interactions in the network and hence its collective dynamics. Depending on the required outcome, the same neural net can give rise to different dynamical networks.

On the other hand, the underlying pattern of synaptic connections may provide a "grammar" and a "vocabulary" for this dynamical behavior. This pattern will prohibit the emergence of certain dynamical networks and make other networks preferential. Through slow evolution, accompanied by plastic changes, the neural net may thus improve its ability to capture important external dynamics.

Obviously, the same mechanisms which are employed by the brain to model external processes can be used by an animal to model the movements of its own body. This self-emulation is necessary to generate signals causing muscle contractions and thus to control motor activity. Since this is a basic function of any animal, the respective mechanisms must have developed already at an early evolutionary stage.

If motorics is controlled through the emergence of dynamical networks, the coordination of the movements of different muscle groups would require interactions between such networks, associated with the generation of different partial movements. These interactions must be relatively weak, so that they do not destroy the desired dynamics of an individual network. They should only coordinate corresponding dynamical processes.

An animal emulates the external world in its brain only as demanded by its physiological functions, such as the necessity to escape danger or pursue food. The brain should therefore extract and internally model only those dynamical aspects of the environment which are relevant for the animal. This means that, effectively, the emergent dynamical networks of the brain would correspond to various *abstractions*, each capturing a particular facet of the reality.

To properly behave, an animal must simultaneously execute many interrelated tasks. Some of them, such as spatial orientation, would require a very rough but global model of the environment. For other purposes, certain dynamical processes must be finely resolved in the internal model. This implies that rough abstractions, forming a skeleton model of the reality, should be allowed to be expanded, revealing their inner structures. Hence, the internal model must represent a dynamical hierarchy built up by its constituent networks.

At high levels of the evolutionary staircase, in humans and, to a less extent, in some animals, the mind appears. This highest function of the brain involves rational or irrational operations with abstract concepts. The concepts form *semantic nets* which are used to describe and predict not only the phenomena in the outside material world, but also the behavior and responses of other individuals. Moreover, self-consciousness develops that allows an individual to construct a model of his or her own person, as perceived by other humans.

The detailed neurophysiological phenomena which underlie concepts and semantic nets are not known. Since concepts are persistent features, they may be associated with plastic modifications of neural nets and evolution of the pattern of synaptic connections. On the other hand, the construction of semantic nets and reasoning, involving operations with them, are relatively fast processes that must proceed at a purely dynamical level.

An individual concept, expressing a meaning, is thus itself a denotation of a certain semantic net. This brings us to a picture of the mind as a space populated with semantic nets that emerge, interact and collectively evolve. Indeed, M. Minsky a long time ago suggested that we talk about the *society* of the mind [465].

In a sense, operations with semantic nets already imply that the mind has its own internal language where "words" are concepts and "sentences" are the networks formed by them. Such "words" and "sentences" are encoded in the pattern of synaptic connections of the neural net and in the dynamical activity patterns of neurons. This internal language has gradually developed in the course of biological evolution in higher animals and humans. From this point of view, the emergence of spoken language represents a natural continuation of such development. Articulating some concepts and links between them allows us to communicate the respective meaning to other individuals. It may be far from accidental that speech is based on the generation of coordinated muscle contractions and thus involves essentially the same neural system which is employed in the control of motor activity.

Consorted actions of many biological cells, forming cooperative groups and giving rise to dynamical networks, are found not only inside the brain. The immune system provides another important example of the cell society. The task of this system is to protect an animal against infections (or cancer) by identifying and destroying alien cells and particles. This is done by antibodies and by special killer cells.

The organism must recognize and combat any invaders, even if it has never encountered them before. The solution of this problem that has been found by nature is based on implementing fast biological evolution. The organism randomly generates lymphocytes bearing receptors that are sensitive to various possible antigens. If a certain lymphocyte has found a target, it begins to rapidly proliferate and triggers a chain of events leading to the production of a large number of specific antibodies or killer cells.

When an infection has been combated, in most cases the organism will maintain a limited population of respective lymphocytes. This immune memory allows it to quickly launch a massive defense against the next invasions of the same bacteria or viruses. The immune system would therefore usually include a large number of various specific lymphocyte populations sensitive to a great variety of already encountered antigens.

Besides fighting against invaders, the immune system should persistently destroy all defect or misfunctioning cells of the organism. However, it should never act against the healthy cells of its host. This means that the immune system must also keep a complete memory of all types of healthy cells in the organism! Whenever a lymphocyte specific against one of such cells appears, its proliferation should be suppressed.

Thus, the immune system effectively represents a society formed by interacting cellular populations. This dynamic network keeps a complex molecular memory and differentially responds to stimuli, suppressing reactions against some of them while generating a strong concerted defense against the others.

Descending the biological hierarchy, we can proceed to considering phenomena inside single living cells. A cell is populated by protein machines representing individual macromolecules or their aggregates (organelles). Some of such machines are found in great numbers, while others have just a few copies inside the whole cell. In certain situations, a massive response of the whole population is triggered by the arrival of just one significant molecule. On the other hand, the response of the entire cell may consist of the generation of only a single molecule.

The machines communicate through messenger molecules and intermediate products with smaller molecular weights. This communication is responsible for the consorted operation of the entire society. As we have seen in Sect. 7.4, coherently operating groups of protein machines may appear in this system. Such synchronous groups form dynamical networks dedicated to the execution of various functional tasks. These spontaneously emerging networks should be distinguished from the reaction scheme. Our analysis has already shown that, depending on the parameters, the same reaction scheme may give rise to various dynamical regimes characterized by different numbers of coherent molecular groups.

The self-organization of dynamical networks and the interactions between such networks and their evolution apparently play a principal role in collective processes which are characterized by a high degree of complexity. If one wants to introduce a "society" as an abstract concept, in terms of the internal organization of this complex system, this can be chosen as its defining property. From such an abstract perspective a society would represent a system consisting of a large number of relatively uniform, active agents which build up fluid networks. Emergent and evolving networks are the basic dynamic entities in this class of complex systems. The elements of the networks can be either individual agents or their coherently operating groups (or associations). The same agent may belong to many different networks.

The performance of a society is determined by the interactions between its constitutive networks. Through their gradual evolution and restructuring, a society can adjust to environmental variations and may increase its competitivity with respect to other societies. A society may even learn to execute certain tasks. In contrast to this, degradation of the network structure and loss of its fluidity may lead to the death of a society.

The ensemble of networks, forming a developed society, would usually have a hierarchical organization. The networks of a lower level would then play the role of individual elements of a higher-level network. As we have generally discussed in Chap. 8, stable dynamics of a complex hierarchically organized system is possible if its different levels are characterized by different timescales.

It might be interesting to analyze from this abstract point of view the properties of various biological and social systems. Obviously, this should rather be done by experts in the respective scientific fields. The framework for this analysis must be provided by general theoretical investigations of networks as dynamical systems. At present, the research focused on the emergence, evolution and interactions of dynamical networks is at an early stage. It is therefore impossible for us to give a systematic exposition of the related topics. Instead we shall provide in this chapter a short summary of the mathematical theory of graphs and consider two examples where the dynamics and evolution of networks are investigated.

9.2 Properties of Graphs

The pattern of connections in a network represents its *graph*. Formally, a graph is a set of dots called *nodes* (or vertices) connected by links called *edges*. Studies of graphs are important for many applications, including transportation logistics and the design of computer chips. A good introduction to the modern mathematical theory of graphs is given by Diestel [466]. Below in this section we follow it to define the basic properties of graphs (the notations are slightly modified here to bring them closer to those used in the physics literature).

Suppose that we have a graph G with N nodes. Then we can enumerate them as $i = 1, 2, \ldots, N$ and construct the adjacency matrix T_{ij}, such that $T_{ij} = 1$ if the graph contains a connection going from node j to node i and $T_{ij} = 0$ otherwise. In *directed* graphs the back and forward connections may be different, so that generally the matrix T_{ij} is not symmetric ($T_{ij} \neq T_{ji}$). Below we consider only simpler graphs where connections are independent of directions. In such graphs the adjacency matrix is symmetric, i.e. $T_{ij} = 1$ if nodes i and j are linked and $T_{ij} = 0$ if the link between them is absent. For definiteness, the diagonal elements of this matrix can be chosen equal to zero ($T_{ii} = 0$).

The description of a graph in terms of its adjacency matrix is not unique. Indeed, by arbitrarily changing the enumeration order of the nodes a different adjacency matrix will generally be obtained. Hence, each graph G of size N has $N!$ isomorphic matrices T_{ij} generated by all possible permutations of N elements. This high degeneracy leads to difficulties in comparing and manipulating graphs that are determined through their adjacency matrices (see the discussion below).

A *subgraph* G' of G is defined as a graph, each node and each edge of which are at the same time one of the nodes and one of the edges of graph G. It can be said that G contains G'. Furthermore, G' is an *induced subgraph of* G if it includes *all* edges that connect the respective nodes in graph G (Fig. 9.1).

On the other hand, a graph G_2 of size N_2 can be *added* to a graph G_1 of size $N_1 \geq N_2$. To do this, we choose N_2 nodes of the graph G_1 and identify them with the nodes of the smaller graph G_2. Then we check whether G_1 already includes all edges connecting the nodes of G_2. If not, the missing edges are added. The resulting graph G will contain as its induced subgraphs both

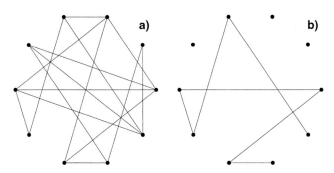

Fig. 9.1. Graph (**a**) and its induced subgraph (**b**)

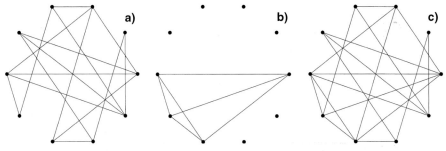

Fig. 9.2. Addition of graph (**b**) to graph (**a**) yields new graph (**c**)

G_1 and G_2 (Fig. 9.2). Note that the addition can be performed in different ways, depending on which nodes of G are identified as the respective nodes of G_2. The *difference* $G - U$ of two graphs G and U is obtained by deleting from G all nodes of U together with their incident edges.

Another operation defined for graphs is their *contraction*. To "contract" a certain edge, we treat as identical the two nodes which are connected by it and replace them by a single node (see Fig. 9.3). The number of nodes and the number of edges in the obtained graph are less by one. Contractions can be sequentially applied to any arbitrarily chosen set of edges in the original graph, yielding a new smaller graph.

The *degree* m_i of a node i is the number of edges connecting it to the other nodes $j \neq i$ of the graph. It is given by

$$m_i = \sum_{j=1}^{N} T_{ij}. \tag{9.1}$$

If all the nodes of a graph have the same degree, this graph is called *regular* (Fig. 9.4). Generally, degrees will be different for different nodes. The *average*

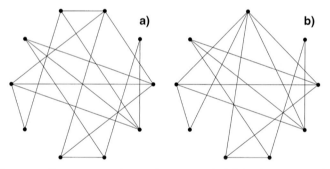

Fig. 9.3. Contraction of one edge in graph (**a**) leads to graph (**b**)

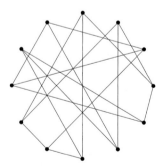

Fig. 9.4. Example of a regular graph of size $N = 12$ and degree $m = 3$

degree $\overline{m}(G)$ of a graph G is the mean number of edges for its nodes,

$$\overline{m}(G) = \frac{1}{N} \sum_{i=1}^{N} m_i. \tag{9.2}$$

The average degree of G can also be expressed as

$$\overline{m}(G) = \frac{1}{N} \sum_{i,j=1}^{N} T_{ij}. \tag{9.3}$$

Furthermore, the minimum and the maximum degrees of a graph are given by the minimum and the maximum degrees of any its nodes, respectively.

The average degree of a graph may be large even when its minimum degree is small. Indeed, a graph may always contain some nodes which are only loosely connected to the rest of vertices. A theorem has been proven [466] which states that every graph G has a subgraph H whose average degree is no less than the average degree of G and whose minimum degree is more than half its average degree. This strongly connected subgraph is obtained by subsequently deleting nodes of small degree from the original graph.

Note that the total number of edges $M = \frac{1}{2}\sum_{ij} T_{ij}$ in graph G can be expressed as $M = N\overline{m}(G)/2$ in terms of its average degree. In a globally coupled graph of size N, where every node is connected to each other (such graphs are also called *complete*), the total number of edges is $M_{max} = N(N-1)/2$. Generally, a large graph $N \to \infty$ is called *dense* if $M \propto N^2$, and *sparse* otherwise.

Many graphs encountered in natural systems are random. The randomness means that an ensemble of graphs should be considered where each graph G is found with some probability $P(G)$. To construct a random graph, various procedures may be employed.

For example, we can independently choose with a certain probability ν each potential connection between the nodes. In this case, matrix elements T_{ij} with $i > j$ represent independent random binary numbers, i.e. $T_{ij} = 1$ with probability ν and $T_{ij} = 0$ with probability $1 - \nu$. According to (9.1), the degree m_i of a node i in such a graph is a sum of N independent random numbers and thus represents itself a random number with some probability distribution $p(m)$. When the size N is large, the central limit theorem of probability theory predicts that this sum should approximately have the Gaussian distribution

$$p(m) = \frac{1}{\sqrt{2\pi N\nu}} \exp\left[-\frac{1}{N\nu}(m - N\nu)^2\right]. \tag{9.4}$$

In the limit $N \to \infty$ the degree m_i of any node in this random graph approaches $N\nu$. Hence, large random graphs generated using this procedure are almost regular, i.e. each node has almost the same number $N\nu$ of connections. These connections are randomly scattered over the graph.

Random graphs, which do not become regular in the limit of large size, can also be easily constructed. Suppose that we have initially generated a random graph G_1 of size N_1 by independently choosing every one of its possible connections with the probability ν_1. Then we generate in the same way a smaller random graph G_2 of size $N_2 < N_1$ with a higher connection probability $\nu_2 > \nu_1$ and *add* it to the first graph. The resulting random graph G will then contain an induced random subgraph G_2 with a higher average degree. By varying the construction parameters ν_2 and N_2, the size and the characteristic average degree of the contained subgraph can be arbitrarily changed. Repeating the same procedure, random graphs that include many different subgraphs with varying average degrees may be obtained. A classical introduction to the properties of random graphs is given by Bollobás [467].

Further fundamental properties of graphs are related to *paths* inside them. A path between two nodes i and j in graph G is a sequence of different adjacent edges that links these two nodes (note that self-intersections of paths are thus excluded). A graph is called *connected* if a path linking any two nodes can always be found. Disconnected graphs consist of components, each representing an individual graph.

The *distance* d_{ij} between nodes i and j is the length of the shortest path between them in the graph. The greatest distance between any two nodes in G defines its *diameter*. The center of graph G is a node whose maximum distance from any other node is as small as possible. This distance represents the *radius* of G.

The diameter (or the radius) of a graph represents one of its most important properties. Indeed, in communication networks the diameter determines the delays in transmitting messages through the network. To increase the speed of a computer chip, the diameter of its graph must therefore be minimized.

Large graphs can nonetheless have short diameters. Indeed, a globally coupled (i.e. a complete) graph of any size N has the unit diameter $D = 1$. It can, moreover, be shown [467] that the diameter D of a random graph with independently chosen connections scales as $D \propto \ln N$ with the graph size. On the other hand, a graph may combine a small minimum degree with large diameter. For example, a regular graph of size $N = L^d$ that represents a cubic d-dimensional lattice (Fig. 9.5) has degree $m = 2^d$ and diameter $D \sim L = N^{1/d}$.

It has been suggested by Milgram [468] that human society may be characterized by a short diameter. He considered a pattern of contacts between individuals and defined the "degree of separation" between any two persons as the minimum number of intermediate acquaintances between them. His analysis has indicated that within the population of the USA there are typically only "six degrees of separation" between any human being and any other. Thus, the world of personal contacts is indeed "small". Other examples

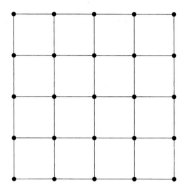

Fig. 9.5. Example of a square lattice of size $N = 25$. The nodes occupying respective positions on the opposite sides are connected by links that are not shown in the figure

of social and economic networks possessing the property of the "small world" are discussed in [469–471].

General estimates for the diameter and the radius of arbitrary graphs are available. For regular graphs with degree m and diameter D the *Moore bound* states [472] that the size N of such a graph cannot exceed

$$N_{\max}(m, D) = 1 + m + \ldots + m(m-1)^{i-1} + \ldots + m(m-1)^{D-1}. \quad (9.5)$$

It can furthermore be proven [466] that *any* graph of radius R and maximum degree $m_{\max}$ should have a sufficiently small size N,

$$N \leq 1 + R m_{\max}^{R}. \quad (9.6)$$

This implies that for a large graph ($N \gg 1$) of radius R the maximum degree cannot be too small, i.e.

$$m_{\max} \geq \left(\frac{N}{R}\right)^{1/R}. \quad (9.7)$$

We can apply this result to roughly estimate the number of acquaintances per person needed to produce the conjectured six-degrees-of-separation law for human society. Assuming that $R = 6$ and taking $N = 6 \times 10^9$, we find that according to (9.7) $m_{\max} > 10^{9/6} \approx 32$. Thus, if each of 6 billion people has about 30 acquaintances, this may indeed explain the emergence of relatively short paths making the world "small".

As we have already noted, large sparse graphs of small diameter are important for many applications, such as the design of computer chips. Generally, the task is to construct so-called *expander graphs*, where any "small" subset of nodes has a relatively "large" neighborhood. The definitions and detailed discussion of various expander graphs, including further references, are given by Chung in the book [472]. Some of the constructed families of graphs come quite close to the upper limit implied by the Moore bound (9.5). For instance, the Ramanujan graphs have a size N which is only by a factor of $(m-1)^{D/2}$ smaller than $N_{\max}(m, D)$ (see [472]).

In 1988 Bollobás and Chung [473] considered graphs which are obtained by adding random matching to a ring. Besides short local connections leading to the left-hand and right-hand neighbours on the ring, each node of such a graph has an additional distant link that connects it to an arbitrarily chosen node. They have proven that the diameter $D(G)$ of such graphs of size N satisfies

$$\log_2 N - c \leq D(G) \leq \log_2 N + \log_2 \log_2 N + c, \qquad (9.8)$$

where c is a constant ($c \leq 10$). Thus, their diameters are almost as small as those of random graphs of the same size and independent of the choice of each connection. However, the total number of edges in the random graph is $M \sim \nu N^2$, whereas the constructed graph has only $M \sim N$ edges. Thus, in contrast to the random graph, it is sparse.

Recently, attention was attracted to a similar simple construction. One takes a d-dimensional lattice and introduces random independent links between distant nodes by assuming that any two non-adjacent nodes of the lattice can be connected with a certain probability p [470, 474]. An important question is: at what critical probability p_c will the diameter of this graph show a crossover from the dependence $D \propto N^{1/d}$, characteristic for lattices, to $D \propto \ln N$, which is typical for completely random graphs. The analysis reveals that the crossover takes place already at a surprisingly low probability of added random connections, i.e. when $p_c \sim N^{-1}$, so that the total number of added connections $p_c N^2$ is comparable with the size N of the graph [474].

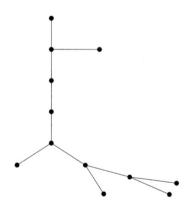

Fig. 9.6. Example of a tree graph

A *cycle* is any closed path in a graph (as for any path, it is not allowed to have self-intersections). A *tree* is a connected graph that contains no cycles (Fig. 9.6). Any two nodes in a tree are linked by a unique path. The minimum length of a cycle contained in graph G is called its *girth* $g(G)$. For example, the girth of a globally coupled graph is three. On the other hand, the girth of any cubic d-dimensional lattice is four. The minimum degree of a graph and its girth are independent: there are graphs combining arbitrarily large minimum degree with arbitrarily large girth [466]. On the other hand, the girth of a graph and its diameter are related. If $D(G)$ is the diameter of a graph G containing a cycle, its girth should be sufficiently small [466], i.e.

$$g(G) \leq 2D(G) + 1. \tag{9.9}$$

An *Euler tour* of a graph is a cycle that contains *all* its edges. In 1736 Euler proved a theorem which states that a connected graph admits such a tour if and only if every its node has even degree [475].

Depending on a particular application, different properties of graphs become essential. In this chapter we are mostly interested in the dynamics of complex systems representing networks. Obviously, the activity of a network should bear some information about its structure. In many situations the pattern of connections of a network (i.e. its graph) is not directly observable and only the *performance* of a graph, expressed in a set of generated signals, can be recorded and analyzed.

An important question is to what extent the performance of a graph can characterize its structure. Can one conclude by measuring generated signals that two networks are similar or identical? Can one reconstruct a graph from its dynamics?

We begin the discussion of these problems by considering networks with simple dynamical properties. Given a graph with N nodes and an adjacency matrix T_{ij}, we construct a network with N dynamical variables $x_i(t)$ described by the system of differential equations

$$\ddot{x}_i = -\sum_{j=1}^{N} T_{ij}(x_i - x_j). \tag{9.10}$$

This dynamical system may be viewed as a one-dimensional "molecule" formed by N identical particles connected by elastic strings. Two particles i and j are connected by a string only if the respective link is present in the graph (i.e. $T_{ij} = 1$).

If we push such a molecule, it starts to vibrate. Because equations (9.10) are linear, the dynamics corresponds to a superposition of $N - 1$ harmonic modes α with different oscillation frequencies ω_α, i.e. we have

$$x_i(t) = C_0 \eta_{i,0} + \sum_{\alpha=1}^{N-1} C_\alpha e^{i\omega_\alpha t} \eta_{\alpha,i}. \tag{9.11}$$

where the constants C_α are determined by the initial conditions. The oscillation frequencies ω_α and the coefficients $\eta_{\alpha,i}$ are given by eigenvalues

$\lambda_\alpha = -\omega_\alpha^2$ and eigenvectors η_α of the eigenvalue equation

$$\widehat{D}\eta_\alpha = \lambda_\alpha \eta_\alpha \qquad (9.12)$$

for the matrix $\widehat{D}$ with elements

$$D_{ij} = T_{ij} - \delta_{ij} \sum_{k=1}^{N} T_{ik}. \qquad (9.13)$$

All eigenvalues λ_α of this matrix are nonpositive. Moreover, one of its eigenvalues is always zero ($\lambda_\alpha = 0$ for $\alpha = 0$) because the dynamical system (9.10) is invariant with respect to arbitrary translations. Note that using the definition (9.1) of the node degree m_i we can also express the matrix elements as

$$D_{ij} = T_{ij} - m_i \delta_{ij}. \qquad (9.14)$$

Thus, any graph generates a complex temporal signal (a certain "melody") which is characterized by a set of its frequencies $\omega_\alpha = \sqrt{-\lambda_\alpha}$ or, equivalently, by the eigenvalue spectrum $\{\lambda_\alpha\}$ for $\alpha = 1, 2, \ldots, N-1$. This spectrum is an imprint of a graph.

An important property of spectra is that they are *invariant*, i.e. do not depend on a particular enumeration of nodes in a given graph. Thus, all isomorphic graphs obtained by any possible permutations of N elements will have the same spectra, though they are described by different adjacency matrices T_{ij}.

A spectrum contains much information about its graph. Since all structural properties of a graph, such as average degrees, diameter and girth, are also invariants, a question arises as to whether they can be quantified in terms of its eigenvalues. A branch of mathematics, known as the *spectral graph theory*, is devoted to studies of such problems. Detailed introductions to this theory have been given by Cvetkovič, Doob and Sachs [476] and by Chung [472].

Different spectra of the same graph can be considered. For instance, a family of spectra specified by a real parameter μ is defined as the eigenvalue set of the matrix

$$A_{ij}(\mu) = T_{ij} - \mu m_i \delta_{ij}. \qquad (9.15)$$

When $\mu = 0$, this is just the adjacency matrix T_{ij}. Speaking of the spectrum of a graph, the spectrum of its adjacency matrix is usually assumed. The matrix (9.14) corresponds to putting $\mu = 1$ in (9.15), so that $D_{ij} = A_{ij}(0)$. The matrix $\widehat{D}$ is called the Laplacian matrix; its eigenvalues define the *Laplacian spectrum* of a graph.

Generally, the eigenvalue family $\lambda_\alpha(\mu)$ of the matrix $A_{ij}(\mu)$ for different parameters μ provides more information about the graph than one of its particular spectra. An exception is the case of regular graphs where degrees of all nodes are equal (i.e. $m_i = m$ for any i). For such graphs, the matrix $A_{ij}(\mu)$ is given by

$$A_{ij}(\mu) = T_{ij} - \mu m \delta_{ij}, \qquad (9.16)$$

and thus it differs from T_{ij} only by the identity matrix. Hence, we have

$$\lambda_\alpha(\mu) = \lambda_\alpha(0) - \mu m, \qquad (9.17)$$

where $\lambda_\alpha(0)$ are the eigenvalues of the adjacency matrix.

The description of a large graph in terms of its adjacency matrix T_{ij} is not convenient because it depends on the order in which all the nodes of the graph are enumerated. By choosing a different enumeration order, another adjacency matrix corresponding to the same graph is obtained. Since the number of different permutations in a sequence of N elements is $N!$, every graph with N nodes is generally described by a whole class of $N!$ adjacency matrices. This number is superexponentially large when graphs with many nodes are considered.

Such great degeneracy makes the comparison between graphs based on their adjacency matrices very time consuming. Indeed, if we are given two matrices T_{ij} and T'_{ij} and asked to find out whether they correspond to the same graph, we should try all possible permutations and each time compare all elements of the two matrices. Alternatively, we can calculate and compare the spectra of matrices T_{ij} and T'_{ij}. If they correspond to the same graph, their spectra must be identical.

The characterization of a graph by its spectrum appears very attractive. In the middle of the 20th century, when spectral investigations of graphs were started, it was believed that a spectrum uniquely defines its underlying graph. Later it was, however, found that in some cases different graphs may possess the same spectrum. Such graphs are called *cospectral*. Figure 9.7 shows two cospectral graphs of size six.

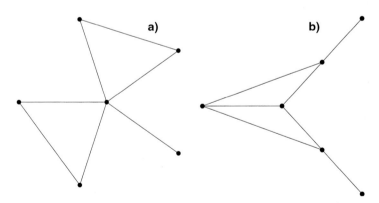

Fig. 9.7. Two cospectral graphs (**a** and **b**) of size $N = 6$

Even if two graphs have the same spectrum of their adjacency matrices, they may be characterized by different spectral families $\lambda_\alpha(\mu)$. In Fig. 9.8 such families for the two cospectral graphs shown in Fig. 9.7 are displayed.

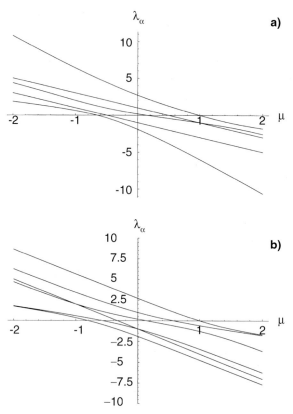

Fig. 9.8. Spectral families $\lambda_\alpha(\mu)$ for the two cospectral graphs shown in Fig. 9.7

The spectra become identical only at $\mu = 0$. Note that the graph shown in Fig. 9.7a has two eigenvalues that coincide at any parameter μ. Therefore we can see only 5 lines in Fig. 9.8a.

Nonetheless, it is not correct to say that a graph can always be identified by examining the spectral family. We have already seen that for any regular graph the parameter dependence of the spectrum is trivial, i.e. $\lambda_\alpha(\mu) = \lambda_\alpha(0) - \mu m$. Therefore, if two regular graphs of degree m have the same spectrum $\lambda_\alpha(0)$ of their adjacency matrices, their spectral families $\lambda_\alpha(\mu)$ are also identical. Figure 9.9 shows the two smallest regular cospectral graphs ($N = 10, m = 4$).

Thus, exact identification of a graph through its spectra is not possible, because regular graphs with the same spectra exist. In practical applications it may, however, be sufficient to identify two graphs, allowing a certain probability of making a mistake. To estimate this mistake, one should study how frequently cospectral regular graphs are found. This can be done by systematically generating all regular graphs of a given size and computing their spectra.

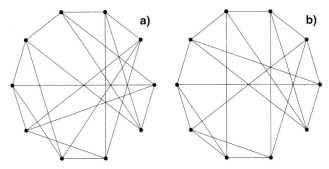

Fig. 9.9. Two regular cospectral graphs (**a** and **b**) of size $N = 10$ and degree $m = 4$

Since the number of possible graphs rapidly grows with the size N, the data are available only for relatively small sizes. Bussemaker et al. [477] have examined *all* connected regular graphs of degree $m = 3$ and sizes $N \leq 14$. The total number of such graphs is 621. Cospectral graphs of size $N < 14$ are absent and only three cospectral graphs of size $N = 14$ have been found. Therefore, the probability of finding cospectral graphs inside this subset is less than 0.5%.

Furthermore, spectra can be used to estimate the similarity between different connected graphs. When a graph is large, changing several of its connections would usually lead to only a small modification of its spectrum. Thus, by comparing their spectra, one can judge how close the two graphs are.

9.3 Clustering and Synchronization in Dynamical Networks

A dynamical network is formed by a population of interconnected elements with simple internal dynamics. The pattern of its connections represents a certain graph. Generally, the connections can be directed and characterized by various strengths. In this section, however, we limit our discussion to networks with symmetric constant connections. Furthermore, we assume that all elements in the considered population are identical.

In Chaps. 6 and 7 we considered the dynamics of globally coupled networks whose connection patterns represented complete graphs. Typically, such networks display synchronization and dynamical clustering. Below we show that this behavior is also characteristic of dynamical networks whose connection patterns are random. As an example, the networks formed by interconnected logistic maps will be investigated following a study [478].

We consider a population of N identical logistic maps $x_i(t+1) = f(x_i(t))$ with $f(x) = 1 - ax^2$ and $i = 1, 2, \ldots, N$. The pattern of connections in the network is specified by a random graph with the adjacency matrix T_{ij} which is obtained by independently generating any possible connection with a fixed probability ν. The collective dynamics of the network is described by the equations

$$x_i(t+1) = \left(1 - \frac{\varepsilon}{\nu(N-1)} \sum_{j=1}^{N} T_{ij}\right) f(x_i(t)) + \frac{\varepsilon}{\nu(N-1)} \sum_{j=1}^{N} T_{ij} f(x_j(t)),$$

$$(9.18)$$

where the coefficient ε determines the coupling strength.

This dynamics can be viewed as involving the "diffusion" of some substance over the graph: every node receives the substance from all other connected nodes of the graph and sends it to any other connected node. Equations (9.18) can also be written in the form

$$x_i(t+1) = f(x_i(t)) + \frac{\varepsilon}{\nu(N-1)} \sum_{j=1}^{N} D_{ij} f(x_j(t)), \qquad (9.19)$$

where D_{ij} is the respective Laplacian matrix (9.14).

We have introduced factors ν in the denominators in (9.18) or (9.19) to simplify the comparison between dynamical behaviors of networks based on graphs with different mean degrees. For the considered random graphs, a typical element would have $m = \nu(N-1)$ connections. When ν is increased, an element will thus receive signals from a larger number of other elements in the network and will be more strongly integrated into its collective dynamics. Our definition ensures that the mean signal received by an element in a network with any connection density will be the same at a fixed coupling strength ε. Note that when $\nu = 1$ equations (9.18) coincide with equations (7.5) that describe the collective dynamics of globally coupled logistic maps.

Numerical simulations of networks (9.18) have been performed [478]. They show that, when the coupling strength ε is gradually increased, these networks experience dynamical clustering and synchronization. However, these phenomena in random networks have significant differences as compared with the globally coupled system (Sect. 7.1).

The most important difference is that exact synchronization and clustering, characterized by identical states of elements, are never observed in the networks. Instead, their elements form compact clouds (or *fuzzy clusters*) whose size depends on the coupling strength and other parameters. Therefore, the employed criteria of clustering and synchronization must be modified.

We shall say that two elements i and j belong at time t with precision δ to the same cluster if the distance $d_{ij}(t) = |x_i(t) - x_j(t)|$ between their states is smaller than δ. Thus defined, clusters depend on the chosen precision. Usually, the finest precision which still identifies the considered cluster will be taken (see, however, also the discussion of the internal structure of fuzzy clusters below in this section).

As a criterion of synchronization in a network, one can require that the mean distance $\bar{d}$ between *any* two elements in the network within a given time interval is less than δ. Using this definition and taking $\delta = 10^{-8}$, synchronization thresholds ε^* were computed at $a = 2$ for various random networks with different sizes N and fixed connection density $\nu = 0.9$ (see Fig. 9.10). Each point in Fig. 9.10 gives the threshold value for one of the randomly generated graphs of the respective size. The open circles show the statistical averages over all generated graphs of that size, and the bars display the statistical dispersion of the data.

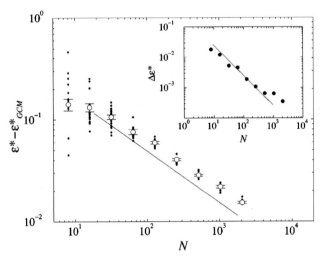

Fig. 9.10. Dependence of the synchronization threshold ε^* on the network size; ε^*_{GCM} is the synchronization threshold of the globally coupled system. The inset show the statistical dispersion of synchronization thresholds. (From [478])

Analyzing Fig. 9.10, we notice that the synchronization thresholds for random graphs are always somewhat higher than the respective threshold $\varepsilon^*_{GCM} = 0.5$ for globally coupled maps (i.e. for complete graphs). In the limit $N \to \infty$ this difference decreases as $(\varepsilon^* - \varepsilon^*_{GCM}) \propto N^{-1/2}$ and the thresholds for random networks approach that of the globally coupled maps.

This conclusion is interesting. It was shown in Sect. 9.2 that large random graphs are approximately regular, i.e. each node has almost exactly νN connections which lead to other, randomly chosen nodes. Now we see that such regular graphs have almost the same synchronization properties as complete graphs, where all possible connections are present. Similar results were obtained in simulations for other edge densities ν and control parameters a.

In globally coupled logistic maps, complete synchronization is preceded by an interval of coupling intensity where clustering is observed and the system behaves as a dynamical glass. This behavior is basically retained when global coupling is replaced by a graph of random connections between the maps. However, dynamical clustering is much more strongly influenced by such quenched disorder than the regimes with complete synchronization.

The precision-dependent order parameter $s(\delta)$ is defined as the fraction of elements having at least one other element at distance less than δ. Figure 9.11 shows this order parameter for precision $\delta = 10^{-3}$ as a function of the coupling strength ε at $a = 2$ for a random graph of size $N = 250$ and connection density $\nu = 0.8$. Complete synchronization sets in at approximately $\varepsilon = 0.56$ and is preceded by partial synchronization (i.e. dynamical clustering) that begins at about $\varepsilon = 0.16$. Small positive values of s which remain below this coupling strength are explained by the fact that δ is relatively large here and therefore the elements separated by such distances may sometimes be found in the asynchronous regime. Note that the order parameter s is maximal in the middle of the dynamical clustering interval and decreases before the transition to complete synchronization takes place.

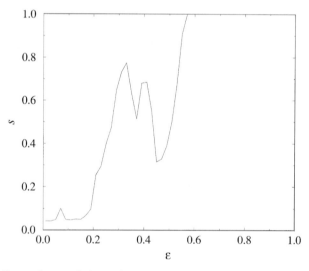

Fig. 9.11. Dependence of the order parameter s on the coupling strength ε for a random network of size $N = 250$ and connection density $\nu = 0.8$. (From [478])

More information about the system can be obtained by examining histograms of distributions over pair distances between elements. Such histograms for a large random network with $N = 1000$ and $\nu = 0.8$ are shown for four different coupling strengths in Fig. 9.12. To construct these histograms, the numbers of pairs with distances falling inside subsequent intervals of width $\Delta d = 0.01$ were counted at a fixed time.

In the turbulent phase (Fig. 9.12a, $\varepsilon = 0.1$) the distribution is almost flat, indicating that the elements are uniformly distributed over the one-element attractor. Inside the partial clustering interval, distributions with different numbers of clusters are observed (e.g. three clusters in Fig. 9.12b for $\varepsilon = 0.25$ and two clusters in Fig. 9.12c for $\varepsilon = 0.35$). Then the form of the distribution changes, signaling the onset of the transition to complete synchronization. Now the distribution has a single maximum at $d = 0$ and a relatively wide shoulder extending to larger pair distances d (see Fig. 9.12d for $\varepsilon = 0.45$). As the coupling strength ε is further gradually increased, the shoulder shrinks until only a single line at $d = 0$ is left, as expected for the complete synchronization.

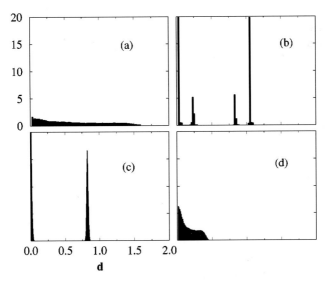

Fig. 9.12. Normalized histograms of distributions over pair distances d for a random network of size $N = 1000$ and connection density $\nu = 0.8$. The coupling strengths are (a) $\varepsilon = 0.1$, (b) $\varepsilon = 0.25$, (c) $\varepsilon = 0.35$ and (d) $\varepsilon = 0.45$. The vertical and horizontal axes have the same scales in all these plots. (From [478])

The distributions in Fig. 9.12 correspond to fixed times and therefore do not yet tell us what is the network dynamics in the respective regimes. Some impression of this dynamics can be gained by looking at the time dependence of the distance $d_{ij}(t)$ between a fixed pair of elements i and j. Figure 9.13 displays such typical dependences for the same system and the same four coupling strengths ε as in Fig. 9.12.

In the absence of synchronization (Fig. 9.13a, $\varepsilon = 0.1$) the pair distance evolves in an irregular way and shows large variations of order unity. In

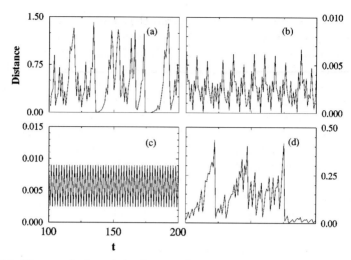

Fig. 9.13. Time evolution of the distance between a pair of elements in the same random network as in Fig. 9.12 at four different coupling strengths (**a**) $\varepsilon = 0.1$, (**b**) $\varepsilon = 0.25$, (**c**) $\varepsilon = 0.35$ and (**d**) $\varepsilon = 0.45$. In parts (**b**) and (**c**) the chosen elements belong to the same cluster. (From [478])

contrast to this, inside the dynamical clustering range the distance between elements belonging to the same (fuzzy) cluster remains small. For example, Fig. 9.13b for $\varepsilon = 0.25$ shows the time-dependent distance between two elements from one of the clusters seen in the histogram Fig. 9.12b (note the much shorter length scale indicated on the right-hand side of this plot!). If one waits a long time, the switching of an element to a different cluster, manifested in a sudden great increase of the pair distance, is found in this case.

Spontaneous self-locking into periodic dynamics was observed at some parameter values in the middle of the clustering interval. Examining the trajectories of individual elements, it was seen that then all of them were periodic, though different for different elements of the network. Thus, the system acquires a rigid internal organization and falls into the state of frozen disorder. An example of this behavior is presented in Fig. 9.13c, which corresponds to the two-cluster regime at $\varepsilon = 0.35$ shown in Fig. 9.12c. The pair distance between two elements in a cluster shows a purely periodic variation here. The elements have fixed affiliations and never switch between the clusters. The distance between the clusters shows similar periodic or quasiperiodic variation, though with a typical magnitude of the order of unity.

Figure 9.13d displays the dynamical behavior found shortly before complete synchronization (for $\varepsilon = 0.45$, as in Fig. 9.12d). We see that the distance between elements shows great variation (note the increase in the vertical scale). The intervals of synchronization alternate with long deviations. Such intermittency explains the appearance of the broad shoulder in the respective histogram. It is formed by the elements that temporally find themselves during a large excursion from the central cluster.

The intermittency is observed inside an interval of coupling strengths, which corresponds to the crossover from robust clustering to complete synchronization in globally coupled logistic maps. As we have noted in Sect. 7.1, extremely long transients are possible in this parameter region. To transform a system of globally coupled maps into a random network, we should randomly delete some connections – and thus introduce quenched disorder. Therefore, one may expect that this would be to a certain extent equivalent to the introduction of relatively strong dynamical noise into the system of globally coupled maps.

To test this conjecture, the dynamics of noisy globally coupled maps was investigated. This system is described by the equations

$$x_i(t+1) = (1 - \varepsilon)f(x_i(t)) + \frac{\varepsilon}{N} \sum_{j=1}^{N} f(x_j(t)) + \mu g(x_i(t))\xi_i(t), \qquad (9.20)$$

where $g(x_i(t)) = 1$ for additive noise and $g(x_i(t)) = x_i(t)$ for multiplicative noise. In these equations, the coefficient μ specifies the noise intensity and $\xi_i(t)$ are independent random numbers (in the considered simulations they were chosen with constant probability density from the segment $[-1, +1]$).

By running the system (9.20), distributions over pair distances were computed in the absence of noise ($\mu = 0$) and with relatively strong additive and multiplicative noise ($\mu = 10^{-3}$). In contrast to the previously shown histograms, time averaging was now performed. The results of such simulations in the parameter region corresponding to the intermittent dynamics ($a = 2$, $\varepsilon = 0.45$, $N = 50$) are shown in Fig. 9.14a–c.

When noise is absent, the averaged distribution over pair distances between elements depends on the initial conditions, as it should be for dynamical glasses. Two such distributions for the same parameter values but different initial conditions are displayed in Fig. 9.14a,b. When sufficiently strong noise is introduced, the dependence on the initial conditions is lost. In this case the average distribution over pair distances is unique (it is shown by the bold line for the additive noise and by the dashed line for the multiplicative noise in Fig. 9.14c).

Applying time averaging to histograms of pair distances for random networks in the intermittent region (Fig. 9.14d), we find that they are also independent of the initial conditions and their shape is similar to that of the histograms for the globally coupled maps with noise.

Under partial synchronization conditions, the networks develop a number of coherent clusters. At a given precision level δ, a cluster is formed by elements that have at least one other element of this group at a distance d shorter than δ. The number and the sizes of clusters, as well as the detailed composition of a cluster, may depend on time.

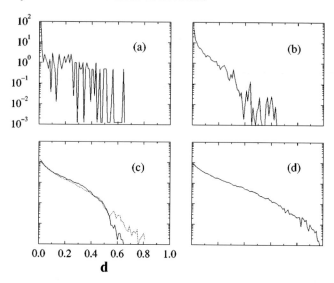

Fig. 9.14. Normalized histograms of distributions over pair distances for $N = 50$ globally coupled maps in the absence of noise (**a** and **b**, for two different initial conditions) and in the presence of additive (**c**, *dashed line*) or multiplicative (**c**, *bold line*) noise of intensity $\mu = 10^{-3}$. The respective histogram for a random network of size $N = 50$ and connection density $\nu = 0.8$ is shown in part (**d**). (From [478])

For a random network with $N = 50$ elements, connection density $\nu = 0.8$ and control parameter $a = 2$, the sizes of all clusters were computed with precision $\delta = 0.1$ at subsequent times for four different coupling strengths ε inside the dynamical clustering interval. The obtained dependences, including the last parts of the transients, are displayed in Fig. 9.15.

At the beginning of the partial synchronization interval, the cluster partitions are very fluid. For instance, at $\varepsilon = 0.25$ a stable cluster with 19 elements is found, while the cluster with 31 elements is unstable and often splits into subgroups of sizes $(17, 14)$, $(18, 13)$, $(20, 11)$, and so on (Fig. 9.15a). As ε is increased, the clusters generally become more stable and their lifetimes grow. For $\varepsilon = 0.28$ the elements are divided into a persistent cluster with 20 elements, while the rest of them irregularly alternate between a single cluster with 30 elements and two clusters with 27 and 3 elements (Fig. 9.15b)

When locking into periodic trajectories takes place at higher coupling strengths, stable clusters are observed. At $\varepsilon = 0.30$ two such clusters with 24 and 26 elements are formed after a transient (Fig. 9.15c). For the coupling strength $\varepsilon = 0.32$, we find a stable cluster with 27 elements and a second group of 23 elements which includes a map that periodically leaves this cluster (Fig. 9.15d).

As the coupling strength is increased, the transients become much longer, the dynamics of the system again becomes chaotic and occasionally a sin-

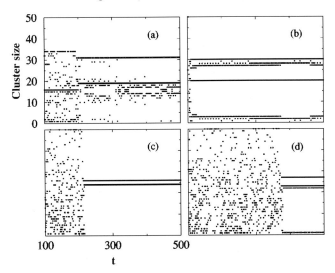

Fig. 9.15. Size and stability of synchronous groups in the clustering phase (see the text)

gle cluster including all elements of the network is found. At $\varepsilon = 0.35$ the intermittent regime begins.

The cluster structure of a network depends on the employed precision δ. If we take a cluster and examine it with a finer precision, we would often find that it consists of a number of smaller and more compact clusters. The hierarchical structure of such a cluster partition is schematically illustrated in Fig. 9.16. Here we show for $\varepsilon = 0.23$ the cluster partitions of the same random network with $N = 50$ elements at a fixed time, as resolved with three different precisions.

When the low precision $\delta = 0.1$ is used, the network is seen forming two clusters with 21 and 29 elements (the size of the cluster is given by the number in parentheses in Fig. 9.16). If the precision is made finer ($\delta = 0.05$), the large cluster with 29 elements is resolved into two smaller clusters of sizes 20 and 9. When $\delta = 0.01$ the cluster with 20 elements splits into two groups with 18 and 2 elements. Moreover, the initial cluster of size 21 is now also divided into two clusters of sizes 16 and 5.

For globally coupled maps, where all elements are equivalent with respect to their connections, the distribution of elements over clusters is arbitrary. If a network is considered, the elements differ in their connection patterns. Can the intrinsic architecture of a network introduce a bias and favor certain cluster partitions?

To investigate this, we introduce relative densities of connections *inside* a cluster and *between* clusters. Suppose that at time t with precision δ the random network separates into M clusters, each containing k_m elements ($m = 1, 2, \ldots, M$). The relative connection density ρ_m inside the cluster C_m is

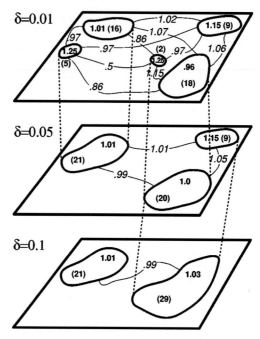

Fig. 9.16. Hierarchical structure of dynamical clusters for a random network of size $N = 50$ and connection density $\nu = 0.8$ at $a = 2$ and $\varepsilon = 0.23$. Numbers in parentheses show the cluster sizes. Numbers without parentheses inside the bounded regions indicate the relative connection density in the respective clusters. Numbers on the *dotted lines* connecting clusters display the relative densities of connections between these clusters. (From [478])

defined as

$$\rho_m = \frac{1}{\nu k_m (k_m - 1)} \sum_{i,j \in C_m} T_{ij}, \qquad (9.21)$$

where the sum is taken over all pairs of elements belonging to this cluster. The relative density of connections $\rho_{l,n}$ between clusters C_l and C_n is given by

$$\rho_{l,n} = \frac{1}{\nu k_l k_n} \sum_{i \in C_l, j \in C_n} T_{ij}, \qquad (9.22)$$

where elements i are chosen from the cluster C_l and elements j from the cluster C_n.

Thus, if $\rho_m > 1$, the density of connections inside the mth cluster is higher than the average connection density for the entire network. If $\rho_{l,n} > 1$ the connections between clusters C_l and C_n are more dense than the average inside the network. As we see from Fig. 9.16, at the low precision $\delta = 0.1$ the density of connections between two existing clusters is indeed smaller than on the average ($\rho_{1,2} = 0.99$), whereas the densities of connections inside the clusters are a little higher ($\rho_1 = 1.01$ and $\rho_2 = 1.03$). For finer precisions $\delta = 0.05$ and $\delta = 0.01$ this is not always true.

The mean relative intracluster connection density ρ_{intra} of the entire network for a partition with M clusters can be defined as

$$\rho_{intra} = \frac{1}{M} \sum_{m=1}^{M} \rho_m, \qquad (9.23)$$

whereas the mean relative intercluster connection density ρ_{inter} is

$$\rho_{inter} = \frac{1}{M(M-1)} \sum_{l,n=1}^{M} \rho_{l,n}, \qquad (9.24)$$

where $l \neq n$.

To statistically analyze the role of the network architecture in determining cluster partitions, an ensemble of 10^4 independently generated random graphs of size $N = 250$ and $\nu = 0.8$ was studied. For each graph, cluster partitions for $\varepsilon = 0.3$ and $a = 2$ at the precision $\delta = 0.1$ were determined at a fixed time and mean intra- and intercluster connection densities were computed [478]. This has allowed us to construct the statistical distributions of these densities, which are shown in Fig. 9.17.

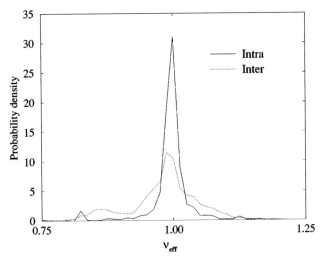

Fig. 9.17. Statistical distributions of mean relative intracluster (bold line) and intercluster (dotted line) connection densities. Averaging over 10^4 different randomly generated networks with $a = 2$ of size $N = 250$ and connection density $\nu = 0.8$ at coupling strength $\varepsilon = 0.3$ and precision $\delta = 0.1$. (From [478])

Before we discuss such distributions, some statistical properties of random graphs should be recalled. We have already seen in Sect. 9.2 that, if each potential connection is independently chosen with the same probability ν, the large graphs ($N \to \infty$) thus generated are approximately regular, i.e. each node has almost exactly νN connections (see (9.4)). The probability of finding subsets of nodes with connection densities significantly different from ν is very small. Therefore, we cannot expect large deviations of mean relative inter- and intracluster connection densities from unity for the clusters which include large numbers of elements.

Indeed, examining Fig. 9.17 we note that both distributions $p(\rho_{\text{intra}})$ and $p(\rho_{\text{inter}})$ have the shape of narrow peaks located near $\rho = 1$. However, these distributions are not identical, as should have been the case if no correlations between the graph architecture and dynamical clustering existed. The centers of these two distributions are slightly shifted in the opposite directions: $< \rho_{\text{intra}} >= 1.013$ and $< \rho_{\text{inter}} >= 0.987$. Moreover, the distribution of intercluster connection densities is significantly broader and has a wide shoulder extending towards lower densities.

Thus, it can be concluded that the elements belonging to the same cluster would typically have a higher density of connections inside this cluster than with the elements belonging to other clusters. This effect is relatively weak for the considered random networks with independently generated connections. It can however become much stronger when non-random networks or random networks generated by more sophisticated procedures (see Sect. 9.2) are taken. Indeed, by analyzing clustering in various small networks with some symmetry properties it was found that certain partitions of elements into clusters are apparently forbidden by the architecture and never observed in simulations [478].

Finally, one more important dynamical property of the considered systems should be discussed. Suppose that within the considered time interval the system exhibits a partition into M clusters C_m of sizes k_m at precision δ. For each cluster we can introduce its dynamical variable X_m as

$$X_m(t) = \frac{1}{k_m} \sum_{i \in C_m} x_i(t), \tag{9.25}$$

so that the states of the elements inside the cluster are given by $x_i(t) = X_m(t) + x'_{i,m}(t)$, where $x'_{i,m}(t)$ are deviations of order δ. By summing equations (9.18) for $i \in C_m$ and dividing by the number k_m of elements in the respective cluster, we obtain

$$X_m(t+1) = \frac{1}{k_m} \sum_{i \in C_m} \left(1 - \frac{\varepsilon}{\nu(N-1)} \sum_{j=1}^{N} T_{ij} \right) f(x_i(t)) \tag{9.26}$$

$$+ \frac{\varepsilon}{\nu k_m(N-1)} \sum_{i \in C_m} \sum_{j=1}^{N} T_{ij} f(x_j(t)).$$

If element j belongs to cluster C_n, we can write

$$f(x_j(t)) = f(X_n(t)) + O(\delta), \tag{9.27}$$

where $O(\delta)$ denotes corrections of order δ. Therefore,

$$\sum_{i \in C_m} \sum_{j=1}^{N} T_{ij} f(x_j(t)) = \sum_{n=1}^{M} \sum_{i \in C_m, j \in C_n} T_{ij} f(X_n(t)) + O(\delta). \tag{9.28}$$

Substituting this into equation (9.26) and neglecting terms of order δ, we obtain

$$X_m(t+1) = \left(1 - \frac{\varepsilon}{\nu k_m(N-1)} \sum_{n \neq m} \sum_{i \in C_m, j \in C_n} T_{ij}\right) f(X_m(t)) \tag{9.29}$$
$$+ \frac{\varepsilon}{\nu k_m(N-1)} \sum_{n \neq m} \sum_{i \in C_m, j \in C_n} T_{ij} f(X_n(t)).$$

Furthermore, taking into account the definition (9.22) of connection densities $\rho_{m,n}$ between different clusters, we can write (9.29) in the final form

$$X_m(t+1) = \left(1 - \sum_{n \neq m} J_{mn}\right) f(X_m(t)) + \sum_{n \neq m} J_{mn} f(X_n(t)), \tag{9.30}$$

where

$$J_{mn} = \frac{\varepsilon k_n \rho_{m,n}}{N-1}. \tag{9.31}$$

This result is remarkable. The system of clusters builds up a *new dynamical network*. This network represents a system of coupled logistic maps $X(t+1) = f(X(t))$ with direction-dependent connections of different weights. The weight of an effective connection leading from map n to map m in this new dynamical network is given by the matrix element J_{mn}.

The connection weights are determined, according to (9.31), by the densities $\rho_{m,n}$ of connections between the respective clusters in the original graph. Moreover, they are proportional to the numbers k_n of elements in clusters C_n. Hence, map n will exercise a stronger influence if it corresponds to a larger cluster C_n which is more densely connected to the cluster C_m.

We have considered in this section the collective dynamics of networks whose individual elements represent chaotic logistic maps. Patterns of connections in these networks were specified by graphs that were randomly generated but remained fixed during the dynamics. The intensity of interactions between connected elements was characterized by the coupling strength parameter.

When the coupling is weak, the dynamics of the elements in the network is effectively independent. If coupling is intensive, strong correlations between elements develop and the dynamics becomes synchronous. These two opposite dynamical regimes are separated by a relatively wide region of coupling strengths where partial synchronization of collective dynamics takes place.

Inside this region the elements condense into a number of coherent fuzzy clusters. These clusters can be viewed as new elementary units. The system of interacting clusters represents a dynamical network characterized by directed connections with varying weights. This effective network *emerges* through partial synchronization.

Depending on the initial conditions, the same original system of coupled logistic maps may produce different cluster partitions. Usually, these cluster partitions slowly change with time because of the splitting and merging of clusters and the switching of individual elements between the clusters. Each cluster partition gives rise to its own effective dynamical network of clusters with a particular set of connection weights. Thus, a great variety of effective dynamic networks can be generated by the same system.

These emergent networks, whose units are coherently operating groups of elements, may have largely different dynamics. Since cluster partitions evolve with time, the emergent networks are *fluid*. As time goes on, one effective network spontaneously transforms into another with a different collective dynamics. The behavior of the system looks, therefore, like slow wandering over a set of dynamic networks corresponding to various cluster partitions.

The lifetimes of particular networks may be relatively long. Moreover, in the middle of the partial synchronization region locking into periodic dynamical regimes characterized by rigid cluster partitions is observed. Such rigid partitions give rise to stable dynamical networks.

The transition from cluster organization to completely synchronous dynamics does not proceed gradually. When the coupling strength is increased, all emergent networks are first destroyed, giving rise to intermittent chaotic dynamics which later converges to the synchronous regime.

The above results were derived using the example of logistic maps. However, similar collective behavior can apparently be expected in populations formed by other chaotic dynamical systems. Indeed, dynamical clustering in systems of globally coupled Rössler oscillators and cross-coupled chaotic neural networks has many properties which are common with the logistic maps.

9.4 Evolution of Graphs

We have seen in the previous section that, through the formation of coherently operating groups, a network with a fixed graph of connections between elements gives rise to a great number of emergent networks with different dynamics. Rearranging the distribution of elements over clusters, such a system may quickly modify its collective behavior and generate a large variety of dynamical patterns. The role of the graph of connections between individual elements is that it yields biases which favor certain dynamical patterns and make other forms of collective behavior improbable or forbidden. Thus, one can say that graphs provide the *grammar* for such complex dynamics.

Over relatively short timescales, the great functional flexibility of dynamical networks is sufficient for many purposes. For instance, emergent short-lived networks of coherently operating neurons in the brain may well be responsible for the rapid emulation of dynamical processes in the external world, for the generation of motor signals and for the short-term memory. On much longer timescales, the patterns of connections between elements (i.e. graphs of dynamical networks) can slowly evolve. In this way the grammar and the dynamical language become optimized for the expression of desirable behavioral patterns. Moreover, through the adaptive evolution of its graph a network also *learns* to recognize the typical, frequently repeated combinations of external stimuli and to appropriately respond to them.

To evolve, a graph must be *plastic*. This means that some connections between elements may disappear and new connections can become established. If the connections are characterized by weights, these weights can change with time. Furthermore, some elements may die out and become replaced by new network units with completely different patterns of connections to the remaining elements. The plasticity is an important and well-known property of the brain and the nervous system. Similar mechanisms can, however, be found in many other systems, from human societies to immune networks.

In the last few decades, the evolution of connection patterns was mainly investigated for neural networks from the viewpoint of information processing. Indeed, through appropriate modification of the weights of synaptic connections between neurons this system may store a large number of different static patterns. This leads to the effect of associative memory: when a new pattern is presented, the system responds by activating the stored pattern which most closely resembles the one presented. Moreover, temporal sequences can also be stored in such networks. A discussion of the related problems by one of the present authors can be found in his earlier book [479].

Below in this section we shall focus our attention on a different aspect. The question which we are going to address is how, through the evolution of their connection patterns, dynamical networks can learn to emulate the dynamics of other systems or other networks. We shall restrict our discussion to an example which has been suggested by Mikhailov [480] and recently analyzed by Ipsen and Mikhailov [481].

Let us consider a dynamical network formed by N identical particles connected by identical elastic strings. The connections are determined by a certain graph G. The dynamics of such a one-dimensional "molecule" is described by the set of linear differential equations $(i = 1, 2, \ldots, N)$

$$\ddot{x}_i + \gamma \, \dot{x}_i + \sum_{j=1}^{N} T_{ij}(x_i - x_j) = \xi_i(t), \tag{9.32}$$

where T_{ij} is the adjacency matrix of graph G. Hence, two particles i and j are connected by a string only if the respective element of the adjacency matrix is $T_{ij} = 1$.

In absence of noise $\xi_i(t)$ this network shows no persistent activity and relaxes to the trivial stationary state. Its response to external noise depends on the pattern of connections between particles. We assume that $\xi_i(t)$ are independent white noises,

$$< \xi_i(t)\xi_j(t') >= 2\gamma\delta_{ij}\delta(t - t'). \tag{9.33}$$

The network (9.32) is used to generate the signal $z(t)$ by taking a weighted sum of velocities $\dot{x}_i(t)$, i.e. as

$$z(t) = \frac{1}{N} \sum_{i=1}^{N} a_i \dot{x}_i(t). \tag{9.34}$$

The weight coefficients a_i are independent random binary numbers that take with equal probability the values $a_i = \pm 1$. For convenience, we assume that the considered network is large, i.e. $N \gg 1$.

The autocorrelation function $f(\tau)$ and the spectral density $f(\omega)$ of the signal $z(t)$ are defined as

$$f(\tau) =< z(t)z(t + \tau) >, \quad f(\omega) = \frac{1}{2\pi} \int_{-\infty}^{\infty} f(\tau)e^{-i\omega\tau}d\tau, \tag{9.35}$$

where the brackets $< ... >$ denote statistical averaging over the noises $\xi_i(t)$ and random weights a_i.

Substituting (9.34) into the definition of $f(\omega)$, we obtain

$$f(\omega) = \frac{1}{2\pi N^2} \sum_{i,j=1}^{N} < a_i a_j > \int_{-\infty}^{\infty} <\dot{x}_i(t)\dot{x}_j(t + \tau) > e^{-i\omega\tau}d\tau. \tag{9.36}$$

Because $a_i = \pm 1$ are independent random numbers, we have $< a_i a_j >= 0$ for $i \neq j$ and $a_i^2 = 1$, so that

$$f(\omega) = \frac{1}{N^2} \sum_{i=1}^{N} f_i(\omega), \tag{9.37}$$

where

$$f_i(\omega) = \frac{1}{2\pi} \int_{-\infty}^{\infty} <\dot{x}_i(t)\dot{x}_j(t + \tau) > e^{-i\omega\tau}d\tau. \tag{9.38}$$

Applying the Fourier transformation to dynamical equations (9.32), we obtain

$$\left(-\omega^2 + i\omega\gamma - \widehat{D}\right) \mathbf{x}(\omega) = \boldsymbol{\xi}(\omega). \tag{9.39}$$

Here vector notations $\mathbf{x}(\omega) \equiv \{x_i(\omega)\}$ and $\boldsymbol{\xi}(\omega) = \{\xi_i(\omega)\}$ are used and $\widehat{D}$ is the Laplacian matrix of the graph, $D_{ij} = T_{ij} - \delta_{ij} \sum_{k=1}^{N} T_{ik}$.

To solve (9.39), we note that the symmetric matrix $\widehat{D}$ has a complete set of orthogonal normalized eigenvectors η_α, defined by the eigenvalue equation

$$\widehat{D}\eta_\alpha = \lambda_\alpha \eta_\alpha. \tag{9.40}$$

The eigenvalues λ_α determine the frequencies $\omega_\alpha = \sqrt{-\lambda_\alpha}$ of the various oscillation modes of the "molecule" (9.32). Note that therefore we always have $\lambda_\alpha \leq 0$. The number of eigenvalues is equal to the number of particles in the system. Since, in the absence of noise, the system is invariant with respect to translations, one of the eigenvalues must be zero. We shall use the ascending order to enumerate frequencies ω_α ($\alpha = 0, 1, 2, \ldots, N-1$), i.e. assign $\alpha = 0$ to the lowest frequency $\omega_0 = 0$.

Decomposing $\mathbf{x}(\omega)$ into a linear superposition of eigenvectors η_α and substituting this into (9.39), the solution is found to be

$$\mathbf{x}(\omega) = \sum_{\alpha=0}^{N-1} \frac{(\xi(\omega) \cdot \eta_\alpha)\, \eta_\alpha}{\omega_\alpha^2 - \omega^2 + i\omega\gamma}. \tag{9.41}$$

Noting that $< \xi_i(\omega)\xi_j^*(\omega') > = (\gamma/\pi)\delta_{ij}\delta(\omega - \omega')$ and using this solution, we obtain

$$f_i(\omega) = \frac{\gamma}{\pi} \sum_{\alpha=0}^{N-1} \frac{\omega^2 \eta_{\alpha,i}\eta_{\alpha,i}^*}{\left(\omega^2 - \omega_\alpha^2\right)^2 + \omega^2\gamma^2}. \tag{9.42}$$

Therefore, the spectral density $f(\omega)$ of the signal $z(t)$ is

$$f(\omega) = \frac{1}{\pi N^2} \sum_{\alpha=0}^{N-1} \frac{\gamma\omega^2}{\left(\omega^2 - \omega_\alpha^2\right)^2 + \omega^2\gamma^2}. \tag{9.43}$$

If we consider the spectral density only for $\omega > 0$ and the damping constant γ is significantly smaller than all eigenfrequencies $\omega_\alpha \neq 0$, this expression can be approximately reduced to

$$f(\omega) = \frac{1}{2\pi N^2} \sum_{\alpha=0}^{N-1} \frac{\gamma_0}{\left(\omega - \omega_\alpha\right)^2 + \gamma_0^2}, \tag{9.44}$$

where $\gamma_0 = \gamma/2$.

Thus, the spectral density $f(\omega)$ is a sum of Lorentz distributions of width γ located at frequencies ω_α which correspond to different oscillation modes of the system (9.32). These frequencies are determined as $\omega_\alpha^2 = -\lambda_\alpha$ in terms of the eigenvalues of the Laplacian matrix $D_{ij} = T_{ij} - \delta_{ij}\sum_{k=1}^{N} T_{ik}$ of the graph G. Hence, the signal $z(t)$ and its spectral density $f(\omega)$ contain information about this graph. Effectively, the considered dynamical network acts as a filter that transforms white applied noise into the signal $z(t)$ with a spectral density $f(\omega)$, whose shape is determined by the pattern of connections in the graph G.

Above, we have shown how the spectral density can be calculated if the graph is known. Next, the *inverse problem* is discussed. Suppose that we receive a signal $z_0(t)$ which is produced by an unknown graph G_0. Can we *reconstruct* this graph G_0 from the spectral density of its signal? We can try to find a solution of this inverse problem by running the stochastic evolution of a test network with plastic connections between elements. The graph G of the test network will then evolve in such a way that the spectral density $f(\omega)$ of its signal $z(t)$ tends to approach the spectral density $f_0(\omega)$ of the received signal $z_0(t)$.

Of course, we already know that cospectral graphs exist. Therefore, different graphs with the same Laplace spectrum and, hence, with the same spectral densities are possible. This means that generally the inverse problem has no unique solution. However, the probability of accidentally finding two graphs with exactly the same spectra is low (see Sect. 9.2). This means that in the great majority of cases the graphs which generate signals with the same spectral density would be indeed identical. Furthermore, if two graphs have only slightly different spectral densities, there is a high probability that these two graphs are themselves similar.

To measure the similarity between two spectral densities $f(\omega)$ and $f_0(\omega)$, we use the distance w defined as

$$w = \int\limits_0^\infty (f(\omega) - f_0(\omega))^2 \, d\omega. \tag{9.45}$$

For convenience, it will be assumed that both spectral densities are always normalized to unity, i.e. $\int_0^\infty f(\omega)d\omega = \int_0^\infty f_0(\omega)d\omega = 1$. The distance w is zero when the two spectral densities coincide.

The test network is again described by equations (9.32). However, its graph G is not fixed now. Instead, it will slowly evolve through random *mutations* and subsequent *selection*. Mutations and selection can be introduced in various ways. We use the following procedure:

Suppose that at the evolutionary step n we have a test graph G with adjacency matrix T_{ij} and spectral density $f(\omega)$, which differs by w from the target spectral density $f_0(\omega)$. To produce a mutation, we choose at random one of the nodes k in the graph G. Now we replace the old connections linking this node k to other nodes in the graph by a new set of connections. To generate the new connections of the node k, we first randomly decide how many such connections there should be after the mutation, i.e. what would be the new degree m of the node k. The number m is chosen at random from 1 to $N - 1$. Then we randomly choose m nodes in the graph and connect them with the node k. As a result, a mutated graph G' is constructed.

The adjacency matrix T'_{ij} of the mutated graph G' is thus different from the matrix T_{ij} only in its kth row and its kth column. The row T_{kj} ($j = 1, 2, \ldots, N$) of the old matrix is replaced by a random sequence that contains

m ones and $N - m$ zeros. Because the matrix is symmetric, the same sequence replaces the kth column of the matrix.

For the mutated graph G', the spectral density $f'(\omega)$ is determined by computing its Laplace spectrum and using (9.44) with subsequent normalization to unity. The distance w' between the new spectral density $f'(\omega)$ and the target spectral density $f_0(\omega)$ is computed.

Next, the decision should be made whether an occurred mutation is accepted or rejected. It is obvious that the mutation decreasing the distance to the target, i.e. with $\Delta w = w' - w < 0$, must always be accepted. However, it would be wise to accept sometimes also a mutation that increases the distance. Otherwise, as in other optimization problems, there is a chance that the evolution ends in a local optimum because, to leave it, a barrier must be overcome. Therefore, the mutation is accepted with a probability $p(\Delta w)$ determined as

$$p(\Delta w) = 1 \text{ for } \Delta w \leq 0 \text{ and } p(\Delta w) = \exp\left(-\frac{\Delta w}{w\theta}\right) \text{ for } \Delta w > 0. \quad (9.46)$$

The parameter θ specifies the characteristic relative distance difference. The mutations leading to $\Delta w/w > \theta$ are accepted only with an exponentially small probability, i.e. extremely rarely.

At the next step $n + 1$ the whole procedure is repeated, taking as G the graph which was selected at the end of the previous step n. The evolution is continued until the spectral density $f(\omega)$ of the test graph approximates with the needed precision the spectral density $f_0(\omega)$ of the target graph.

Numerical simulations of evolving graphs, using the described algorithm, have been performed [481]. In these simulations both the target and the test graphs had sizes $N = 50$ and their spectral densities were determined assuming the same damping constant γ.

A random graph G_0 with connection density $\nu_0 = 0.9$ is chosen in the first simulation. Its spectral density $f_0(\omega)$ has for $\gamma = 0.08$ the form of a narrow peak shown by the solid line in Fig. 9.18. As the initial evolving graph we also take a random graph, but with a much lower connection density $\nu = 0.1$. Its spectral density for $\gamma = 0.08$ is shown by the dashed line in Fig. 9.18. This distribution is much broader and its center is located at a lower frequency.

The evolution, employing the above mutation scheme and the selection rule (9.46) with $\theta = 0.01$, leads to rapid convergence of the spectral density $f(\omega)$ of the evolving graph G to the spectral density $f_0(\omega)$ of the target graph. The spectral density $f(\omega)$ which was obtained after 10000 steps is displayed by a dotted line in Fig. 9.18. It is so close to the target spectral density $f_0(\omega)$ that the two plots almost coincide.

Examining the corresponding dependence of the distance $w(n)$ between two spectral densities (Fig. 9.19), we observe that it indeed rapidly falls down within the first 1000 steps. Then, however, the evolution becomes very slow and is characterized by the appearance of increasingly long intervals of

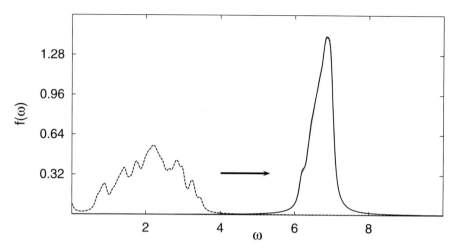

Fig. 9.18. Spectral densities for the target random graph with $\nu = 0.9$ (*solid line*), for the initial graph (*dashed line*) and for the final graph after 10^4 evolution steps (*dotted line*); $N = 50$, $\gamma = 0.08$ and $\theta = 0.01$

time over which all mutations are rejected and the distance $w(n)$ remains constant. This slowing down of the evolution process takes place when the target is nearly reached (the distance between the final and the target spectral densities after 10 000 steps in Fig. 9.18 is $w = 0.07$).

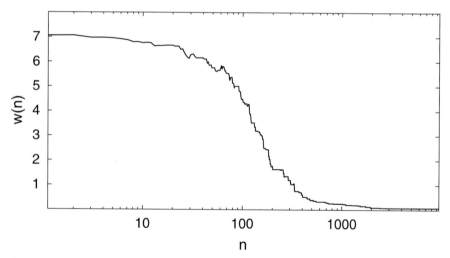

Fig. 9.19. Time dependence of the distance $w(n)$ between the two spectral densities corresponding to Fig. 9.18. Note the logarithmic timescale

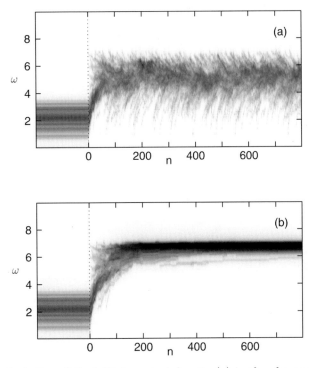

Fig. 9.20. Evolution of the initial spectral density (**a**) in the absence of selection ($\theta = \infty$) and (**b**) when selection is included ($\theta = 0.01$). The spectral density $f(\omega)$ at subsequent moments n is shown in a gray scale. The parameters are the same as in Fig. 9.18

To illustrate the evolution process and the role of selection, we first show in Fig. 9.20a the evolution of the initial spectral density when *no* selection is applied, i.e. the result of any mutation is accepted. The spectral density $f(\omega)$ is shown in a gray scale as a function of the evolutionary time n. The mutations are introduced starting from the moment $n = 0$. They lead to the development of a strongly fluctuating broad spectrum.

When the selection rule (9.46) is in operation (Fig. 9.20b), the evolution is different. The spectrum quickly shifts to the position corresponding to the center of the target spectrum and then slowly evolves, finely tuning to the details of the target spectral density.

The target graph used in this example was relatively uniform, since it was constructed by independently choosing any potential connection with the same probability ν. Therefore, its spectral density $f_0(\omega)$ had the shape of a single narrow peak. To further investigate the efficiency of the evolution process, in the second considered example we employ a more elaborate construction of the target graph G_0.

To generate it, we begin with a sparse random graph G_1 of size $N = 50$ and low connection density $\nu = 0.15$. Then we generate two different small random graphs g_1 and g_2 of equal sizes $N = 15$ and high connection density $\nu = 0.9$. These two small graphs are added to the sparse graph G_1. To do this, we choose two non-overlapping subsets of nodes of this graph, each containing 15 elements, and identify them with the nodes of the two small dense graphs. The graph G_0 is then obtained by retaining all connections that are present either in G_1 or in g_1 and g_2. Hence, it would contain two dense small regions on the sparse background (Fig. 9.21). The spectral density of the target graph thus constructed is shown for $\gamma = 0.08$ by the solid line in Fig. 9.22. It has two main maxima and several smaller peaks.

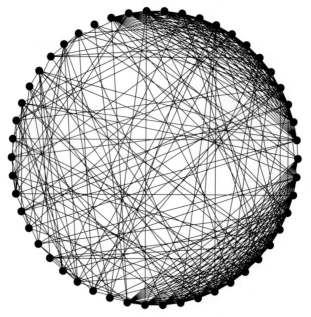

Fig. 9.21. A sparse random graph of size $N = 50$ with two dense random subgraphs of size $N = 15$

To start the evolution process, we now choose a random graph with high connection density $\nu = 0.9$ (the same as the target graph in the previous example). The spectral density of this graph is shown for $\gamma = 0.08$ by the dashed line in Fig. 9.22. Running the evolution with $\theta = 0.01$, we find that the graph again tends to approach the target spectrum. Its spectral density after 10^5 evolution steps is displayed by the dotted line. Though some differences are apparent, the final density $f(\omega)$ closely reproduces the profile of the target spectral density $f_0(\omega)$. The initial stage of the evolution process is illustrated

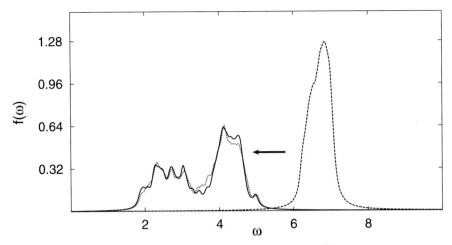

Fig. 9.22. Spectral densities for the target graph shown in Fig. 9.21 (*solid line*), for the initial graph (*dashed line*) and for the final graph after 10^5 evolution steps (*dotted line*); $N = 50$, $\gamma = 0.08$ and $\theta = 0.01$

in Fig. 9.23b. For comparison, we also show the evolution process in the absence of selection (Fig. 9.23a).

The time dependence of the distance w between two spectral densities, shown in Fig. 9.24, is similar to that found in the previous example. The transition from the rapid decrease to the final slow evolution stage is again observed after approximately 1000 iterations. The convergence rate is significantly lower now, so that even after 10^5 steps the distance is $w = 0.37$.

To estimate the efficiency of evolutionary selection, we should remember that the total number M_N of different graphs of size N is astronomically large even for relatively small sizes. Indeed, to define a graph we should specify its $N(N-1)/2$ possible edges, which can be done in $2^{N(N-1)/2}$ different ways. The actual number is a little smaller, because the graphs corresponding to the various permutations are identical. Since the number of permutations is $N!$, we have $M_N = 2^{N(N-1)/2}/N!$. Taking, for example, $N = 50$ yields $M_{50} = 2^{50 \cdot 49/2}/50! = 1.9 \times 10^{304}$, which is by many orders of magnitude larger than the total number of atoms in the universe (usually estimated as 10^{80}). Hence, the probability that, by randomly generating test graphs, we would encounter a certain given graph or even come close to it is practically zero. It is, therefore, remarkable that the evolution process allows us to approach the target graph within just a hundred thousand iterations.

On the other hand, the final stage of the evolution is very slow. Suppose, for instance, that the test graph differs from the target graph only in the connection pattern of a single node, i.e. it is separated from the target graph by a single mutation. Obviously, this would be that particular mutation which converts the test graph into the target. It can, therefore, be found only by

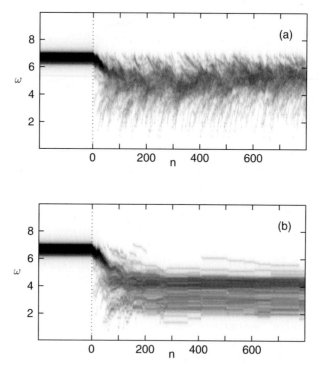

Fig. 9.23. Evolution of the initial spectral density (**a**) in absence of selection ($\theta = \infty$) and (**b**) when selection is included ($\theta = 0.01$). The spectral density $f(\omega)$ at subsequent moments n is shown in a gray scale. The parameters are the same as in Fig. 9.22

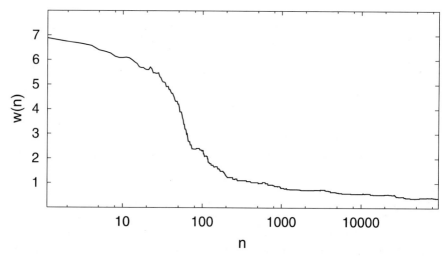

Fig. 9.24. Time dependence of the distance $w(n)$ between the two spectral densities, corresponding to Fig. 9.22

trying all possible one-node mutations. The total number of such mutations is $2^{N-1}N$ and this gives the estimate of the number of iterations which would generally be needed. When $N = 50$, this number is $2.8 \cdot 10^{16}$, which is not astronomical but still too large.

We see that the exact reconstruction of a graph from its spectrum through evolutionary selection is not feasible even for relatively small graphs (even if the existence of cospectral graphs is neglected). However, within a short time the evolution can yield a graph that would be quite similar to the target in terms of its spectrum. This behavior is typical of evolutionary techniques in many complex combinatorial optimization problems (see [479]).

In the above examples the spectral densities $f_0(\omega)$ were produced by target graphs. However, the same approach remains applicable when $f_0(\omega)$ is just the power spectrum of some external signal. Through subsequent mutations and selection an evolving plastic network would then tune itself to reproduce this spectrum with sufficiently high precision.

We have earlier noted that the dynamical network (9.32) is essentially a filter which transforms white noise into a stochastic signal with a certain autocorrelation function $f(\tau)$ and spectral density $f(\omega)$. Our analysis suggests that such plastic networks may tune themselves to reproduce a large variety of spectral densities. Thus, they can emulate signals generated by various stochastic systems.

The nature of the great emulation capacity of plastic dynamical networks is clear. By rearranging its connections, such a network can exhibit a vast number of different behavioral patterns, corresponding to different graphs. Hence, almost any function can be effectively "written" into the structure of the connections between the elements of a sufficiently large dynamical network.

Our discussion in this section has been limited to stochastic networks (9.32) that possess no intrinsic dynamics and only respond in various ways to external noise. The advantage of this analysis was that we could directly relate the behavior of such networks to the spectral properties of the underlying graphs. Therefore, instead of actually simulating the dynamics of networks, we could numerically compute the eigenvalues of the Laplacian matrices of the respective graphs and use these eigenvalues to estimate the efficiency of subsequent mutations.

The above evolutionary algorithm can, however, be easily extended to dynamical networks with intrinsical chaotic dynamics. Such networks would generate chaotic signals even in absence of noise. The autocorrelation function and the spectrum of these signals would again depend on the pattern of connections in the network, i.e. on its graph. The evolution of the graph can be designed in such a way that the network eventually learns to actively generate a prescribed signal or actively emulate the behavior of a different dynamical system.

9.5 Further Reading

The dynamics and evolution of networks play a fundamental role in the functioning of the immune system. The picture of the immune system given in Sect. 9.1 corresponds to the clonal selection theory proposed by Jerne [482] and Burnet [484]. The bone marrow randomly produces a great diversity of lymphocyte clones, which are integrated into the dynamics of the immune system. Most of these clones will die within a few days if they do not participate in immune reactions. When the immune system is warding off an infection, selected fitting idiotypes are not only expanded but also refined in a form of rapid evolution producing idiotypes with slightly differing fit which are in turn expanded if they provide a better fit with the antigen. For models of various aspects of clonal selection theory, see [485].

In order to efficiently recognize foreign antigens, the immune system has to be complete, meaning that it has to recognize virtually all chemical shapes that a (hitherto) unknown antigen might display. Perelson and Oster estimated the minimal size of a complete immune system [486]. Their analysis was based on the concept of shape space spanned by various parameters that determine the fit between antibody receptors and the antigen (see also [487–489]). Jerne [483] noted that in a complete immune system the antibodies would also react with other antibodies of the immune system. Idiotypes and anti-idiotypes consequently form a network. This raises the problem of how the immune system avoids mounting an immune response against itself, i.e. how to avoid auto-immunity. Another problem in relation to idiotypic networks is how the immune system avoids the excitation of one clone percolating through the network, leading to an excitation of large parts of the immune system, a question raised in early immune network models that were developed in the 1970s (see [490, 491]). These models did not yet take into account the network structure. Perelson et al. investigated the connectivity aspects of the immune network and described phase transitions suggesting that parts of different connectivity could be linked to different functional aspects [492].

Rapid proliferation of network models in the early 1990s has produced significant insights into the operation of the immune system [489, 493]. In these models the dynamics and meta-dynamics of networks have been emphasized. The models have led to an organism-centered view of the immune system, where it was seen as being primarily responsible for the somatic identity and only secondarily to the defensive role [493–495]. The conditions under which percolation can be avoided and immunological memory can be maintained by stimulatory network interactions have been investigated. These studies used percolation theory and random graph theory. Weisbuch et al. have shown that both localized memories and percolation are possible as part of the dynamical repertoire of model immune systems [496].

The mechanisms of natural tolerance, assuring that the immune system does not react against molecules of its own organism, have been actively discussed. The network's interaction with an external antigen has been studied, viewing the antigen as an infectious agent which is controlled by the network dynamics [497–500]. An important question is whether tolerance towards auto-antigens can be achieved through the diversity of dynamical configurations and regimes (i.e. through the dynamical repertoire) [498–501].

In the above-mentioned models tolerance was obtained only at parameter values corresponding to stable steady states of an idiotypic network. When antibody levels fluctuate, which is a more realistic situation, the introduction of a constant auto-antigen in significant concentration always produced an unbounded immune response. Calenbuhr et al. [503, 504] have shown that in a chaotic regime, characterized by dynamical equivalence of clones, a persistently present and constant antigen causes degeneration of the chaotic attractor into an attractor in which the clones are no longer equivalent. Instead of launching an unbounded immune response, the system becomes stabilized. Hence, in the presence of constant antigen, the system accomodates the stimulating molecule into its dynamics by shifting from an attractor of high complexity to another of lower complexity. Stewart and Varela [501] and Detours et al. [502] also took into account the changes of the immune network structure due to supply of new clones from the bone marrow. They have studied pattern formation in the shape space and showed that the immune networks could develop partitions corresponding to tolerant and suppressive zones.

A different important example of evolving networks is provided by computer networks. During the 1980s, computers became increasingly interconnected both within and between institutions. These large evolving networks are characterized by the absence of central control. Many features make computer networks reminiscent of biological and social forms of organization. Consequently, the term "computational ecologies" has been coined by Huberman [505]. The World Wide Web (WWW) has developed into a network which is currently growing at an exponential rate with roughly a million pages added every day. Huberman and Adamic [506] have investigated the WWW growth dynamics and found that web pages are distributed among sites according to a universal power law, i.e. many sites have only a few pages, whereas very few sites have hundreds or thousands of pages. They have studied the dynamics of web site visits and found universal power laws similar to the Pareto law of income distributions. This implies that a small number of sites command the traffic of a large segment of the web population, a signature of winner-take-all markets. The web can be looked at as a large directed graph, with documents being the vertices and links (URLs) representing the edges. Albert et al. [507] have pointed out that the topology of this graph is essential for finding information. They have analyzed the local connectivity of a sufficiently large subset of the web and determined the probabilities that a document has a certain number of outgoing and incoming links. Both probabilities followed a power law over several orders of magnitude.

Industrial enterprises operate within evolving network structures determined by their relationships with customers and suppliers. Richter and Wakuta [508] analyzed from this perspective the European and Japanese car industries. In Europe, car components were supplied until the early 1990s to manufacturers by a huge number of relatively autonomous firms. Suppliers did not tend to have exclusive links to car manufacturers, which were rather selected for each transaction according to cost and quality criteria. Therefore, there were many partner changes, which inhibited synergy effects and joint project development. The Japanese situation was different because there the suppliers of parts often had exclusive delivery contracts with a small number of customers. Consequently, the suppliers were strongly dependent on their customers. Richter and Wakuta proposed to describe these two extreme situations as open and closed networks, respectively. Open networks, like the European ones, are characterized by a continuous change of the companies concerned, implying linkages with suppliers being continuously dissolved, while new ones enter in. Closed networks, like the Japanese ones, are characerized by tighter relations of longer duration, which make it difficult for other companies to enter into existing structures.

As technologies progress, networks evolve. Not only are additional connections between existing partners established, but also links to entirely new players emerge. A quantitative analysis of networks can be used to determine the position of a company in the changing environment. Companies that do not form part of the right network may be underperforming [509].

In Luhmann's autopoietic theory of social systems [510], a society is built on a network of abstract ideas. It evolves by producing new ideas and decisions. The general autopoietic theory of living systems was proposed by Maturana and Varela in the beginning of the 1970s [511–513]. Based on the theories of Maturana and Varela and of Luhmann, Richter and Calenbuhr [514] have analysed and compared evolving biological and economic systems. Corporate structures, their networks, growth and evolution have recently been compared to biological structures and processes [514–519]. Knowledge management can make use of the society concept to deal with the corporate form of the learning process [520].

Ecological systems are formed by species that are linked through predator–prey relationships or compete for the same food. Hence, they represent predator–prey networks and food webs. Such networks are described by various multi-species versions of the Lotka–Volterra equations. Following May [521], the question has been discussed of how the complexity of such networks is related to the stability of the respective ecosystems [522, 523]. The extinction and introduction of new species lead to the evolution of ecological networks. Mathematical models of such evolving networks have been investigated by Bak and Sneppen [524], Sneppen et al. [525], Solé and Manrubia [526, 527] and Hunding and Engelhardt [528, 529].

Another motivation for studies of evolving networks comes from neurophysiology. It is generally assumed that the long-term memory is related to the modification of the synaptic connections between neurons. Neural plasticity is also involved in learning processes. There is a large number of publications on plastic neural networks responsible for associative memory and learning (e.g. see [479]). A learning mechanism for networks of coupled oscillatory neurons, based on synchronization effects and employing the modifications of connection weights and delay times, has been proposed [530].

Ito and Kaneko [531] have recently investigated the dynamics of plastic networks formed by chaotic oscillators. Each oscillator represented a circle map. Initially, the connection weights between the maps were chosen at random. In parallel to the system's dynamics, the pattern of connection weights could slowly change, so that a connection weight was increased or decreased depending on the degree of synchronization between the two units. Furthermore, a constant input was applied to one of the units. The numerical simulations have shown that in the parameter region, characterized by weakly chaotic and desynchronized dynamics, the units spontaneously develop a hierarchically organized, layered structure of connections.

A problem which is closely related to the evolution of networks is the evolution of *functions*. Indeed, a mathematical function is a map that puts some values into a correspondence to its arguments and thus establishes a certain binary network. Kataoka and Kaneko [532, 533] have considered the dynamics of functions, defined through the evolution equation

$$f_{n+1}(x) = (1 - \epsilon)f_n(x) + \epsilon f_n(f_n(x)), \tag{9.47}$$

where ϵ is a control parameter. Starting from different initial functions $f_0(x)$, this equation generates a sequence of functions $f_1(x), f_2(x), \ldots, f_n(x), \ldots$. The asymptotic behavior of this sequence in the limit $n \to \infty$ has many interesting properties, including convergence to certain attractors. Kataoka and Kaneko [532] justify their study by referring to the problems involving the origin and evolution of languages (including the genetic DNA code).

References

Chapter 1

1. E. Schrödinger, *What is Life? The Physical Aspect of a Living Cell* (Cambridge University Press, Cambridge 1944)
2. A.M. Turing, Phil. Trans. Roy. Soc. **237**, 37 (1952)
3. G. Nicolis, I. Prigogine, *Self-Organization in Nonequilibrium Systems* (Wiley, New York 1977)
4. H. Haken, *Synergetics. An Introduction* (Springer, Berlin 1977)
5. E. Lorenz, J. Atm. Sci. **20**, 130 (1963)
6. Ya.G. Sinai, Dokl. Akad. Nauk SSSR **153**, 1261 (1963)
7. H.R. Maturana, F.J. Varela, *Autopoesis and Cognition* (Reidel, Boston 1980)
8. Y. Kuramoto, *Chemical Oscillations, Waves and Turbulence* (Springer, Berlin 1984)
9. A.T. Winfree, *When Time Breaks Down* (Princeton University Press, Princeton, NJ 1987)
10. J.D. Murray, *Mathematical Biology* (Springer, Berlin 1989)
11. J. Casti, *Alternate Realities: Mathematical Models of Nature and Man* (Wiley, New York 1989)
12. A.S. Mikhailov, *Foundations of Synergetics I. Distributed Active Systems* (Springer, Berlin 1990; 2nd revised edn. 1994)
13. A.S. Mikhailov, A.Yu. Loskutov, *Foundations of Synergetics II. Chaos and Noise* (Springer, Berlin 1991; 2nd revised edn. 1996)
14. H. Haken, A.S. Mikhailov (eds.), *Interdisciplinary Approaches to Nonlinear Complex Systems* (Springer, Berlin 1993)
15. S.A. Kauffman, *The Origins of Order. Self-Organization and Selection in Evolution* (Oxford University Press, New York 1993)
16. E. Bonabeau, M. Dorigo, G. Theraulaz, *Swarm Intelligence: From Natural to Artificial Systems* (Oxford University Press, Oxford 1999)
17. C.G. Langton, *Artificial Life: An Overview* (MIT Press, Cambridge, Ma., 1995)
18. S. Camazine, J.L. Deneubourg, N.R. Franks, J. Sneyd, G. Theraulaz, E. Bonabeau, *Self-Organization in Biological Systems* (Princeton University Press, Princeton 2001)
19. W.B. Arthur, S.N. Durlauf, D.A. Lane (eds.), *The Economy as an Evolving Complex System II* (Addison-Wesley, Reading, MA 1997)
20. K. Kaneko, I. Tsuda, *Complex Systems: Chaos and Beyond* (Springer, Berlin 2001)

Chapter 2

21. C.J. Romanes, "Preliminary observations on the locomotor system of meduse", Phil. Trans. Roy. Soc. **CLXVI**, 269 (1876)
22. A.G. Mayer, "Rhythmical pulsation in scyphomeduse", Papers from the Tortugas Laboratory, Washington, **I**, 115–131 (1908)
23. C.R. Mines, "On dynamic equilibrium in the heart", J. Physiol. **XLVI**, 349 (1913)
24. Th. Lewis, *The Mechanism and Graphic Registration of the Heart Beat*, 3rd edn. (Shaw and Sons, London 1925)
25. W. Garrey, "The nature of fibrillary contraction of the heart. Its relation to tissue mass and form", Amer. J. Physiol. **23**, 397 (1914)
26. N. Wiener, A. Rosenblueth, "The mathematical formulation of the problem of conduction of impulses in a network of connected excitable elements, specifically in cardiac muscle", Arch. Inst. Cardiol. Mexico **16**, 205 (1946)
27. A.S. Mikhailov, V.S. Zykov, Sov. Phys. - Dokl. **31**, 51 (1986)
28. A.S. Mikhailov, *Foundations of Synergetics I. Distributed Active Systems*, 2nd edn. (Springer, Berlin, Heidelberg 1994)
29. D.A. Kessler, H. Levine, Phys. Rev. E **48**, 4801 (1993)
30. H. Levine, Phys. Rev. Lett. **66**, 2400 (1991)
31. B.N. Vasiev, P. Hogeweg, A.V. Panfilov, Phys. Rev. Lett. **73**, 3173 (1994)
32. Th. Höfer, Ph.K. Maini, Phys. Rev. E **56**, 2074 (1997)
33. A. Huth, Ch. Wissel, J. Theor. Biol. **156**, 365 (1992)
34. A. Huth, Ch. Wissel, Comm. Theor. Biol. **3**, 169 (1993)
35. A. Huth, Ch. Wissel, Ecol. Modeling **75/76**, 135 (1994)
36. E.O. Wilson, *The Social Insects* (Belknap Press, Cambridge, MA 1971)
37. C.J. Lumsden, B. Hoelldobler, J. Theor. Biol. **100**, 81(1983)
38. J.L. Deneubourg, S. Goss, Ethology, Ecology and Evolution **1**, 295 (1989)
39. J.L. Deneubourg, G. Theraulaz, R. Beckers, "Swarm Made Architectures", In *Towards a Practice of Autonomous Systems*, Proc. First European Conf. on Artificial Life, eds. F.J. Varela, P. Bourgine (MIT Press/Bradford Books, Cambridge, MA 1992) p. 123
40. J.L. Deneubourg, S. Goss, N. Franks, A. Sendova-Franks, C. Detrain, L. Chretien. "The Dynamics of Collective Sorting: Robot-like Ants and Ant-like Robots", In *Simulation of Animal Behaviour: From Animals to Animats*, eds. J.A. Meyer, S. Wilson (MIT Press, Cambridge, MA 1991) p. 356
41. J.L. Deneubourg, S. Aron, S. Goss, J.M. Pasteels, J. Ins. Behav. **3**, 159 (1990)
42. S. Goss, S. Aron, J.L. Deneubourg, J.M. Pasteels, Naturwissenschaften **76**, 579 (1989)
43. V. Calenbuhr, J.L. Deneubourg, J. Theor. Biol. **158**, 359 (1992)
44. V. Calenbuhr, J.L. Deneubourg, J. Theor. Biol. **158**, 394 (1992)
45. V. Calenbuhr *Collective Behaviour in Social and Gregarious Insects: Chemical Communication and Self-Organization.* Ph.D. Thesis, Free University of Brussels (1992)
46. V. Braitenberg, *Vehicles – Experiments in Synthetic Psychology* (MIT Press, Cambridge, MA 1984)
47. R.A. Brooks, AI Journal **47**, 139 (1991)
48. H. Asama, T. Fukuda, T. Arai, I. Endo (eds.), *Distributed Autonomous Robotic Systems 2* (Springer, Berlin 1996)
49. T. Lueth, R. Dillmann, P. Dario, H. Wörn (eds.), *Distributed Autonomous Robotic Systems 3* (Springer, Berlin 1998)
50. N.R. Franks, Am. Sci. **77**(2), 138 (1989)
51. R. Jander, K. Daumer, Insectes Sociaux **21**, 45, (1974)

52. W. Hangartner, Z. vergl. Physiol. **57**, 103 (1967)
53. R.H. Leuthold, "Orientation Mediated by Pheromones in Social Insects", In *Pheromones and Defensive Secretions in Insects,* eds. C. Noirot, P.E. Howse, J. Le Masne (University of Dijon Press, Dijon 1975) p. 197
54. G.F. Oster, E.O. Wilson, *Caste and Ecology in the Social Insects* (Princeton University Press, Princeton, NJ 1978)
55. B. Hölldobler, E.O. Wilson, *The Ants* (Springer, Berlin, Heidelberg 1990)
56. M.H. Hansell, *Animal Architecture and Building Behaviour* (Longman, London 1984)
57. P.-P. Grasse, Insectes Sociaux **6**, 41 (1959)
58. J.D. Murray, *Mathematical Biology* (Springer, Berlin, Heidelberg 1989)
59. I.S. Balakhovskii, Biophysics **10**, 1175 (1965)
60. S.A. MacKay, J. Cell Sci. **33**, 1 (1978)
61. G.K. Moe, W.C. Rheinboldt, J.A. Abildskov. Am. Heart J. **67**, 200 (1964)
62. O. Selfridge, Arch. Inst. Cardiol. Mexico **18**, 177 (1948)
63. J.M. Smith, R.J. Cohen, Proc. Nat. Acad. Sci. USA **81**, 233 (1984)
64. M. Markus, B. Hess, Nature **347**, 56 (1990)
65. J.R. Weimar, J.J. Tyson, L.T. Watson, Physica D **55**, 302 (1992)
66. J.R. Weimar, J.J. Tyson, L.T. Watson, Physica D **55**, 328 (1992)
67. E.F. Codd, *Cellular Automata* (Academic Press, New York 1966)
68. S. Wolfram, Physica D **10**, 1 (1984)
69. C.G. Langton, Physica D **10**, 135 (1984)
70. E.R. Berlekamp, J.H. Conway, R.K. Guy, *Winning Ways* (Academic Press, London 1982)
71. M. Gardner, Sci. Am. **223**, 120 (1970); Sci. Am. **224**, 112 (1971); Sci. Am. **224**, 117 (1970)
72. J. von Neumann, In *The Theory of Self-Reproducing Automata*, edited and completed by A.W. Burks, (University of Illinois Press, Champaign, IL 1966)
73. A.W. Burks (ed.), *Essays on Cellular Automata* (University of Illinois Press, Champaign, IL 1968)
74. M. Gerhardt, H. Schuster, J.J. Tyson, Physica D **46**, 392 (1990)
75. D. Young, Math. Biosciences **72**, 51 (1984)
76. C. Burks, D. Farmer, Physica D **10**, 157 (1984)
77. A. Lindenmayer, J. Theor. Biol. **30**, 455 (1967)
78. R. Axelrod, *The Complexity of Cooperation: Agent Based Models of Competition and Collaboration Intelligence* (Princeton University Press, Princeton, NJ 1997)
79. I. Aoki, Bull. Ocean. Res. Inst. University Tokyo **12**, 1 (1980)

Chapter 3

80. H.C. Berg, S.A. Anderson, Nature **245**, 380 (1973)
81. M. Silverman, M. Simon, Nature **249**, 73 (1974)
82. S.H. Larson, R.W. Reader, E.N. Kort, W.-W. Tso, J. Adler, Nature **249**, 74 (1974)
83. H.P. Greenspan, J. Fluid Mech. **84**, 125 (1978)
84. W.D. Bascom, R.L. Cottington, C.R. Singleterry, Adv. Chem. **43**, 112 (1964); **43**, 341 (1964)
85. J.L. Anderson, M.E. Lowell, D.C. Prieve, J. Fluid Mech. **117**, 107 (1982)
86. S.S. Dukhin, B.V. Derjagin, In *Surface and Colloidal Science*, Vol. 7, ed. E. Matievic (Wiley, New York 1974) p. 322

87. N.O. Young, J.S. Goldstein, M.J. Block, J. Fluid Mech. **6**, 350 (1959)
88. A.S. Mikhailov, D. Meinköhn, "Self-motion in Physico-chemical Systems far from Thermal Equilibrium", In *Stochastic Dynamics*, eds. L. Schimansky-Geyer, Th. Pöschel (Springer, Berlin, Heidelberg 1997) p. 334
89. L.D. Landau, E.M. Lifshitz, *Fluid Mechanics* (Pergamon, Oxford 1959)
90. A.S. Mikhailov, A.Yu. Loskutov, *Foundations of Synergetics II. Chaos and Noise*, 2nd edn. (Springer, Berlin, Heidelberg 1996)
91. K. Krischer, A.S. Mikhailov, Phys. Rev. Lett. **73**, 3165 (1994)
92. M. Or-Guil, M. Bode, C.P. Schenk, H.-G. Purwins, Phys. Rev. E **57**, 6432 (1998)
93. M. Bode, "Pattern Formation in Complex Dissipative Systems: a Particle Approach", Habilitationsschrift, (Westfälische Wilhelms-Universität, Münster 1998)
94. G. Wolf (ed.), *Encyclopedia Cinematographica: Microbiology* (Inst. für den Wissenschaftl. Film, Göttingen 1967)
95. R. Ben-Jacob, O. Stochet, A. Tenenbaum, I. Cohen, A. Czirók, T. Viscek, Phys. Rev. E **53**, 1835 (1996)
96. J. Toner, Y. Tu, Phys. Rev. Lett. **75**, 4326 (1995)
97. J. Toner, Y. Tu, Phys. Rev. E **58**, 4828 (1998)
98. Z. Csahók, A. Czirók, Physica A **243**, 304 (1997)
99. M.J. Lighthill, G.B. Whitham, Proc. Roy. Soc. A **229**, 281 (1955)
100. G.B. Whitham, *Linear and Nonlinear Waves* (Wiley, New York 1974)
101. J. Treiterer, Ohio State University Technical Report No. PB 246 094, Columbus, OH (1975)
102. B.S. Kerner, P. Konhäuser, Phys. Rev. E **48**, 2335 (1993)
103. B.S. Kerner, S.L. Klenov, P. Konhäuser, Phys. Rev. E **56**, 4200 (1997)
104. W. Alt, "Cell Motion and Orientation", In *Towards the Frontiers of Mathematical Biology*, eds. W. Alt, S. Levin (Springer, Berlin 1997) p. 45
105. H.C. Berg, D.A. Brown, Nature **239**, 500 (1972)
106. W. Alt, J. Math. Biol. **9**, 147(1980)
107. B.E. Farell, R.P. Daniele, D.A. Lauffenburger, Cell Motil. Cytoskeleton **16**, 279(1990)
108. H.C. Berg, Sh. Khan, "A Model for the Flagellar Rotary Motor", In *Mobility and Recognition in Cell Biology*, eds. H. Sund, C. Veeger (de Gruyter, Berlin 1983) p. 485
109. F. Oosawa, Sh. Hayashi, Adv. Biophys. **22**, 151 (1986)
110. P. Laenger, Biophys. J. **53**, 53 (1988)
111. W. Junge, H. Lill, S. Engelbrecht, Trends Biochem. Sci. **22**, 420 (1997)
112. S.B. Vik, B.J. Antonio. J. Biol. Chem. **269**, 30364 (1994)
113. T.C. Elston, G. Oster, Biophys. J. **73**, 703 (1997)
114. T.C. Elston, H. Wang, G. Oster, Nature **391**, 510 (1998)
115. P. Dimroth, H. Wang, M. Grabe, G. Oster, Proc. Natl. Acad. Sci. USA **96**, 4924 (1999)
116. G. Oster, H. Wang, M. Grabe, Proc. Roy. Soc. **355**, 523 (2000)
117. G. Oster, H. Wang, "ATP Synthase: Two Rotary Molecular Motors Working Together", In *Encyclopedia of Molecular Biology*, ed. T.E. Creighton (Wiley, New York 1998), p. 211
118. P. Dimroth, G. Kaim, U. Matthey, Biochimica et Biophysica Acta **1365**, 87(1998)
119. G. Kaim, P. Dimroth, FEBS Lett. **434**, 57(1998)
120. G. Kaim, P. Dimroth, EMBO J. **17**, 5887(1998)
121. G. Kaim, P. Dimroth, EMBO J. **18**, 4118 (1999)
122. G. Oster, H. Wang, Structure **7**, R67 (1999)

123. A. Okubo, *Diffusion and Ecological Problems* (Springer, Berlin 1980)
124. W.J. Bell, T.R. Tobin, Biol. Rev. **57**, 219 (1982)
125. H.C. Berg, E.M. Purcell, Biophys. J. **20**, 193 (1977)
126. W. Alt, BioSystems **34**, 11 (1995)
127. D.W. Stevens, J.R. Krebs, *Foraging Theory* (Princeton University Press, Princeton, NJ 1986)
128. D. Grünbaum, J. Math. Biol. **38**, 169(1999)
129. B. Alberts, D. Bray, J. Lewis, M. Raff, K. Roberts, J.D. Watson, *The Cell* (Garland, New York 1994)
130. C.J. Brokaw, Math. BioSci. **90**, 247 (1988)
131. C.J. Brokaw, "Descriptive and Mechanistic Models of Flagellar Motiliy", In *Biological Motion*, eds. W. Alt, G. Hoffmann (Springer, Berlin 1990) p. 128
132. H. Machemer, "Cilia and Flagella", In *Biological Motion,* eds. W. Alt, G. Hoffmann (Springer, Berlin 1990) p. 121
133. Y. Mogami, J. Pernberg, H. Machemer, J. Math. Biol. **30**, 215 (1992)
134. S. Camalet, F. Jülicher, J. Prost, Phys. Rev. Lett. **82**, 1590 (1999)
135. W. Alt, "Mathematical Models and Analysing Methods for the Lamellipodial Activity of Leukocytes", In *Biomechanics of Active Movement and Deformation of Cells*, ed. N. Akkas (Springer, Berlin, Heidelberg 1990) p. 403
136. W. Alt, H.W. Kaiser, "Observation, Modeling and Simulation of Keratinocyte Movement", In *Biomechanics of Active Movement and Division of Cells*, ed. N. Akkas (Springer, Berlin, Heidelberg 1994) p. 445
137. W. Alt, A. Deutsch, G. Dunn (eds.), *Dynamics of Cell and Tissue Motion* (Birkhäuser, Basel 1997)
138. P.D. Dale, J.A. Sherratt, P.K. Maini, Bull. Math. Biol. **59**, 1077 (1997)
139. L. Olsen, P.K. Maini, J.A. Sherratt, B. Marchant, J. Theor. Med. **1**, 175 (1998)
140. I. Durand, P. Jönson, Ch. Misbah, A. Valance, K. Kassner, Phys. Rev. E **56**, R3776 (1997)
141. H.-U. Keller, M. Bessis, Nouv. Rev. Fr. Hematol. **15**, 439 (1975)
142. S.E. Malawista, A.D. Chevance, J. Cell. Biol. **95**, 960 (1982)
143. H. Gruler, Liquid Crystals **24**, 49 (1998)
144. M. Dupeyrat, E. Nakache, Bioelectrochem. Bioenergetics **5**, 134 (1987)
145. E. Nakache, M. Dupeyrat, J. Colloid. Interface Sci. **94**, 187 (1983)
146. C. Reynolds, Computer Graphics **21**, 25 (1987)
147. T. Vicsek, A. Czirók, E. Ben-Jacob, J. Cohen, O. Stochet, Phys. Rev. Lett. **75**, 1226 (1995)
148. Z. Csahók, T. Vicsek, Phys. Rev. E **52**, 5297 (1995)
149. A. Czirók, H.E. Stanley, T. Vicsek, J. Phys. A **30**, 1375 (1997)
150. A. Czirók, A.-L. Barabási, T. Vicsek, Phys. Rev. Lett. **82**, 209 (1999)
151. A. Czirók, T. Vicsek, "Collective Motion", In *Statistical Mechanics of Biocomplexity*, eds. D. Reguera, J.M.G. Villar, J.M. Rubi (Springer, Berlin 1999) p. 152
152. M. Schienbein, H. Gruler, Bull. Math. Biol. **55**, 585 (1993)
153. H. Gruler, "Self-Organized Molecular Machines", In *Chaos and Complexity*, eds. Trân Thanh Vân, P. Bergé, R. Conte, M. Dubois (Editions Frontieres, Gif-sur-Yvette 1995) p. 173
154. L. Edelstein-Keshet, J. Watmough, D. Grunbaum, J. Math. Biol. **36**, 515 (1998)
155. A. Mogilner, L. Edelstein-Keshet, J. Math. Biol. **38**, 534 (1999)
156. D. Grünbaum, J. Math. Biol. **33**, 139 (1994)
157. D. Grünbaum, SIAM J. Appl. Math. **61**, 43 (2000)
158. G. Flierl, D. Grünbaum, S. Levin, D. Olsen, J. Theor. Biol. **196**, 397 (1999)

159. D. Chao, S. Levin, "Herding Behavior: the Emergence of Large-Scale Phenomena from Local Interactions", In *Differential Equations with Applications to Biology*, eds. S. Ruan, C.S.K. Wolkowicz, J. Wu, Fields Institute Communications, Vol. 21 (Am. Math. Soc., Providence 1999) p. 81
160. F. Schweitzer, W. Ebeling, B. Tilch, Phys. Rev. Lett. **80**, 5044 (1998)
161. W. Ebeling, F. Schweitzer, B. Tilch, BioSystems, **49**, 17 (1999)
162. U. Erdman, W. Ebeling, L. Schimansky-Geier, F. Schweitzer, Eur. Phys. J. B **15**, 105 (2000)
163. R.E. Chandler, R. Herman, E.W. Montroll, Oper. Res. **6**, 165 (1958)
164. E. Komentani, T. Sasaki, J. Operat. Research Japan **2**, 11 (1958)
165. I. Prigogine, R. Herman, *Kinetic Theory of Vehicular Traffic* (Elsevier, New York 1971)
166. W. Leutzbach, *Introduction to the Theory of Traffic Flow* (Springer, Berlin 1988)
167. A.D. May, *Traffic Flow Fundamentals* (Prentice-Hall, Englewood Cliffs, NJ 1990)
168. D. Helbing, *Verkehrsdynamik* (Springer, Berlin 1997)
169. K. Nagel, M. Schreckenberg, J. Phys. I (France) **2**, 2221 (1992)
170. S. Lübeck, M. Schreckenberg, K.D. Usadel, Phys. Rev. E **57**, 1171 (1998)
171. D. Helbing, Phys. Rev. E **57**, 6176 (1998)
172. F. Schweitzer (ed.), *Self-Organization of Complex Structures: From Individual to Collective Dynamics* (Gordon and Breach, London 1997)
173. H. Meersman, E. van de Voorde, W. Winkelmans (eds.), *World Transport Research*, Selected Proceedings of the 8th World Conference on Transport Research, Vol. 3: *Transport Modelling/Assessment* (Pergamon Press, London 1999)

Chapter 4

174. R. Feynman, R. Leighton, M. Sands, *The Feynman Lectures in Physics*, Vol. 1 (Addison-Wesley, Reading, MA 1963)
175. H. Risken, *The Fokker–Planck Equation* (Springer, Berlin, Heidelberg 1989)
176. A.S. Mikhailov, A.Yu. Loskutov, *Foundations of Synergetics II. Chaos and Noise*, 2nd revised edn. (Springer, Berlin, Heidelberg 1991, 1996)
177. M.O. Magnasco, Phys. Rev. Lett. **71**, 1477 (1993)
178. A. Ajdari, D. Mukamel, L. Peliti, J. Prost, J. Phys. I (France) **4**, 1551 (1994)
179. R. Bartussek, P. Hänggi, J.G. Kissner, Europhys. Lett. **28**, 459 (1994)
180. R.D. Astumian, M. Bier, Phys. Rev. Lett. **72**, 1766 (1994)
181. C.R. Doering, W. Horsthemke, J. Riordan, Phys. Rev. Lett. **72**, 2984 (1994)
182. J. Luczka, R. Bartussek, P. Hänggi, Europhys. Lett. **31**, 431 (1995)
183. A.F. Huxley, Prog. Biophys. Chem. **7**, 255 (1957)
184. A.F. Huxley, R.M. Simmons, Nature **233**, 533 (1971)
185. M. Meister, S.C. Caplan, H.C. Berg, Biophys. J. **55**, 905 (1989)
186. R.D. Vale, F. Oosawa, Adv. Biophys. **26**, 97 (1990)
187. S.M. Simon, C.S. Peskin, G.F. Oster, Proc. Natl. Acad. Sci. USA **89**, 3770 (1992)
188. G. Adam, M. Delbrück, In *Structural Chemistry and Molecular Biology*, eds. A. Rich, N. Davidson (Freeman, San Francisco 1968) p. 198
189. K. Hall, D.G. Cole, Y. Yeh, J.M. Scholey, R.J. Baskin, Nature **364**, 457 (1993)
190. X. Michalet, D. Bensimon, B. Fourcade, Phys. Rev. Lett. **72**, 168 (1994)
191. A.S. Mikhailov, B. Hess, J. Theor. Biol. **176**, 185 (1995)

192. E.M. Purcell, Am. J. Phys. **45**, 3 (1977)
193. E.M. Lifshitz, L.P. Pitaevskii, *Statistical Physics*, Part 2 (Pergamon, London 1985)
194. V. Pareto, *Cours d'Economique Politique* (Université de Lausanne, Lausanne 1897)
195. E.W. Montroll, M.S. Shlesinger, Proc. Natl. Acad. Sci. USA **79**, 3380 (1982)
196. A.J. Lotka, J. Wash. Acad. Sci. **16**, 317 (1926)
197. A.J. Lotka, J. Phys. Chem. **14**, 271 (1910)
198. A.J. Lotka, *Elements of Physical Biology* (Williams and Wilkins, Baltimore 1925)
199. A.N. Kolmogorov, Dokl. Akad. Nauk. USSR **30**, 301 (1941); **32**, 16 (1941)
200. L.D. Landau, E.M. Lifshitz, *Statistical Physics*, Part 1 (Pergamon, London 1980)
201. P. Bak, C. Tang, K. Wiesenfeld, Phys. Rev. Lett. **59**, 381 (1987)
202. H. Kesten, Acta Math. **131**, 207 (1973)
203. R. Engelhardt, *Emergent Percolating Nets in Evolution*, Ph.D. Thesis, University of Copenhagen (1998)
204. J.D. Murray, *Mathematical Biology* (Springer, Berlin, Heidelberg 1989)
205. A.S. Mikhailov, Dokl. Akad. Nauk SSSR **243**, 786 (1978)
206. A.S. Mikhailov, Phys. Lett. A **73**, 143 (1979)
207. A.S. Mikhailov, Z. Phys. B **41**, 277 (1981)
208. B.B. Mandelbrot, *Fractals: Form, Chance, and Dimension* (Freeman, San Francisco 1983)
209. C.E. Shannon, Bell Syst. Tech. J. **30**, 50 (1951)
210. W. Ebeling, J. Freund, F. Schweitzer, *Komplexe Strukturen: Entropie und Information* (Teubner, Stuttgart, Leipzig 1998)
211. S. Manrubia, D.H. Zanette, Phys. Rev. E **58**, 295 (1998)
212. W. Horsthemke, R. Lefever, *Noise-Induced Transitions* (Springer, Berlin, Heidelberg 1984)
213. A. Schenzle, H. Brand, Phys. Rev. A **20**, 1628 (1979)
214. R. Graham, A. Schenzle, Phys. Rev. A **25**, 1731 (1982)
215. D. Sornette, Physica A **250**, 295 (1998)
216. T. Hill, M. Kirshner, Intl. Rev. Cytol. **78**, 1 (1982)
217. P. Janmey, C. Cunningham, G.F. Oster, T. Stossel, In *Swelling Mechanics: From Clays to Living Cells and Tissues*, ed. T. Karaliss (Springer, Berlin, Heidelberg 1992) pp. 333–346
218. H. Miyamoto, H. Hotani, In *Taniguchi International Symposium on Dynamics of Microtubules*, ed. H. Hotani (The Taniguchi Foundation, Taniguchi, Japan 1988) pp. 220–242
219. C.S. Peskin, G.M. Odell, G.F. Oster, Biophys. J. **65**, 316 (1993)
220. D. Bray, *Cell Movements* (Garland, New York 1992)
221. B. Alberts, D. Bray, J. Lewis, M. Raff, K. Roberts, J.D. Watson, *The Cell* (Garland, New York 1994)
222. T.J. Mitchison, Ann. Rev. Cell Biol. **4**, 527 (1988)
223. S. Inoue, E.D. Salmon, Mol. Biol. Cell **6**, 1148 (1995)
224. M. Dogterom, B. Yurke, Science **278**, 856 (1997)
225. A. Mogilner, G.F. Oster, Biophys. J. **71**, 3030 (1996)
226. A. Mogilner, G.F. Oster, Eur. Biophys. J. **25**, 47 (1996)
227. A. Mogilner, G.F. Oster, Eur. Biophys. J. **28**, 235 (1999)
228. R. Lill, F.E. Nargang, W. Neupert, Curr. Opin. Cell Biol. **8**, 505 (1996)
229. S.T. Hwang, G. Schatz, Proc. Natl. Acad. Sci. USA **86**, 8432 (1989)
230. C. Wachter, G. Schatz, B.S. Glick, Mol. Biol. Cell. **5**, 465 (1994)

231. P.J. Kang, J. Ostermann, W. Shilling, E. Neupert, E.A. Craig, N. Pfanner, Nature **348**, 137 (1990)
232. P.E. Scherer, U.C. Krieg, S.T. Hwang, D. Vestweber, G. Schatz, EMBO J. **9**, 4315 (1990)
233. W. Neupert, F.-U. Hartl, E.A. Craig, N. Pfanner, Cell **63**, 447 (1990)
234. S.M. Simon, C.S. Peskin, G.F. Oster, Proc. Natl. Acad. Sci USA, **89**, 3770 (1992)
235. B.S. Glick, Cell **80**, 11 (1995)
236. N. Pfanner, M. Meijer, Curr. Biol. **5**, 132 (1995)
237. J.-F. Chauwin, G.F. Oster, B.S. Glick, Biophys. J. **74**, 1732 (1998)
238. C.S. Peskin, G.B. Ermentrout, G.F. Oster, In *Cell Mechanics and Cellular Engineering*, eds. V.C. Mow et al. (Springer, New York 1993), p. 479
239. D.A. Erie, T.D. Yager, P. von Hippel, Ann. Rev. Biophys. Biomol. Struct. **21**, 379 (1992)
240. A. Polyakov, E. Severinova, S.A. Darst, Cell **83**, 365 (1995)
241. T.D. Yager, P. von Hippel, In *Escherichia Coli and Salmonella Typhimurium*, ed. F. Neidhardt (American Society for Microbiology, Washington 1987) p. 1241
242. H. Yin, M.D. Wang, K. Svoboda, R. Landick, S.M. Block, J. Gelles, Science **270**, 1653 (1995)
243. H.-Y. Wang, T. Elston, A. Mogilner, G.F. Oster, Biophys. J. **74**, 1186 (1998)

Chapter 5

244. R. May, Nature **261**, 459 (1976)
245. A.S. Mikhailov, A.Yu. Loskutov, *Foundations of Synergetics II. Chaos and Noise*, 2nd revised edn. (Springer, Berlin 1991, 1996)
246. M.C. Mackey, L. Glass, Science **197**, 287 (1977)
247. L. Glass, M.C. Mackey, Ann. New York Acad. Sci. **316**, 214 (1979)
248. M.C. Mackey, U. An Der Heiden, Funk. Biol. Med. **1**, 156 (1982)
249. M.C. Mackey, J.G. Milton, Ann. New York Acad. Sci. **504**, 16 (1988)
250. G. Benchetrit, P. Baconnier, J. Demongeot (eds.), *Concepts and Formalizations in the Control of Breathing* (Manchester University Press, Manchester 1987)
251. L. Rensing, U. An Der Heiden, M.C. Mackey (eds.), *Temporal Disorder in Human Osillatory Systems* (Springer, Berlin, Heidelberg, New York 1987)
252. C. Haurie, D.C. Dale, M.C. Mackey, Blood **92**, 2629 (1998)
253. K. Ikeda, H. Daido, O. Akimoto, Phys. Rev. Lett. **45**, 709 (1980)
254. P. Nardone, P. Mandel, R. Kapral, Phys. Rev. A **33**, 2465 (1986)
255. N. Khrustova, G. Veser, A.S. Mikhailov, R. Imbihl, Phys. Rev. Lett. **75**, 3564 (1995)
256. N. Khrustova, A.S. Mikhailov, R. Imbihl, J. Chem. Phys. **107**, 2096 (1997)
257. J.-M. Grandmont, P. Malgrange, J. Econ. Th. **40**, 3 (1986)
258. R. Deneckere, S. Pelikan, J. Econ. Th. **40**, 13 (1986)
259. M. Boldrin, L. Montrucchio, J. Econ. Th. **40**, 26 (1986)
260. R.A. Dana, L. Montrucchio, J. Econ. Th. **40**, 40 (1986)
261. J.-M. Grandmont, G. Laroque, J. Econ. Th. **40**, 138 (1986)
262. K. Shell, Cited in J.-M. Grandmont, P. Malgrange, J. Econ. Th. **40**, 3 (1986)
263. D. Cass, K. Shell, J. Polit. Econ. **91**, 193 (1983)
264. C. Azariadis, J. Econ. Th. **25**, 380 (1981)

265. C. Azariadis, R. Guesnerie, "Sunspots and Cycles". CARESS Working Paper 8322R, University of Pennsylvania (1983; revised 1984)
266. R. Guesnerie, J. Econ Th. **40**, 103 (1986)
267. R.H. Day, W. Huang, J. Econ. Beh. Org. **14**, 299 (1990)
268. V.L. Smith, G.L. Suchanek, A.W. Williams, Econometrica **5**, 1119 (1988)
269. P. Anderson, K.J. Arrow, D. Pines (eds.), *The Economy as an Evolving Complex System* (Addison-Wesley, Reading, MA 1988)
270. W.B. Arthur, S.N. Durlauf, D.A. Lane (eds.), *The Economy as an Evolving Complex System II* (Addison-Wesley, Reading, MA 1997)
271. W.B. Arthur, Am. Econ. Rev. **84**, 406 (1994)
272. W.B. Arthur, Complexity **1**, 20 (1995)
273. W.B. Arthur, J.H. Holland, B. LeBaron, R. Palmer, P. Taylor, In W.B. Arthur, S.N. Durlauf, D.A. Lane (eds.), *The Economy as an Evolving Complex System II* (Addison-Wesley, Reading, MA 1997), p. 15
274. R. Axelrod, W.D. Hamilton, Science **211**, 1390 (1981)
275. R. Axelrod, *The Evolution of Cooperation* (Basic Books/Harper Collins, New York 1984)
276. R. Axelrod, D. Dion, Science **242**, 1385 (1988)
277. N.S. Glance, B.A. Huberman, J. Math. Soc. **17**, 281 (1993)
278. N.S. Glance, B.A. Huberman, Phys. Lett. A **165**, 432 (1992)
279. B.A. Huberman, N.S. Glance, In *Modelling Rational and Moral Agents*, ed. P. Danielson (Oxford University Press, Oxford 1996), p. 210
280. B.B. Mandelbrot, J. Business **36**, 394 (1963)
281. R.N. Mantegna, Physica A **179**, 232 (1991)
282. R.N. Mantegna, H.E. Stanley, Nature **376**, 46 (1995)
283. J.-P. Bouchaud, Physica A **263**, 415 (1999)
284. L. Laloux, P. Cizeau, J.-P. Bouchaud, M. Potters, Phys. Rev. Lett. **83**, 1467 (1999)
285. V. Plerou, P. Gopikrishnan, B. Rosenow, L.A.N. Amaral, H.E. Stanley, Phys. Rev. Lett. **83**, 1471 (1999)
286. J.-P. Bouchaud, M. Potters, *Theorie des Risques Financiéres* (Alea-Sacley, Eyrolles 1997), English translation in preparation
287. R.N. Mantegna, H.E. Stanley, *Introduction to Econophysics: Correlations and Complexity in Finance* (Cambridge University Press, Cambridge 1999)
288. H. Takayasu, H. Miura, T. Hirabayashi, K. Hamada, Physica A **184**, 127 (1992)
289. M. Levy, S. Solomon, Int. J. Mod. Phys. C **7**, 595 (1996)
290. G. Caldarelli, M. Marsili, Y.-C. Zhang, Europhys. Lett. **40**, 479 (1997)
291. A.-H. Sato, H. Takayasu, Physica A **250**, 231 (1998)
292. J.-P. Bouchaud, R. Cont, Eur. Phys. J. B **6**, 543 (1998)

Chapter 6

293. N. Wiener, *Nonlinear Problems in Random Theory* (M.I.T. Press, Cambridge, MA 1958)
294. N. Wiener, *Cybernetics* (M.I.T. Press, Cambridge, MA 1961)
295. A.T. Winfree, J. Theor. Biol. **16**, 15 (1967)
296. D.H. Zanette, A.S. Mikhailov, Phys. Rev. E **60**, 4571 (1999)
297. J.W. Rayleigh, *Theory of Sound*, Vol. 1, 2nd edn. (Dover, New York 1945) p. 81

298. A.T. Winfree, *The Geometry of Biological Time* (Springer, Berlin, Heidelberg 1980)
299. Y. Kuramoto, *Chemical Oscillations, Waves and Turbulence* (Springer, Berlin, Heidelberg 1984)
300. J.D. Murray, *Mathematical Biology* (Springer, Berlin 1989)
301. A. Goldbeter, *Biochemical Oscillations and Cellular Rhythms: The Molecular Bases of Periodic and Chaotic Behaviour* (Cambridge University Press, Cambridge 1996)
302. D. Battogtokh, A. Preusser, A.S. Mikhailov, Physica D **106**, 327 (1997)
303. P. Hartman, *Ordinary Differential Equations* (Wiley, New York 1964)
304. K. Alligood, T. Sauer, J.A. Yorke, *Chaos: An Introduction to Dynamical Systems* (Springer, New York 1997)
305. H. Fujisaka, T. Yamada, Prog. Theor. Phys. **69**, 32 (1983)
306. A.S. Mikhailov, A.Yu. Loskutov, *Foundations of Synergetics II. Chaos and Noise*, 2nd revised edn. (Springer, Berlin 1996)
307. R.M. Coleman, *Wide Awake at 3:00 A.M.* (Freeman, New York 1986)
308. J. Buck, Nature **211**, 562 (1966)
309. J. Buck, Quart. Rev. Biol. **63**, 265 (1988)
310. T.J. Walker, Science **166**, 891 (1969)
311. R.D. Traub, R. Miles, R.K.S. Wong, Science **243**, 1319 (1989)
312. D.C. Michaels, E.P. Matyas, J. Jalife, Circ. Res. **61**, 704 (1987)
313. B. Hess, Quart. Rev. Biophys. **30**, 2 (1997)
314. A.V.M. Herz, J.J. Hopfield, Phys. Rev. Lett. **75**, 1222 (1995)
315. K. Wiesenfeld, P. Colet, S.H. Strogatz, Phys. Rev. E **57**, 1563 (1998)
316. Y. Aizawa, Prog. Theor. Phys. **56**, 703 (1976)
317. Y. Yamaguchi, K. Kometani, H. Shimizu, J. Stat. Phys. **26**, 719 (1981)
318. G.B. Ermentrout, Physica D **41**, 219 (1990)
319. P.C. Matthews, S.H. Strogatz, Phys. Rev. Lett. **65**, 1701 (1990); P.C. Matthews, S.H. Strogatz, Physica D **52**, 293 (1991)
320. V. Hakim, W.J. Rappel, Phys. Rev. A **46**, R7347 (1992)
321. N. Nakagawa, Y. Kuramoto, Physica D **75**, 74 (1994)
322. K. Kaneko, Physica D **41**, 137 (1990)
323. L.M. Pecora, T.L. Carroll, Phys. Rev. Lett. **64**, 821 (1990)
324. L.M. Pecora, T.L. Carroll, Phys. Rev. A **44**, 2374 (1991)
325. J.F. Heagy, T.L. Carroll, L.M. Pecora, Phys. Rev. E **50**, 1874 (1994)
326. L.O. Chua, L. Kocarev, K. Eckart, M. Itoh, Int. J. Bifurcation Chaos Appl. Sci. Eng. **2**, 705 (1992)
327. T.L. Carroll, Phys. Rev. E **53**, 3117 (1996)
328. T.L. Carroll, G.A. Johnson, Phys. Rev. E **57**, 1555 (1998)
329. L.M. Pecora, T.L. Carroll, G.A. Johnson, D. Mar, Chaos **7**, 520 (1997)
330. W.L. Ditto, K. Showalter, Chaos **7**, 4 (1997)
331. D.H. Zanette, A.S. Mikhailov, Phys. Rev. E **57**, 276 (1998)
332. M.G. Rosenblum, A.S. Pikovsky, J. Kurths, Phys. Rev. Lett. **76**, 1804 (1996); **78**, 4193 (1997)
333. A. Pikovsky, M. Rosenblum, J. Kurths, *Synchronization: A Universal Concept in Nonlinear Science* (Cambridge University Press, Cambridge 2002)
334. S. Hayes, C. Grebogi, E. Ott, A. Mark, Phys. Rev. Lett. **73**, 1781 (1994)
335. K.M. Cuomo, A.V. Oppenheim, Phys. Rev. Lett. **71**, 65 (1993)
336. N. Gershenfeld, G. Grinstein, Phys. Rev. Lett. **74**, 5024 (1995)
337. Lj. Kocarev, U. Parlitz, Phys. Rev. Lett. **74**, 5028 (1995)
338. J.H. Peng, E.J. Ding, M. Ding, W. Yang, Phys. Rev. Lett **76**, 904 (1996)
339. S. Boccaletti, A.A. Farini, F.T. Arecchi, Phys. Rev. E **55**, 4979 (1997)
340. V. Petrov, V. Gaspar, J. Masere, K. Showalter, Nature **361**, 240 (1993)

341. R. Roy, T.W. Murphy Jr., T.D. Maier, Z. Gills, E.R. Hunt, Phys. Rev. Lett. **68**, 1259 (1992)
342. R. Meucci, W. Gadomski, M. Ciofini, F.T. Arecchi, Phys. Rev. E **49**, R2528 (1994)
343. R. Meucci, M. Ciofini, R. Abbate, Phys. Rev. E **53**, R5537 (1996)
344. E.R. Hunt, Phys. Rev. Lett. **67**, 1953 (1991)
345. D. Auerbach, P. Cvitanovic, J.-P. Eckmann, G. Gunaratne, I. Procaccia, Phys. Rev. Lett. **58**, 2387 (1987)
346. E. Ott, C. Grebogi, J.A. Yorke, Phys. Rev. Lett. **64**, 1196 (1990)
347. K. Pyragas, Phys. Rev. A **170**, 421 (1992)
348. S. Boccaletti, F.T. Arecchi, Europhys. Lett. **31**, 127 (1995)
349. S. Boccaletti, J. Bragard, F.T. Arecchi, Phys. Rev. E **59**, 6574 (1999)
350. A. Amengual, E. Hernández-García, R. Montagne, M. San Miguel, Phys. Rev. Lett. **78**, 4379 (1997)
351. L.G. Morelli, D.H. Zanette, Phys. Rev. E **58**, R8 (1998)
352. P. Grassberger, Phys. Rev. E **59**, R2520 (1999)
353. M.K.S. Yeung, S.H. Strogatz, Phys. Rev. Lett. **82**, 648 (1999)
354. E.M. Izhikevich, Phys. Rev. E **58**, 905 (1998)

Chapter 7

355. R. May, Nature **261**, 459 (1976)
356. S. Grossmann, S. Thomae, Z. Naturforsch. **32a**, 1353 (1977)
357. M. Feigenbaum, J. Stat. Phys. **19**, 25 (1978); **21**, 669 (1979)
358. K. Kaneko, Physica D **41**, 137 (1990)
359. H. Fujisaka, T. Yamada, Prog. Theor. Phys. **69**, 32 (1983)
360. J. Milnor, Comm. Math. Phys. **99**, 177 (1985); **102**, 517 (1985)
361. K. Kaneko, Physica D **124**, 322 (1998)
362. K. Kaneko, Physica D **77**, 456 (1994)
363. S. Manrubia, A.S. Mikhailov, Europhys. Lett. **50**, 580 (2000)
364. O.E. Rössler, Phys. Lett. A **57**, 97 (1976)
365. D.H. Zanette, A.S. Mikhailov, Phys. Rev. E **57**, 276 (1998)
366. D.H. Zanette, A.S. Mikhailov, Phys. Rev. E **62**, R7571 (2000)
367. W.C. McCulloch, W. Pitts, Bull. Math. Biophys. **5**, 115 (1943)
368. A.S. Mikhailov, *Foundations of Synergetics I. Distributed Active Systems* (Springer, Berlin 1990)
369. D.H. Zanette, A.S. Mikhailov, Phys. Rev. E **58**, 872 (1998)
370. B. Hess, A.S. Mikhailov, J. Theor. Biol. **176**, 181 (1995)
371. B. Hess, A.S. Mikhailov, Science **264**, 223 (1994)
372. P. Stange, A.S. Mikhailov, B. Hess, J. Phys. Chem. B **102**, 6273 (1998)
373. K. Wiesenfeld, P. Hadley, Phys. Rev. Lett. **62**, 1335 (1989)
374. K.Y. Tsang, R.E. Mirollo, S.H. Strogatz, K. Wiesenfeld, Physica D **48**, 102 (1991)
375. D. Golomb, D. Hansel, B. Shraiman, H. Sompolinsky, Phys. Rev. A **45**, 3516 (1992)
376. K. Okuda, Physica D **63**, 424 (1993)
377. V. Hakim, W.J. Rappel, Phys. Rev. A **46**, R7347 (1992)
378. N. Nakagawa, Y. Kuramoto, Physica D **75**, 74 (1994)
379. N. Nakagawa, Y. Kuramoto, Prog. Theor. Phys. **89**, 313 (1993)
380. T. Shibata, K. Kaneko, Phys. Rev. Lett. **81**, 4116 (1998)
381. T. Shibata, K. Kaneko, Physica D **124**, 177 (1998)

382. T. Shibata, T. Chawanya, K. Kaneko, Phys. Rev. Lett. **82**, 4424 (1999)
383. G. Perez, S. Sinha, H.A. Cerdeira, Physica D **63**, 341 (1993)
384. F. Xie, H.A. Cerdeira, Phys. Rev. E **54**, 3235 (1996)
385. W. Just, Phys. Rep. **290**, 101 (1997)
386. K. Kaneko, Physica D **54**, 5 (1991)
387. S.C. Manrubia, A.S. Mikhailov, Int. J. Bifurc. and Chaos **10**, 2465 (2000)
388. A. Hampton, D.H. Zanette, Phys. Rev. Lett. **83**, 2179 (1999)
389. W. Wang, I.Z. Kiss, J.L. Hudson, Chaos **10**, 248 (2000)
390. C.M. Gray, P. König, A.K. Engel, W. Singer, Nature **338**, 334 (1989)
391. G. Buzsaki, Z. Horvath, R. Urioste, J. Hetke, K. Wise, Science **256**, 1025 (1992)
392. W. Gerstner, "What is different with spiking neurons?", In *Plausible Neural Networks for Biological Modelling*, eds. H. Masterbroek, J.E. Vos (Kluwer, Dordrecht 2001) pp. 23–48
393. S. Thorpe, D. Fize, C. Marlot, Nature **381**, 520 (1991)
394. C.S. Peskin, *Mathematical Aspects of Heart Physiology* (Courant Inst. Math. Sci., New York 1975)
395. R.E. Mirollo, S.H. Strogatz, SIAM J. Appl. Math. **50**, 1645 (1990)
396. Y. Kuramoto, Physica D **50**, 15 (1991)
397. E.M. Izhikevich, IEEE Trans. Neural Networks **10**, 508 (1999)
398. H.R. Wilson, J.D. Cowan, Biophys. J. **12**, 1 (1972)
399. B.W. Knight, J. Gen. Physiology **59**, 734 (1972)
400. W. Gerstner, J.L. van Hemmen, "Coding and Information in Neural Networks", In *Models of Neural Networks II*, eds. E. Domany, J.L. van Hemmen, K. Schulten (Springer, Berlin 1994) p. 1
401. W. Gerstner, Neural Comput. **12**, 43 (2000)
402. A. Treves, Int. J. Neural Syst. **3** (Suppl.), 115 (1992)
403. J.J. Hopfield, A.V.M. Herz, Proc. Natl. Acad. Sci. USA **92**, 6655 (1995)
404. V.M. Tsodyks, T. Sejnowski, Network **6**, 111 (1995)
405. C. van Vreeswijk, H. Sompolinsky, Science **274**, 1724 (1996)
406. M. Tsodyks, I. Mitkov, H. Sompolinsky, Phys. Rev. Lett. **71**, 1280 (1993)
407. U. Ernst, K. Pawelzik, T. Geisel, Phys. Rev. Lett. **74**, 1570 (1995)
408. X.J. Wang, J. Rinzel, Neuroscience **53**, 899 (1993)
409. D. Golomb, J. Rinzel, Physica D **72**, 259 (1994)
410. H.D. Abarbanel, M.I. Rabinovich, A. Selverston, M.V. Bazhenov, R. Huerta, M.M. Sushik, L.L. Rubchinskii, Phys.-Usp. **39**, 337 (1996)
411. K. Kaneko, T. Yomo, Bull. Math. Biol. **59**, 139 (1997)
412. K. Kaneko, T. Yomo, J. Theor. Biol. **1999**, 243 (1999)
413. B. Alberts, Cell **92**, 291 (1998)
414. C.W.F. McClare, J. Theor. Biol. **30**, 1 (1971)
415. L.A. Blumenfeld, A.N. Tikhonov, *Biophysical Thermodynamics of Intracellular Processes: Molecular Machines of a Living Cell* (Springer, Berlin 1994)
416. H. Gruler, D. Müller-Enoch, Eur. Biophys. J. **19**, 217 (1991)
417. M. Schienbein, H. Gruler, Phys. Rev. E **56**, 7116 (1997)
418. A.S. Mikhailov, B. Hess, J. Phys. Chem. **100**, 19059 (1996)
419. P. Stange, A.S. Mikhailov, B. Hess, J. Phys. Chem. B **103**, 6111 (1999)
420. P. Stange, A.S. Mikhailov, B. Hess, J. Phys. Chem. B **104**, 1844 (2000)

Chapter 8

421. R. Rammal, G. Toulouse, M.A. Virasoro, Rev. Mod. Phys. **58**, 765 (1986)
422. H.A. Ceccatto, B.A. Huberman, Physica Scripta **37**, 145 (1988)
423. D. Sherrington, S. Kirkpatrick, Phys. Rev. Lett. **35**, 1972 (1975)
424. N. Metropolis, A.W. Rosenblueth, M.N. Rosenblueth, A.H. Teller, J. Chem. Phys. **6**, 1087 (1953)
425. V. Dotsenko, *An Introduction to the Theory of Spin Glasses and Neural Networks* (World Scientific, Singapore 1994)
426. K.H. Fischer, J.A. Hertz, *Spin Glasses* (Cambridge University Press, Cambridge 1991)
427. S.F. Edwards, P.W. Anderson, J. Phys. F **5**, 965 (1975)
428. G. Parisi, J. Phys. A **13**, 1887 (1980)
429. G. Parisi, Phys. Rev. Lett. **50**, 1946 (1983)
430. M. Mezard, G. Parisi, M. Virasoro, *Spin-Glass Theory and Beyond* (World Scientific, Singapore 1987)
431. A.P. Young, Phys. Rev. Lett. **51**, 1206 (1983)
432. R.N. Bhatt, A.P. Young, J. Magn. Magn. Mater. **54–57**, 191 (1986)
433. S.C. Manrubia, A.S. Mikhailov, Europhys. Lett. **50**, 580 (2000)
434. S.C. Manrubia, U. Bastolla, A.S. Mikhailov, Eur. Phys. J. B **23**, 497 (2001)
435. A.N. Kolmogorov, C. R. Acad. Sci. USSR **30**, 301 (1941); **32**, 16 (1941)
436. L.D. Landau, E.M. Lifshitz, *Fluid Mechanics* (Pergamon Press, New York 1987)
437. U. Frisch, *Turbulence: The Legacy of A.N. Kolmogorov* (Cambridge University Press, Cambridge 1995)
438. A.S. Mikhailov, D.H. Zanette, unpublished
439. J.D. Murray, *Mathematical Biology* (Springer, Berlin 1989)
440. R.J. Britten, E.H. Davidson, Science **165**, 349 (1969)
441. E. Davidson, R.J. Britten, Science **204**, 1052 (1976)
442. J. Monod, F. Jacob, Cold Spring Harbor Symp. Quant. Biol. **26**, 389 (1961)
443. B.C. Goodwin, Adv. in Enzyme Regulation **3**, 425 (1965)
444. B. Alberts, D. Bray, J. Lewis, M. Raff, K. Roberts, J.D. Watson, *Molecular Biology of the Cell* (Garland, New York 1994)
445. S.A. Kauffman, *The Origins of Order – Self-Organization and Selection in Evolution* (Oxford University Press, New York 1993)
446. R.M. May, *Stability and Complexity in Model Ecosystems* (Princeton University Press, Princeton, NJ 1973)
447. J. Hofbauer, K. Sigmund, *Evolutionstheorie und dynamische Systeme – Mathematische Aspekte der Selektion* (Paul Parey, Berlin 1984)
448. S.A.L.M. Kooijman, M.P. Boer, *Dynamics Energy Budgets in Biological Systems: Theory and Applications in Ecotoxicology* (Cambridge University Press, Cambridge 1993)
449. L. Edelstein-Keshet, *Mathematical Models in Biology* (Random House, New York 1988)
450. H.L. Smith, P. Waltman, *The Theory of the Chemostat* (Cambridge University Press, Cambridge 1994)
451. D. Ludwig, SIAM J. Appl. Math. **55**, 564 (1995)
452. M. Giampietro, Agriculture, Ecosystems and Environment **65**, 201 (1997)
453. R.U. Ayres, *Resources, Environment and Economics: Applications of the Materials/Energy Balance Principle* (John Wiley & Sons, New York 1978)
454. R.U. Ayres, P.M. Weaver, *Eco-restructuring: Implications for Sustainable Development* (United Nations University Press, Tokyo 1998)

455. C.S. Holling, "The Resilience of Terrestrial Ecosystems: Local Surprise and Global change", In *Sustainable Development of the Biosphere*, eds. W.C. Clark, R.E. Munn (1986)
456. B.A. Huberman, T. Hogg, "The Behaviour of Computational Ecologies" In *The Ecology of Computation*, ed. B.A. Huberman (Elsevier, North-Holland 1988) p. 77–115
457. A.S. Mikhailov, *Foundations of Synergetics I. Distributed Active Systems*, 2nd revised edn. (Springer, Berlin 1990; 1995)
458. S. Galluccio, J.-P. Bouchaud, M. Potters, Physica A **259**, 449 (1998)
459. A. Gábor, I. Kondor, Physica A **274**, 229 (1999)
460. R. Axelrod, D.S. Bennett, British J. Political Sci. **23**, 211 (1993)
461. S. Galam, J. Math. Phychol. **30**, 426 (1986)
462. S. Galam, J. Stat. Phys. **61**, 943 (1990)
463. S. Galam, Physica A **274**, 132 (1999)
464. R.N. Mantegna, Eur. Phys. J. B **11**, 193 (1999)

Chapter 9

465. M. Minsky, *The Society of Mind* (Simon and Schuster, New York 1988)
466. R. Diestel, *Graph Theory* (Springer, Berlin 1997)
467. B. Bollobás, *Random Graphs* (Academic Press, New York 1985)
468. S. Milgram, Psychol. Today **2**, 60 (1967)
469. M. Kochen (ed.), *The Small World* (Ablex, Norwood, NJ 1989)
470. D.J. Watts, S.H. Strogatz, Nature **393**, 440 (1998)
471. D.J. Watts, *Small Worlds: The Dynamics of Networks between Order and Randomness* (Princeton University Press, Princeton, NJ 1999)
472. F.R.K. Chung, *Spectral Graph Theory* (American Mathematical Society, Providence 1997)
473. B. Bollobás, F.R.K. Chung, SIAM J. Discrete Math. **1**, 328 (1988)
474. M.E.J. Newman, D.J. Watts, Phys. Rev. E **60**, 7332 (1999)
475. L. Euler, Comment. Acad. Sci. I. Petropolitanae **8**, 128 (1736)
476. D.M. Cvetkovič, M. Doob, H. Sachs, *Spectra of Graphs. Theory and Applications* (Academic Press, New York 1980)
477. F.C. Bussemaker, S. Čobelič, D.J. Cvetkovič, J.J. Seidel, J. Comb. Theory B **23**, 234 (1977)
478. S.C. Manrubia, A.S. Mikhailov, Phys. Rev. E **60**, 157 (1999)
479. A.S. Mikhailov, *Foundations of Synergetics I. Distributed Active Systems* (Springer, Berlin 1990)
480. A.S. Mikhailov, J. Phys. A **21**, L487 (1988)
481. M. Ipsen, A.S. Mikhailov, "Evolutionary reconstruction of networks". Preprint arXiv:nlin. AO/0111023 v1, 9 November 2001
482. N.K. Jerne, Proc. Natl. Acad. Sci. (USA) **41**, 848 (1955)
483. N.K. Jerne, Ann. Immunol. (Inst. Pasteur) **124 C**, 373 (1974)
484. M.F. Burnet, Aust. J. Sci. **20**, 67 (1957)
485. A.S. Perelson (ed.), *Theoretical Immunology, Parts I and II* (Addison-Wesley, Redwood City, CA 1988)
486. A.S. Perelson, G.F. Oster, J. Theor. Biol. **81**, 645 (1979)
487. L.A. Segel, A.S. Perelson, "Shapespace Analysis of Immune Networks", In *Cell to Cell Signalling: From Experiments to Theoretical Models"*, ed. A. Goldbeter (Academic Press, New York 1989) p. 273
488. L.A. Segel, A.S. Perelson, Immunology Lett. **22**, 91 (1989)

489. R.J. De Boer, A.S. Perelson, J. Theor. Biol. **149**, 381 (1990)
490. P.H. Richter, Eur. J. Immunol. **5**, 350 (1975)
491. P.H. Richter, "The Network Idea and the Immune Response", In *Theoretical Immunology*, eds. G.I. Bell, A.S. Perelson, G.H. Pimbley (Dekker, New York 1978), p. 539
492. R.J. Bagley, J.D. Farmer, N. Packard, A.S. Perelson, I. Stadnyk, BioSystems **23**, 113 (1989)
493. F.J. Varela, A. Coutinho, Immunology Today **12**, 159 (1991)
494. F.J. Varela, F.J. Stewart, A. Coutinho, "What is the Immune System For?", In *Thinking about Biology*, eds. W. Stein, F.J. Varela (Addision-Wesley, Reading, MA 1993)
495. F. Tauber, *The Immune Self. Theory or Metaphor. A Philosophical Inquiry* (Cambridge University Press, Cambridge 1994)
496. G. Weisbuch, R.J. De Boer, A.S. Perelson, J. Theor. Biol. **146**, 483 (1990)
497. R.J. De Boer, P. Hogeweg, Bull. Math. Biol. **51**, 381 (1989)
498. A.U. Neumann, G. Weisbuch, Bull. Math. Biol. **54**, 21 (1992)
499. A.U. Neumann, G. Weisbuch, Bull. Math. Biol. **54**, 699 (1992)
500. G. Weisbuch, R.M. Zorzenon Dos Santos, A.U. Neumann, J. Theor. Biol. **163**, 237 (1993)
501. J. Stewart, F.J. Varela, J. Theor. Biol. **153**, 477 (1991)
502. V. Detours, H. Bersini, J. Stewart, F.J. Varela, J. Theor. Biol. **170**, 401 (1994)
503. V. Calenbuhr, H. Bersini, J. Stewart, F.J. Varela, J. Theor. Biol. **177**, 199 (1995)
504. V. Calenbuhr, F.J. Varela, H. Bersini, Intl. J. Bif. Chaos **6**(9), 1691 (1996)
505. B.A. Huberman, T. Hogg, "The Behavior of Computational Ecologies", In *The Ecology of Computation* ed. B.A. Huberman (Elsevier, North Holland 1988), p. 77–115
506. B.A. Huberman, L.A. Adamic, Nature **401**, 131 (1999)
507. R. Albert, H. Jeong, A.-L. Barabasi, Nature **401**, 130 (1999)
508. F.J. Richter, Y. Wakuta, Eur. Management J. **11**, 262 (1993)
509. R.W. Rycroft, D.E. Kash, Research - Technology Management, **42**, 13 (1999)
510. N. Luhmann, *Social Systems* (Stanford University Press, Stanford 1995)
511. F.J. Varela, H.R. Maturana, R. Uribe, BioSystems **5**, 187 (1974)
512. H.R. Maturana, Int. J. Man-Machine Stud. **7**, 313 (1975)
513. H.R. Maturana, F.J. Varela, *Autopoiesis and Cognition* (Reidel, Boston 1980)
514. F.J. Richter, V. Calenbuhr, Human Systems Management **19**, 11 (2000)
515. A. de Geeus, *The Living Company* (Harvard Business School Press, Boston 1997)
516. M. Zeleny, *Autopoiesis, Dissipative Structures, and Spontaneous Social Orders* (Westview Press, Boulder 1980)
517. M. Zeleny, Human Systems Management **9**, 57 (1990)
518. M. Zeleny, Soziale Systeme **1**, 179 (1995)
519. M. Zeleny, "Beyond the Network Organization: Self-Sustainable Web Enterprises", In *Business Networks in Asia: Promises, Doubts, and Perspectives* ed. F.J. Richter (Quorum, Westport 1999)
520. W.R. Bukowitz, R.L. Williams, *The Knowledge Management Fieldbook* (Prentice Hall, Englewood Cliffs, NJ 1999)
521. R.M. May, *Stability and Complexity in Model Ecosystems* (Princeton University Press, Princeton, NJ 1973)
522. W. Jansen, J. Math. Biol. **25**, 411 (1987)
523. R. Law, J.C. Blackford, Ecology **73**, 567 (1992)
524. P. Bak, K. Sneppen, Phys. Rev. Lett. **71**, 4083 (1993)

525. K. Christensen, R. Donagelo, B. Koiler, K. Sneppen, Phys. Rev. Lett. **81**, 2380 (1998)
526. R.V. Solé, S.C. Manrubia, Phys. Rev. E **54**, R42 (1996)
527. R.V. Solé, J. Bascompte, S.C. Manrubia, Proc. Roy. Soc. (London) B **263**, 1407 (1996)
528. R. Engelhardt, *Emergent Percolating Nets in Evolution*, Ph.D. Thesis, University of Copenhagen (1998)
529. A. Hunding, R. Engelhardt, Origins of Life **30**, 439 (2000)
530. E.M. Izhikevich, IEEE Trans. Neural Netw. **10**, 508 (1999)
531. J. Ito, K. Kaneko, Neural Netwroks **13**, 275 (2000)
532. N. Kataoka, K. Kaneko, Physica D **138**, 225 (2000)
533. N. Kataoka, K. Kaneko, Physica D **149**, 174 (2000)

Subject Index

Springer Series in Synergetics

Springer Series in Synergetics

Springer Series in Synergetics